Practical Reliability Analysis

Ken Neubeck

Upper Saddle River, New Jersey
Columbus, Ohio

Library of Congress Cataloging-in-Publication Data
Neubeck, Ken.
　Practical reliability analysis / Ken Neubeck.
　　p. cm.
　Includes bibliographical references and index.
　ISBN 0-13-042020-4
　1. Reliability (Engineering) I. Title.

TA169.N48 2004
620'.00452—dc21
2002034595

Editor in Chief: Stephen Helba
Executive Editor: Debbie Yarnell
Associate Editor: Kimberly Yehle
Production Editor: Louise N. Sette
Production Supervision: Carlisle Publishers Services
Design Coordinator: Diane Ernsberger
Cover Designer: Mark Shumaker
Cover art: Concept by Ken Neubeck
Production Manager: Brian Fox
Marketing Manager: Jimmy Stephens

This book was set in Times by Carlisle Communications, Ltd. It was printed and bound by R. R. Donnelley & Sons Company. The cover was printed by Phoenix Color Corp.

Copyright © 2004 by Pearson Education, Inc., Upper Saddle River, New Jersey 07458. Pearson Prentice Hall. All rights reserved. Printed in the United States of America. This publication is protected by Copyright and permission should be obtained from the publisher prior to any prohibited reproduction, storage in a retrieval system, or transmission in any form or by any means, electronic, mechanical, photocopying, recording, or likewise. For information regarding permission(s), write to: Rights and Permissions Department.

Pearson Prentice Hall™ is a trademark of Pearson Education, Inc.
Pearson® is a registered trademark of Pearson plc
Prentice Hall® is a registered trademark of Pearson Education, Inc.

Pearson Education Ltd.　　　　　　　　　　　Pearson Education Australia Pty. Limited
Pearson Education Singapore Pte. Ltd.　　　　Pearson Education North Asia Ltd.
Pearson Education Canada, Ltd.　　　　　　　Pearson Educación de Mexico, S.A. de C.V.
Pearson Education—Japan　　　　　　　　　　Pearson Education Malaysia Pte. Ltd.

10 9 8 7 6 5 4 3 2 1
ISBN: 0-13-042020-4

To my wonderful wife, Fran, for all of her encouragement and support in my writing projects. To my parents, Ray and Claire, who taught me determination when growing up.

About the Author

Ken Neubeck has accrued over two decades of reliability engineering experience in the aerospace and commercial fields. After obtaining his B.S. and M.S. degrees in applied math from SUNY at Stony Brook, he started employment with the Fairchild Republic Company and worked on the A-10 Close Air Support Aircraft program as a reliability engineer for over twelve years. After the company closed in 1987, he continued as a reliability engineer for electronics companies and other commercial businesses. Ken has written books on the A-10 and F-105 aircraft as well as on amateur radio topics where he holds the call sign WB2AMU. He continues to conduct extensive research in the area of radio propagation.

Preface

WHY THIS BOOK?

Actually, the question stated above is best answered by another question: Why isn't the subject of reliability taught more often as a regular subject in more universities? Reliability is perhaps one of the better fields for engineering and applied math students to apply their academic background in practical ways.

But as a number of writers have pointed out in some of the industry reliability newsletters over the years, many newly graduated engineers from U.S. universities seem to have had little exposure to the measure of reliability nor are they equipped to design components with reliability in mind. One of the reasons for this has been brought out by Professor Marvin Roush in his article "Reliability: Where Are the Universities?" (*RAC Newsletter,* September 1988). He states, "It is my premise that there are few university engineering faculty who would be comfortable teaching reliability considerations within their particular special discipline. With their limited knowledge about reliability, they would find it difficult to select homework problems and design problems which incorporate reliability, let alone grade the solutions."

Reliability is occasionally offered as a basic course in either the applied mathematics or engineering curriculum in a number of U.S. universities. Sometimes it is listed as an elective, but it is rarely offered on a regular basis when listed in this manner in the college catalog. It appears that the ideal way to bridge this problem is to bring engineers from the reliability field to teach and develop textbooks that suitably discuss this subject. With this basic goal in mind, I undertook this book project—a reliability textbook that is suitable for teaching at the university level, as well as being a useful guide to engineers who confront reliability problems in their field of work. Hence, this is my answer to "Why this book?"

Reliability engineering received a major boost as a distinct field in engineering when the U.S. space program took off during the 1960s. The government needed a program to ensure high reliability for components that are used in the space environment and this resulted in the birth of the reliability engineering field. Military programs followed suit shortly thereafter, and many reliability tasks were required to be performed and submitted formally as a data item to the customer on new aircraft programs. It is incredible to think that on some programs, such as the Grumman LEM program, there were over 100 engineers dedicated to performing many reliability-type tasks!

In recent years, the number of military programs has declined significantly and reliability requirements have been relaxed, yet reliability remains a very important aspect of engineering. It still must be addressed in various ways for both military and commercial products. Poor reliability can still undermine a new product as well as adversely impact a company's reputation.

In 1961, the classic book ***Reliability Theory and Practice*** (Prentice Hall) by Igor Bazovsky was published. This book provided the mathematical background for the basic aspect of reliability theory and has been used as an important reference by many reliability engineers during their careers. Since this classic book came out, a number of similar books have dealt with the numerical aspect of reliability theory.

Unfortunately, several of the newer reliability books tend to get caught up with the bookkeeping tasks of describing methods to generate failure rate prediction values rather than exploring the thinking behind it. At the time of this writing, PRISM is one of the current prediction methods used for calculating individual component failure rate. The danger of tailoring a reliability book around a current prediction method is that methods are revised constantly and the book can be quickly outdated. Thus, the book in your hands will not spend an inordinate amount of time on specific methods for developing failure-rate predictions. Instead, the focus is on promoting practical methods of thinking in solving reliability problems.

As any engineer involved with developing new designs can testify, the exercise of number-crunching does not guarantee high reliability performance. There is a practical side, and this has rarely been addressed fully in any reliability text that has been written previously. No current book serves as a practical guide for all types of engineers to resolve the various reliability issues that may come up. As a mathematician, I have an inherent love for numbers and analysis, but through years of experiences, I have found that many of the calculated reliability values had little bearing on how well a product will really perform during field service. A practical side needs to be addressed that provides a grounded reality for the analysis that needs to be performed.

I was able to live the reliability engineering experience firsthand. After graduating with B.S. and M.S. degrees in applied math, I did not immediately see many opportunities for direct applications in the workplace for the math skills that I had learned. After my first five years as a quality analyst for the Fairchild Republic Company, a former aircraft manufacturing company, I finally got a break with an opportunity to transfer into the company's engineering department as a reliability engineer. In this area, I finally saw a chance to use much of my math background, and I understood that reliability engineering was an ideal field for students in math and engineering to explore.

I was very fortunate to have two excellent senior reliability engineers in the department at that time who had a wealth of experience and education in the field. Dave Conroe and Sam Sobel had the expertise to handle any theoretical or practical problem that came to the department. Sam used to write up numerous reliability and math exercises for me to do on a weekly basis, while Dave would refer me to various sources for additional learning. I learned more about applied math during my time in this department than from all of the college math courses that I had taken.

The book includes true-life examples of reliability problems that, in most cases, were based on experiences during my professional work career. It uses the "lessons learned" approach, and readers can take the concepts presented and use them when dealing with reliability-related issues that they encounter.

This book is not just for reliability engineers; it is also for project, electrical, and mechanical engineers, as well as engineering managers to use as a guide when developing a design or in troubleshooting an existing design. It provides hardcore methods towards solving problems by providing a useful set of tools, both in analytical and narrative form for helping any engineer when dealing with reliability issues involving product design. These issues may involve developing corrective action for problems that arise from a unit undergoing testing or field service, as well as working up an analysis to prove a hypothesis. Reliability engineering may very well have had its roots in the aerospace field, but it is applicable for all products that require engineering to make them.

This book provides the necessary caution when using mathematical analysis for a reliability problem. The old saying "figures don't lie, but liars figure" is especially true in the politics of many aerospace and electronic companies. This is particularly true in the area of reliability predictions and publishing MTBF values.

There is a dichotomy of sorts when discussing reliability engineering. There are both qualitative and quantitative aspects to this field. It is my experience that applied math analysis and many reliability concepts go hand and hand. For this book, it is helpful to have some background in applied mathematics, preferably in the area of probability and statistics. The main thing is not to have a fear of numbers or a fear of troubleshooting equipment. The numerical examples in this book assume minimal prior knowledge in the field of statistics.

It is my belief that it is impossible to teach reliability without teaching a reasonable amount of mathematical theory. Teaching reliability without the mathematical fundamentals is like playing music without the basic rhythm that guides it structure. Reliability has no meaning without the math theory behind it to provide the beat.

On the other side, it is important to be able to present mathematical concepts that are fundamental to reliability, some of them complex, in a way that is both interesting and helpful to the reader's understanding. Some previous approaches in teaching reliability and related subjects may have placed perhaps too much emphasis on the math and the approach comes off very dry. In this case, it is like playing music without the emotion because it is weighed down with too much technique.

Thus, a balance is needed between presenting both quantitative math concepts and the qualitative analysis that is presented in case studies. Chapters 1, 2, 3, 10, 11, and 13 are the reliability math-based chapters which would be used in a traditional reliability mathematics course. The remaining chapters involve qualitative examples of reliability engineering from which selected topics could be presented to the student at the discretion of the teacher. However, these chapters would be of particular interest to engineers and program managers in industry. Specifically, Chapters 5 through 9 focus on reliability issues involving the basic design in terms of mechanical, temperature, and humidity factors—these are primarily the types of issues that would be addressed in a physics of failure approach that is a major trend in the reliability field.

There have been very few books on reliability written in the past that have discussed the qualitative portion of the design to the level that it is covered in this book. With the concept that reliability covers many things, each chapter in this book exists as a stand-alone essay. Thus, a teacher may be able to select the appropriate chapters of this book that are suitable for the specific approach being taken when constructing and tailoring a reliability course for a school's curriculum. Chapters 1, 2, and 3 are recommended for the baseline material for any course.

This book uses case studies to illustrate the main message or approach that is being presented in a particular section or chapter. This approach will make it easier for students and engineers from various disciplines to grasp and understand reliability concepts as well as ideas in related fields. I had appreciated only a handful of math courses and textbooks over the years, with those that presented real-life applications as my favorites.

There is a bit of humor when actual life experiences show that designs made by human beings can often be less than perfect. But human beings generally have the ability to bounce back and learn from their mistakes and develop better designs the next time around. The axiom "One must experience failure before achieving success" is appropriate. Some of the examples presented here were obvious situations that reflected a lack of understanding by design engineering. Other case study examples show that limitations in technology at the time had made it impossible to achieve a reliable design that was desired. The reliability engineer has to be able to identify all of the potential pitfalls that lie ahead as a new design is developed and help keep members of the engineering team informed.

This book is designed for use as either an undergraduate reliability-engineering course or as part of a one-week reliability seminar course. It is also meant to be a practical book that could be an important reference book in the working engineer's bookcase.

ACKNOWLEDGMENTS

A project of this size requires a support staff of people to make it happen and I want to acknowledge as well as thank a number of people for helping me with this project.

I want to thank my wife, Fran, for typing the mathematical tables for me and for her overall support to me.

In addition to Dave Conroe (who also reviewed the math chapters) and Sam Sobel, I want to recognize two individuals who I had worked with over the years in different companies, Mark Schmidt and John Golden. John provided a tremendous amount of background material on reliability as well as edited certain chapters. Mark provided editing in the chapters concerning various engineering concepts and helped me construct some of the charts. Mark, John, and I represented the Sagittarius A-team.

Additional technical support in constructing figures was provided by Alan Wyman.

I want to thank my father, Ray, for teaching me over the years the various troubleshooting tips in the area of fixing electronic equipment. Our common experiences in the amateur radio hobby have given me a lot of insights into reliability issues.

I am also grateful to my former college professor, Dr. David Adler from Dowling College, who was able to teach me some valuable methods of mathematical analysis in approximation theory that I have used throughout my professional life in solving problems.

I would like to acknowledge the reviewers of this text during the preliminary review process of this book: Val Hawks—Brigham Young University; Eric Constans—Rowan University; Bart P. Hamilton—University of Akron; Matthew P. Stephens—Purdue University; Bilal M. Ayyub—University of Maryland, College Park; Peter J. Hardro—Naval Undersea Warefare Center, Newport–University of Massachusetts, Dartmouth; Tunc Aldemir—The Ohio State University; and Harry F. Cullen—Palomar College.

OTHER BOOKS BY KEN NEUBECK

- *Six Meters, A Guide to the Magic Band* (1994 1st edition, 1998 2nd edition, 2003 3rd edition, Worldradio Books)
- *A-10 Warthog in Action* (Mini series) (1995, Squadron/Signal Publications)
- *A-10 Warthog Walk Around* (1999, Squadron/Signal Publications)
- *F-105 Thunderchief Walk Around* (2000, Squadron/Signal Publications)
- *F-105 Thunderchief in Action* (2002, Squadron/Signal Publications)

Contents

Chapter 1   RELIABILITY TERMS AND DEFINITIONS   1

    Introduction   1
1.1 Definitions and Terms Used in Reliability Analysis   2
    1.1.1 Reliability   2
    1.1.2 Failure Rate   2
    1.1.3 Mean Time Between Failures (MTBF)   3
    1.1.4 Life History Curve (The Reliability Bathtub Curve)   3
1.2 Reliability Distributions   4
    1.2.1 The Exponential Reliability Distribution   5
    1.2.2 The Binomial Reliability Distribution   7
    1.2.3 The Poisson Reliability Distribution   7
    1.2.4 The Weibull Reliability Distribution   9
1.3 System Reliability Models   10
    1.3.1 Serial Reliability Model   11
    1.3.2 Parallel Reliability Model   14
    1.3.3 Standby Reliability Model   16
    1.3.4 k of n Configuration Reliability Model   19
    1.3.5 Combination Reliability Model   21
1.4 Summary   22
1.5 Exercises   23

Chapter 2   CONFIDENCE LIMITS AND THEIR USE IN RELIABILITY ANALYSIS   25

    Introduction   25
2.1 Confidence Limits   25
2.2 Using the Chi-Square Distribution to Calculate Confidence Limits   31
2.3 One-Sided Binomial Confidence Limits   36
2.4 Summary   38
2.5 Exercises   39

Chapter 3 RELIABILITY PROGRAM TASKS 40

     Introduction 40
- 3.1 Electronic Parts Stress Analysis 41
- 3.2 Reliability Prediction Techniques 45
    - 3.2.1 Failure Rate Prediction Method for Electronic Parts 47
    - 3.2.2 Non-Electronic Parts Failure Rate Prediction Methods 47
- 3.3 Reliability Apportionment 48
- 3.4 Failure Modes and Effects Analysis (FMEA) 49
- 3.5 Failure Modes, Effects, and Criticality Analysis (FMECA) 54
- 3.6 Fault-Tree Analysis (FTA) 57
- 3.7 Summary 62
- 3.8 Exercises 63

Chapter 4 THE BENEFITS OF RELIABILITY TESTING TO PRODUCT DESIGN 66

     Introduction 66
- 4.1 Reliability Development Test Objectives 67
- 4.2 Thermal Cycling 67
- 4.3 Vibration Cycling 70
- 4.4 Test Documentation and Other Items Needed for Reliability Testing 72
    - 4.4.1 Test Articles 72
    - 4.4.2 Test Equipment 74
    - 4.4.3 Reliability Test Procedure 74
    - 4.4.4 Test Log Book 74
    - 4.4.5 Failure Reporting 74
    - 4.4.6 Implementing Corrective Action 76
    - 4.4.7 Final Test Report 76
- 4.5 The Different Types of Reliability Development Tests 76
- 4.6 Reliability Test Case Studies 77
    - 4.6.1 Reliability Test Case Study #1 78
    - 4.6.2 Reliability Test Case Study #2 79
    - 4.6.3 Reliability Test Case Study #3 81
    - 4.6.4 Reliability Test Case Study #4 85
    - 4.6.5 Reliability Test Case Study #5 86
    - 4.6.6 Reliability Test Case Study #6 87
    - 4.6.7 Reliability Test Case Study #7 89
- 4.7 Highly Accelerated Life Test (HALT) 89
- 4.8 Production Reliability Tests 90
    - 4.8.1 Environmental Stress Screening (ESS) Tests or Burn-In 91
    - 4.8.2 Production Reliability Acceptance (PRAT) Tests 92
- 4.9 Troubleshooting Tips for Reliability Test Failures 92
    - 4.9.1 Thermal Testing Failures 92
    - 4.9.2 Vibration Testing Failures 93

4.10 Guidelines for the Timely Completion of Reliability Tests   93
4.11 Summary   94
4.12 Exercises   94

Chapter 5   MECHANICAL DESIGN IMPACT ON ELECTRONIC COMPONENT RELIABILITY   96

Introduction   96
5.1 Basic Concepts of Structural Reliability   97
5.2 Mechanical Failure Modes of Electrical Components   97
    5.2.1 Potentiometers   98
    5.2.2 Resistor Networks   100
    5.2.3 Tubular Capacitors   101
    5.2.4 Lamps   102
    5.2.5 Transformers and Large Inductors   102
    5.2.6 Wiring   104
    5.2.7 Printed Circuit Boards   105
    5.2.8 Flexible Circuit Boards   109
    5.2.9 Connector and Sockets   110
    5.2.10 Relays and Switches   113
    5.2.11 Fans, Motors, and Meters   114
5.3 The Reliability Engineer's Role in Mechanical Design   114
5.4 Mechanical Reliability Design Case Studies   115
    5.4.1 Case Study #1   115
    5.4.2 Case Study #2   118
5.5 Summary   121
5.6 Exercises   121

Chapter 6   THERMAL FACTORS AND RELIABILITY   123

Introduction   123
6.1 The Thermal Environment   124
6.2 Thermal Analysis   124
6.3 Thermal Survey   125
    6.3.1 Purpose of Conducting a Thermal Survey   125
    6.3.2 Thermal Survey Ground Rules and Setup   125
    6.3.3 Thermal Survey Example   126
6.4 Components Affected by Cold Temperatures   127
    6.4.1 Liquid Crystal Displays (LCDs)   127
    6.4.2 O-Rings   128
    6.4.3 Plastic Parts   128
6.5 Components Affected by Hot Extremes   129
    6.5.1 Transistors and Integrated Circuits   129

6.6 Components Affected by Thermal Variation   129
    6.6.1 Circuit Boards and Solder Connections   129
6.7 Remedies for Thermal Relief   130
    6.7.1 Heat Sinks   131
    6.7.2 Fans   131
    6.7.3 Heaters   134
6.8 Thermal Factors Case Studies   134
    6.8.1 Thermal Factors Case Study #1   134
    6.8.2 Thermal Factors Case Study #2   135
    6.8.3 Thermal Factors Case Study #3   136
6.9 Summary   137
6.10 Exercises   137

Chapter 7 THE IMPACT OF WATER ON PRODUCT RELIABILITY   139

Introduction   139
7.1 How the Different States of Water Can Impact a Design   139
7.2 Remedies for Exposure to Water   141
7.3 Case Studies on the Effects of Water   142
    7.3.1 Effects of Water Case Study #1   142
    7.3.2 Effects of Water Case Study #2   144
    7.3.3 Effects of Water Case Study #3   146
    7.3.4 Effects of Water Case Study #4   146
    7.3.5 Effects of Water Case Study #5   148
    7.3.6 Effects of Water Case Study #6   150
    7.3.7 Effects of Water Case Study #7   150
7.4 Summary   152
7.5 Exercises   152

Chapter 8 FAILURE ANALYSIS AND TROUBLESHOOTING METHODS   153

Introduction   153
8.1 Basic Types of Failures   154
8.2 Basic Troubleshooting Methods and Tools   154
    8.2.1 Use of the Multimeter for Electrical Troubleshooting   155
    8.2.2 Troubleshooting Charts   155
    8.2.3 Troubleshooting Chart Case Study   156
8.3 Fault Isolation Troubleshooting Methods for Hard Failures   158
    8.3.1 Fault Isolation Troubleshooting of Hard Failures Case Study   158
8.4 Latent Failure Troubleshooting Methods   159
    8.4.1 Latent Failure Case Study   160

8.5 Secondary Failure Troubleshooting Methods 161
    8.5.1 Secondary Failure Troubleshooting Case Study 162
8.6 Intermittent Failure Troubleshooting Methods 162
    8.6.1 Intermittent Failure Case Study #1 164
    8.6.2 Intermittent Failure Case Study #2 167
    8.6.3 Intermittent Failure Case Study #3 168
    8.6.4 Intermittent Failure Case Study #4 169
    8.6.5 Intermittent Failure Case Study #5 170
    8.6.6 Intermittent Failure Case Study #6 170
8.7 Repetitive Failures 172
    8.7.1 Repetitive Failure Case Study 172
8.8 Summary 173
8.9 Exercises 174

Chapter 9 THE DESIGN CONCEPT'S IMPACT ON RELIABILITY PERFORMANCE 175

Introduction 175
9.1 Case Studies on Design Concepts 176
    9.1.1 Unproven Design Concept Case Study 176
    9.1.2 Current Technology Not Able to Meet Design Requirements Case Study 179
    9.1.3 Lack of Bulletproofing in Design Case Study 180
    9.1.4 Failure to Anticipate Differences Case Study 183
    9.1.5 Needless Complications of Design Case Study 184
    9.1.6 Allowing for a Margin of Error in the Design Case Study 185
9.2 Summary 193
9.3 Exercises 195

Chapter 10 RELIABILITY GRAPHS AND DUANE GROWTH CURVES 196

Introduction 196
10.1 MTBF Graphs 196
    10.1.1 MTBF Graph Case Study #1 197
    10.1.2 MTBF Graph Case Study #2 198
10.2 The Duane Reliability Growth Curve 204
    10.2.1 Reliability Growth Curve Case Study #1 207
    10.2.2 Reliability Growth Curve Case Study #2 210
    10.2.3 Reliability Growth Curve Case Study #3 212
10.3 Growth Curve Theory in Other Fields 215
10.4 Summary 217
10.5 Exercises 218

Chapter 11 USING HYPOTHESIS TESTING TO SOLVE RELIABILITY PROBLEMS 220

Introduction 220
11.1 The Concept of Hypothesis Testing 221
11.2 The F-Test of Significance 221
    11.2.1 F-Test Case Study #1 223
    11.2.2 F-Test Case Study #2 225
    11.2.3 F-Test Case Study #3 227
11.3 The Chi-Square (χ^2) Test of Independence 231
    11.3.1 Chi-Square Test Case Study #1 235
    11.3.2 Chi-Square Test Case Study #2 236
    11.3.3 Chi-Square Test Case Study #3 237
11.4 Summary 240
11.5 Exercises 241

Chapter 12 SOFTWARE RELIABILITY CONCEPTS 243

Introduction 243
12.1 Basic Reliability Software Concepts 244
    12.1.1 The Three Different Types of Software 244
    12.1.2 Reliability Issues with Software 244
12.2 Software in Generic Customer Terminal Design 247
    12.2.1 Generic Customer Terminal Software Case Study #1 249
    12.2.2 Generic Customer Terminal Software Case Study #2 250
12.3 Software in Aircraft Applications 251
    12.3.1 Aircraft Software Case Study #1 252
    12.3.2 Aircraft Software Case Study #2 253
12.4 Medical Equipment and Software 254
    12.4.1 Medical Equipment Software Case Study #1 255
    12.4.2 Medical Equipment Software Case Study #2 255
12.5 Automobile and Software 256
    12.5.1 Automobile Software Case Study 256
12.6 Amateur Radio Transceiver and Software 257
    12.6.1 Amateur Radio Software Case Study 257
12.7 Handheld Electronic Units 258
    12.7.1 GPS Handheld Unit Case Study 258
12.8 Summary 258
12.9 Exercises 259

Chapter 13 MAINTAINABILITY CONCEPTS   261

    Introduction 261
- 13.1 Maintainability Terms 261
  - 13.1.1 Mean Time to Repair (MTTR) 261
  - 13.1.2 Mean Downtime (MDT) 262
  - 13.1.3 Mean Time Between Maintenance (MTBM) 263
  - 13.1.4 Inherent Availability 263
  - 13.1.5 Operational Availability 263
  - 13.1.6 Mission Capable Rate 264
- 13.2 Maintenance Concept 265
  - 13.2.1 Organizational Level 265
  - 13.2.2 Intermediate Level 266
  - 13.2.3 Depot Level 266
- 13.3 Maintainability Documentation 266
  - 13.3.1 Discrepancy Reports and Failure Tags 266
  - 13.3.2 Component Maintenance Manuals and Parts Lists 267
  - 13.3.3 Repair History Sticker 268
- 13.4 Dispatch Reliability 268
- 13.5 Other Maintenance Engineering Tasks 269
  - 13.5.1 Maintainability Demonstration 269
  - 13.5.2 Maintainability Demonstration Case Study 270
  - 13.5.3 Testability or Test Point Analysis 270
  - 13.5.4 Bit Demonstration 270
  - 13.5.5 Bit Demonstration Case Study 271
- 13.6 Preventative Maintenance Concepts 271
  - 13.6.1 Preventative Maintenance Case Study #1 272
  - 13.6.2 Preventative Maintenance Case Study #2 273
- 13.7 Designing for Maintainability 275
  - 13.7.1 Designing for Maintainability Case Study #1 276
  - 13.7.2 Designing for Maintainability Case Study #2 277
  - 13.7.3 Designing for Maintainability Case Study #3 278
  - 13.7.4 Designing for Maintainability Case Study #4 280
- 13.8 Life Cycle Cost 281
- 13.9 Summary 281
- 13.10 Exercises 282

Chapter 14 RELIABILITY EVALUATIONS AND PROTOTYPES   283

    Introduction 283
- 14.1 General Guidelines for a Reliability Evaluation 283
- 14.2 Qualitative Reliabilty of Mechanical Design 284
  - 14.2.1 Mechanical Reliability Evaluation Checklist 284

14.3 Structural Issues 285
    14.3.1 Sharp Corners 285
    14.3.2 Material 285
14.4 Qualitative Reliability Review of Electrical Design 285
    14.4.1 Wiring 286
    14.4.2 Noise Bypass Protection 288
    14.4.3 Isolation or Protection Circuitry 289
    14.4.4 Printed Circuit Board Design 289
    14.4.5 Part Selection 291
14.5 Miscellaneous Design Review Items 292
    14.5.1 Software 292
    14.5.2 Continuity of Design 292
14.6 The Straw Man Concept for a New Design 293
    14.6.1 Computer Modeling or Virtual Prototyping 293
14.7 Construction of a Prototype Unit 293
14.8 Case Studies 294
    14.8.1 Engineering, Reliability and Maintainability Review (ER & MR) 294
    14.8.2 Reliability and Maintenance Evaluation of New Product Case Study 295
    14.8.3 Reliability Evaluation of a Replacement Technology 297
    14.8.4 Competition Between Prototypes-Case Study 298
14.9 Evaluation Techniques Used in Making a More Robust Design 299
    14.9.1 Taguchi's Loss Function 299
    14.9.2 Design of Experiments (DOE) 299
    14.9.3 Evolutionary Operation (EVOP) 300
    14.9.4 Benchmarking 300
    14.9.5 Best Practices 300
14.10 Summary 301
14.11 Exercises 301

Chapter 15 RELIABILITY MANAGEMENT 302

Introduction 302
15.1 Benefits of an Integrated Reliability Program 302
15.2 Managing Reliability Program Tasks 303
    15.2.1 Reliability Management for New Programs 305
    15.2.2 Reliability Management for Mature Programs 306
15.3 Decisions Regarding Reliability Task Management 306
    15.3.1 Cost Analysis 306
    15.3.2 Manpower and Schedule 307
    15.3.3 Outsourcing Reliability Tasks 308
    15.3.4 Tools Selections 308

15.4 Reliability Program Case Study   308
15.5 Summary   311
15.6 Exercises   311

Chapter 16   EPILOGUE: OTHER USEFUL SKILLS   312

Introduction   312
16.1 Non-Engineering Skills   312
16.2 Case Studies and Discussions   313
    16.2.1 Knowing What the Customer Wants   313
    16.2.2 Ability to Multi-Task   316
    16.2.3 Maintaining a Distrustful Nature   317
    16.2.4 Expecting the Unexpected   319
    16.2.5 Having a Natural Curiosity in Other Areas   320
    16.2.6 Writing Skills   322
    16.2.7 Not Being Afraid to Break Things During Testing   323
    16.2.8 A Realization That There Is No One Magic Bullet   324
    16.2.9 Dispelling the Myth: "Anyone Can Build These Parts (or Do This Job)"   325
16.3 Summary of Book   326
16.4 Exercises   328

GLOSSARY   329

Appendix A   STATISTICAL TABLES   330

Chi-Squared Distribution—Table 1   331
F Distribution Values—Table 2   332
F Distribution Values—Table 3   333

Appendix B   REFERENCES AND SOURCE MATERIAL   334

INDEX   336

CHAPTER 1

Reliability Terms and Definitions

INTRODUCTION

The term reliability is used amazingly often in everyday life. One example is the coined phrase "dependable service and high reliability," often seen in newspaper ads as a major selling point for home appliances. How surprised most people would be if they knew that there was an actual field in engineering known as reliability engineering, a field that got its start during the United States space programs of the 1960s and continues to exist today.

Reliability has two aspects: qualitative analysis and quantitative analysis. Over the years, the quantitative aspects of the field have been emphasized as skills for reliability engineers, particularly in the military and commercial aircraft programs. Several quantitative tools may be used by the engineer as part of the tasks for a new program. However, qualitative analysis is necessary too, and often needed in the area of troubleshooting field failures and test failures. Thus, practical reliability analysis is the knowledge of both areas, and this book will cover both aspects.

This first chapter will introduce many of the terms used in the quantitative aspect of reliability. A good deal of this aspect uses elements of probability and statistics, and a basic understanding of these elements is required for anyone studying reliability. In this chapter, the reader will be introduced to the various terms and numerical representations that are used as the foundation of numerical reliability analysis. Descriptions of the various numerical expressions and models that are standard in the reliability field will be provided in this chapter and will provide a foundation of understanding for the reader. All of these will be useful tools in our approach for practical reliability analysis.

1.1 DEFINITIONS AND TERMS USED IN RELIABILITY ANALYSIS

A number of terms need to be defined when we discuss reliability. It is necessary to establish a convention or common meaning for these terms so that there are no misunderstandings when they are used. These terms may be used in reliability analysis reports, engineering specifications, and in design review meetings.

1.1.1 Reliability

How do we define reliability? Reliability can be defined as the probability of a device performing its purpose adequately for the period of time intended under the specified conditions encountered. A device can be anything, ranging from a single component to an electronic device to an entire aircraft. The concept of reliability is not just confined to military or commercial aircraft. It also applies to automobiles, railroads, ATMs, consumer products, and equipment used by public utilities.

1.1.2 Failure Rate

We can measure reliability performance by tracking the different types of failures along with the number of failures that is accrued by a product during its service in the field. In the field of engineering, we need a way to measure this performance by introducing the concept of failure rate. In addition, by having a basic unit of measurement for reliability, we will be able to determine the number of spare parts that will be needed to maintain sufficient support for a product design during the life of the program. The failure rate is the cornerstone unit of measurement used in the field of reliability engineering.

The basic term that we use to measure reliability is known as the failure rate of an item, or how often does it fail during a certain amount of time. For example, we may observe a number of failures against the operating hours or cycles of an item during testing or field service. Failure rate is expressed as the ratio of the total number of failures to the total operating time. This can be expressed mathematically as:

$$\lambda = K/T$$

where λ is the failure rate and K is the number of failures and T is the total operating time.

Failure rate may be expressed in terms of failures per hour and is often represented by the symbol lambda, λ. As this failure rate can be quite small for many products, a number of professional organizations use the convention in expressing lambda in terms of failures per one million hours or in terms of the scientific notation of 10^6 hours. Also, failure rate may be expressed in terms of the number of operating cycles or the number of transactions in lieu of operating time. A very important fact to note is that this failure rate calculation represents an average value.

FAILURE RATE EXAMPLE

Calculate the failure rate for a commercial product such as a washing machine that has accumulated five failures that resulted in five service calls during 1,200 hours of operation.

This is calculated as:

5 failures/1,200 hours = .00417 failures per hour

In terms of scientific notation, the answer calculated above is expressed as:

4170 failures/10^6 hours or $\lambda = 4{,}170 \times 10^{-6}$ [failures per hour]

1.1.3 Mean Time Between Failures (MTBF)

Another way to express reliability in quantitative terms is by taking the reciprocal of the failure rate. This is known as the Mean Time Between Failures (MTBF), and it is the ratio of total operating time to the total number of failures. It can be expressed mathematically as:

$$MTBF = T/K$$

where T is the total operating time and K is the total number of failures. The MTBF is the reciprocal of the failure rate and can be expressed as:

$$MTBF = 1/\lambda$$

Most equipment specifications on new products will have the reliability specified in terms of MTBF. The MTBF that is calculated, like the failure rate, is an average value.

MTBF EXAMPLE

If we use the previous example of the washing machine, the MTBF can be calculated either by dividing the number of hours 1,200 by 5 failures to yield a MTBF value of 240 or by taking the reciprocal of λ (i.e., $1/\lambda$) or:

$$1/(.00417) = 240 \text{ hours}$$

This number becomes useful in that we can expect the machine to break down after 240 hours of use, and this information can be used to establish spares requirements.

1.1.4 Life History Curve (The Reliability Bathtub Curve)

One of the common representations for the reliability performance of components is the life history curve. This representation involves observing the reliability performance of a very large sample of homogeneous components that is entering field service or testing at the same start time, $T = 0$. If we were to observe these components over a lifetime, or $T = T_m$ (where m represents the time at the end of life), and not replace components when they fail, we would see three basic periods of failure performance:

1. Infant mortality (or early failures)
2. Random failures (constant failure rate)
3. Wearout failures (or end of life failures)

When these three portions are combined and presented on a graph, they form a life history or "bathtub" curve as shown in Figure 1–1.

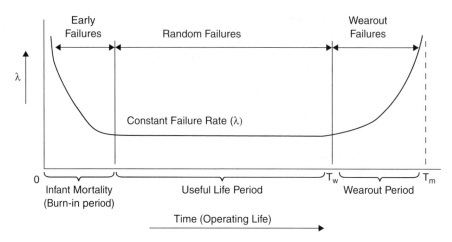

Figure 1-1 Reliability Bathtub Curve

In this model, the population of components will initially experience a high failure rate known as infant mortality. This period of time represents the burn-in or debugging period where weak components, as well as failures caused by design mistakes or process errors, are weeded out. When this initial phase has passed with the weak components weeded out and the design mistakes corrected, the remaining component population reaches a relatively constant failure rate period, known as the useful life period.

Most reliability determinations are usually concerned with this useful life or random failure period. A number of reliability distributions such as the exponential that are used to describe this phase of constant failure rate will be discussed shortly. The idea of a constant failure rate is generally an oversimplification, and failures can result from complex, uncontrollable and often unknown causes. After the useful life period ends, wearout failures begin to occur (at time T_w as shown in the figure) until the end of life is reached at T_m.

As stated before, components are not replaced when they fail for this model. However, from the experience of the bathtub curve, a policy can be developed where during the useful life period, components are replaced when they fail. Thus, before components fail as they reach the end of their projected useful life, they should be replaced as part of preventative maintenance. In addition, other methods of preventative maintenance such as lubrication or cleaning can help extend the useful life. The concept of preventative maintenance will be discussed later in Chapter 13.

1.2 RELIABILITY DISTRIBUTIONS

The reliability performance of components can often be described by mathematical expressions that are known as reliability distributions. These reliability distributions are derived from the same frequency distributions that are used in the area of probability where they provide an analytical representation of all possible outcomes. The term probability can be defined as the percentage or likelihood that a specific event will occur and that the frequency distribution represents an analytical description of the probabilities of all possible outcomes.

Reliability can be viewed in the context of probability when the reliability of a component is equal to the probability that the component does not fail during the interval $[0, t]$. This can be expressed in another way where the reliability is equal to the probability that the component is still functioning at time t. For example, the probability of the variable x occurring between time a and time b can be expressed as:

$$P(a < x < b) = \int_a^b f(x)\, dt$$

The function $f(x)$ is called the probability density function of x or the distribution function of x. The type of probability density function depicted here is a continuous function, as it is based on a continuous interval from time a to time b, where the variable may take on any values in this interval.

The other major probability frequency distribution that is used in the field of reliability is the discrete probability function. This function is based on an event-type situation, where the outcome can take on any of a series of finite values, $x_1, x_2, x_3 \ldots x_n$, (where n represents a positive integer). The discrete probability function then assigns probability values to these outcome values, and it is called the probability mass function. The probability associated with each of the possible outcomes, x_i, can be represented as:

$$P(x_i) \geq 0 \text{ for all values of } i$$

And the sum of these probabilities must satisfy the following:

$$\sum_{i=1}^{\infty} P(x_i) = 1$$

An example of a discrete probability function is the action of flipping a penny, where there are only two outcomes, heads or tails, and thus $i = 2$. The discrete probability of obtaining either heads or tails is 50 percent each, and the sum of the probabilities of these two possible outcomes is 1.

When either the element of time or cycles is introduced into these two types of probability functions (discrete or continuous), two major reliability distributions are the result. Continuous reliability distributions typically use time as the variable over the interval being measured. An example of this type is the reliability performance of a light bulb population over time. Discrete reliability distributions are an event-type or specific cycle-type function where either the event is successful or not. An example of this would be the reliability of a single rocket launch being successful or not. The discrete reliability distributions make use of two specific outcomes, where reliability is expressed as R, and unreliability is expressed as Q.

We will now discuss some of the different reliability distributions that fall into either of the two basic categories of reliability distributions: continuous or discrete reliability distributions.

1.2.1 The Exponential Reliability Distribution

When a device is subject only to failures that occur at random intervals, and that expected number of failures is the same for equally long operating periods (i.e., the failure rate λ is constant), its reliability is exponentially based and is expressed as:

$$R(t) = e^{-\lambda t}$$

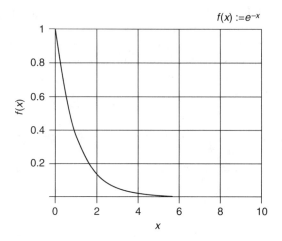

Figure 1-2 Exponential Reliability Distribution

This is the exponential reliability distribution, and it can be applied during the random failure period of the bathtub curve. The value e is the base of the natural logarithm, and λ is a constant that is called the chance failure rate. The value t is an arbitrary operating time for which we are measuring the reliability or R of the device that is known as the exponential case of chance failure. The exponential distribution is a continuous reliability function over an interval of time.

The exponential can be graphed as shown in Figure 1–2.

The term unreliability is often used in the area of the exponential reliability distribution, and this is generally expressed using the letter Q. For the continuous reliability distribution, unreliability is based on the function of time, and it can be expressed in relation to reliability as:

$$Q(t) = 1 - R(t)$$

$Q(t)$ for the exponentially based reliability distribution derived above is

$$Q(t) = 1 - e^{-\lambda t}$$

A simple approximation procedure can be developed through the expansion of $e^{-\lambda t}$ into an infinite series as follows:

$$R(t) = e^{-\lambda t} = 1 - \lambda t + \lambda^2 t^2/2! - \lambda^3 t^3/3! + \ldots$$
$$Q(t) = 1 - e^{-\lambda t} = \lambda t - \lambda^2 t^2/2! + \lambda^3 t^3/3! - \ldots$$

For very small values of $\lambda +$ the term $\lambda^2 t^2/2!$ and subsequent terms of the series can be neglected as they are very small, leaving:

$$Q(t) = 1 - e^{-\lambda t} \cong \lambda t$$

as the approximation for very small values of λt (where $\lambda t < .1$). Thus:

$$R(t) \cong 1 - \lambda t$$

EXPONENTIAL RELIABILITY DISTRIBUTION EXAMPLE

What is the reliability of an electrical device that has an exponential reliability distribution with a failure rate (λ) of 9×10^{-6} over time $t = 1{,}000$ hours?

Using the exponential reliability formula we get:

$$R = e^{-\lambda t} = e^{-(.000009)(1000)} = .99104$$

In this example, we can see that $Q(t)$ is:

$$Q(t) = 1 - .099104 = .000896$$

and if the approximation of λt is used, $Q(t)\ (.000009)\ (1000) = .009$

The exponential is an important reliability distribution that is used a lot in the reliability field, particularly in the area of component testing and subassembly testing over a continuous period of time.

1.2.2 The Binomial Reliability Distribution

The binomial reliability distribution is a discrete distribution and can be represented by:

$$(R + Q)^n = 1$$

where n represents the total number of trials conducted. For example, in the case of flipping coins for heads and tails, if we state that obtaining heads is a success, while obtaining tails is a failure, and $R = Q = .50$, we can see the following expansion for $n = 2$ (two trials) by using the basic equation:

$$(R + Q)^2 = R^2 + 2RQ + Q^2 = 1$$

It can be applied to two coin flips where the probability of obtaining heads or tails is 50 percent:

$$(.5 + .5)^2 = (.5)^2 + 2(.5)(.5) + (.5)^2 = 1$$

The first term represents the combination where both flips are heads, the second term represents the combination where one flip is heads and the other flip is tails, and the third term is the combination where both flips are tails.

When comparing the exponential and binomial reliability distributions, the exponential reliability distribution has one advantage over the binomial reliability distribution in that actual probability does not need to be known. However, because of the way that the binomial distribution is constructed, it is useful for certain applications in the area of system reliability. For example, the binomial reliability distribution becomes very useful in the area of calculating the system reliability for groups of components that are operating in parallel, as will be shown later in Section 1.3.4. The use of the binomial reliability distribution is also important in the area of confidence limits that is discussed in Chapter 2.

1.2.3 The Poisson Reliability Distribution

The Poisson reliability distribution is a discrete distribution that provides a useful tool in the use of the binomial distribution. One of the weaknesses involving the binomial distribution is that it cannot be directly applied to the calculation of event probabilities in the time domain. This is because the total number of favorable and unfavorable events (such as failures) is usually unknown.

The Poisson distribution can be used to overcome this situation by linking event probabilities to the time domain. Thus, for the binomial expression $(R + Q)^n = 1$, an equivalent expression using the number e (the base of the natural logarithm) is:

$$e^{-x} e^x = 1.$$

The term e^x can be expanded into an infinite series such as:

$$e^x = 1 + x + x^2/2! + x^3/3! + \ldots x^m/m!$$

By using the equation, $e^{-x} e^x = e^0 = 1$, we substitute into the equation above and get:

$$e^{-x}(1 + x + x^2/2! + x^3/3! + \ldots) = 1$$
$$e^{-x} + xe^{-x} + e^{-x} x^2/2! + e^{-x} x^3/3! + \ldots = 1$$

By substituting $x = \lambda t$, we get the Poisson reliability distribution:

$$e^{-\lambda t} + \lambda t e^{-\lambda t} + e^{-\lambda t}(\lambda t)^2/2! + e^{-\lambda t}(\lambda t)^3/3! + \ldots = 1$$

By moving terms to the other side of the equation, we get the following:

$$e^{-\lambda t} = 1 - \lambda t e^{-\lambda t} + e^{-\lambda t}(\lambda t)^2/2! + e^{-\lambda t}(\lambda t)^3/3! + \ldots)$$

The right side of the equation is the same formula as the exponential reliability distribution, and we can express this equation in terms of reliability and unreliability (using the symbols R and Q), as follows:

$$R = 1 - (Q_1 + Q_2 + Q_3 + \ldots)$$

Thus, we have reliability expressed in terms of individual unreliability values that are based on discrete number of failures (where Q_i is the probability of exactly i failures occurring during time period t). So for a probability of exactly one failure during time period t, we get:

$$Q_1(t) = (\lambda t) e^{-\lambda t}$$

For a probability of exactly two failures, we get:

$$Q_2(t) = (\lambda t)^2 e^{-\lambda t}/2!$$

The probability of zero failure is the reliability term on the right side of the above equation, which is the same term as the exponential reliability distribution:

$$R_0(t) = e^{-\lambda t}$$

POISSON RELIABILITY DISTRIBUTION EXAMPLE

What is the combined probability of one or two failures occurring during a time period of 1,000 hours for an electronic device with a failure rate of 100×10^{-6}? What is the probability of three or more failures occurring for this device during this time period?

We need to calculate the probabilities for the two cases for the first question:

A. We first calculate the probability of exactly one failure occurring during time t:

$$Q_1(t) = (\lambda t)\, e^{-\lambda t} = (1000 \times 100 \times 10^{-6})\, e^{-(.0001)(1000)} = .09048$$

B. The probability of exactly two failures occurring during time t is:

$$Q_2(t) = (\lambda t)^2\, e^{-\lambda t}/2! = (1000 \times 100 \times 10^{-6})^2\, e^{-(.0001)(1000)}/2! = .004524$$

C. Thus the probability of both one or two failures occurring during this time is the sum or:

$$.09048 + .004524 = .095004$$

The second part of this problem is to determine what is the probability of three or more failures occurring during this time period and this can be done in a simple manner by using the equation:

$$R = 1 - (Q_1 + Q_2 + Q_3 + \ldots)$$

By moving terms, we get:

$$(Q_3 + \ldots) = 1 - R - Q_1 - Q_2$$
$$(Q_3 + \ldots) = 1 - .90499 - .09048 - .004524 = .000006$$

It can be seen that there is similarity between the exponential reliability distribution and the Poisson distribution. Because of this relationship, the Poisson reliability distribution is particularly useful in simplifying some applications of the binomial reliability distribution as will be seen later in the K out of N system model as shown in Section 1.3.4.

1.2.4 The Weibull Reliability Distribution

One of the more complicated continuous reliability distributions is the Weibull reliability distribution, which is frequently used in life data analysis. The Weibull reliability function takes into account that failure rates are not usually constant throughout the life of a component or system. The Weibull distribution is widely used in certain areas (such as manufacturing) because of its versatility and the fact that the Weibull distribution can assume different shapes depending on the specific parameter values chosen. The Weibull is unique in that it can best fit the data that is collected as opposed to trying to fit the data to a normal curve distribution.

The Weibull is very similar to the exponential reliability function with the addition of a shaping parameter. The basic formula for the Weibull reliability function is:

$$R(T_1) = e^{-[(T_1 - \gamma)/\eta]^\beta} \quad \text{or for clarity, written as: } \exp - (T_1 - \gamma)/\eta)^\beta$$

For most reliability applications, we consider $\gamma = 0$ and $\alpha = 1/\eta$. This leaves the simplified form of:

$$R(t) = e^{-(\alpha t)^\beta} \quad \text{or for clarity, } R(t) = \exp - (\alpha t)^\beta$$

β is called the shape parameter, and it indicates whether the failure rate is increasing or decreasing as shown as follows:

if $\beta < 1.0$, the failure rate is decreasing,
if $\beta = 1.0$, the failure rate is constant,
and if $\beta > 1.0$, the failure rate is increasing.

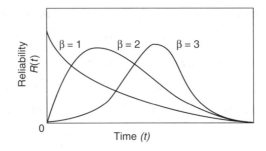

Figure 1-3 Weibull Distribution for $\beta = 1, 2$ and 3 ($\lambda = 0, \alpha = 1$)

If $\beta = 1.0$, it can be seen that the equation for $R(t)$ above becomes:

$R(t) = e^{-(\alpha t)}$, which is the same as the exponential distribution.

Likewise, the failure rate is also simplified when $\beta = 1$. The general formula is:

$$\lambda(t) = (\beta/\alpha)(t/\alpha)^{\beta-1}$$

Substituting $\beta = 1$ yields $\lambda(t) = (1/\alpha)(t/\alpha)^0 = 1/\alpha$. Figure 1-3 shows some samples of Weibull reliability curves with different values for β and $\alpha = 1$.

As described before, when $\beta = 1$, we get the exponential distribution as seen in the figure. When we have the $\beta = 2$, we get a curve that is skewed to the left that shows rapid reliability growth initially and then a gradual decline which can simulate the wearout stage.

WEIBULL RELIABILITY DISTRIBUTION EXAMPLE

What is the equivalent failure rate for $\beta = .8$ and $\alpha = 1$ for $t = 100$ and $t = 1000$ hours?
Using the equation for $\lambda(t)$, we get:

For $t = 100$ hours, $\lambda(t) = (.8/\alpha)(100/\alpha)^{-.2} = .318$

For $t = 1000$ hours, $\lambda(t) = (.8/\alpha)(1000/\alpha)^{-.2} = .201$

It can be seen from above that with a $\beta < 1.0$, the failure rate is decreasing as hours increase.

The parameters of the Weibull are difficult to calculate without using graphical methods. The graphing methods are explained in significant details in *Reliability for Technology, Engineering, and Management* (Prentice Hall, 1998) by Paul Kales. The important fact to understand about the use of the Weibull distribution is that it accommodates the concept of a nonconstant failure rate.

1.3 SYSTEM RELIABILITY MODELS

In the real world, we typically deal with multiple components that make up a system. We have been introduced to a number of reliability distributions; now we will be able to use some of these distributions in evaluating the reliability of a system that uses more than one component. How the components are connected to each other determines what type of system reliability model is used and, ultimately, the reliability value for the system.

There are different types of system reliability models. These are typically used to analyze items such as an aircraft completing its flight successfully or a string of individual electronic components working properly in a circuit board. These system reliability models may sometimes be referred to as the mission reliability of a system (referred to here as R_s)

1.3.1 Serial Reliability Model

The simplest reliability model is the serial reliability model. In this model, each component of the system needs to be working for overall system success. Even though these components may not physically be connected in a serial fashion, the serial model for these components are depicted graphically as elements that are arranged in a series string shown in Figure 1–4.

Figure 1-4 Serial Reliability Model

Reliability (R_s) for this serial model is calculated by multiplying the individual component reliability values. For n components in the system, this is represented as:

$$R_s = R_1 \times R_2 \times R_3 \times R_4 \ldots \times R_n$$
$$= (1 - Q_1)(1 - Q_2)(1 - Q_3)(1 - Q_4) \ldots (1 - Q_n)$$

For the case where the reliability of each component is based on the exponential reliability distribution, the reliability of the ith component is expressed $R_i = e^{-\lambda_i t}$ and thus the system reliability, R_s can be expressed as:

$$R_s = (1 - e^{-\lambda_1 t})(1 - e^{-\lambda_2 t})(1 - e^{-\lambda_3 t}) \ldots (1 - e^{-\lambda_{nt}})$$

The equivalent hardware failure rate for a serial model is simply the addition of the individual failure rates:

$$\lambda_1 + \lambda_2 + \lambda_3 + \ldots + \lambda_n = \lambda_{sys}$$

Thus, MTBF = $1/\lambda_{sys}$

SERIAL RELIABILITY MODEL EXAMPLE #1

What is the reliability of an ATM that consists of the following components with associated failure probability percentage? Assume that for this example, all ATM components are required to be working as a serial reliability model with each component being represented by a failure probability percentage based on field data.

1. Card Reader   5% failure rate
2. Monitor   0.1% failure rate
3. Printer   1% failure rate
4. Deposit Unit   1% failure rate
5. Cash Dispenser   2% failure rate

The schematic diagram for this problem is represented as shown in Figure 1–5.

Figure 1-5 Serial Reliability Example (ATM)

In order to calculate the overall reliability for this series model, the components are expressed in terms of unreliability, $Q = (1 - R)$. Thus:

$$R_s = (1 - .05)(1 - .001)(1 - .01)(1 - .01)(1 - .02)$$
$$= (.95)(.999)(.99)(.99)(.98) = 91.2\%$$

SERIAL RELIABILITY MODEL EXAMPLE #2

Quite often, the term mission reliability comes up in describing an aircraft mission or flight. This is a serial model that can either be expressed in terms of its individual phases of the mission or in terms of the complete mission.

For example, suppose we have a training aircraft that flies a two-hour training mission as shown in Figure 1–6. The mission consists of four distinct submissions of:

A. 5 minutes (.083 hours) of loiter time at sea level
B. 1.5 hours of the main mission

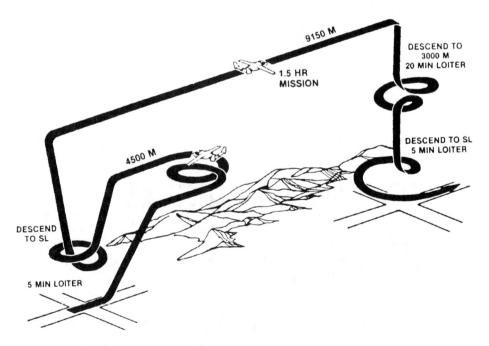

Figure 1-6 Mission Reliability Example
An aircraft mission reliability is a serial model that consists of individual portions of the flight as pictured above.

C. 20 minutes (.33 hours) of loiter time at 3,000 M
D. 5 minutes (.083 hours) of loiter time at sea level

If the aircraft has an overall MTBF of 10 hours (which equates to a failure rate of .1 failures per hour), we can calculate the mission reliability either for the whole two hours or in terms of its individual times because we are dealing with a serial model.

If we use the individual times we get

$$R_s = (e^{-.083 \times .1})(e^{-1.5 \times .1})(e^{-.33 \times .1})(e^{-.083 \times .1})$$
$$= (.9917)(.8607)(.9675)(.9917). = .819$$

R_s can also be expressed as $e^{-(.083 + 1.5 + .33 + .083)(.1)}$.

If we use the total mission time of 2 hours and failure rate of .1 failures per hour, we get:

$$R_s = e^{-.1 \times 2} = .819$$

Therefore, because we have a serial model for the mission model, we can compute the overall mission either way in terms of the overall time or individual times where the individual reliability values are multiplied together.

SERIAL RELIABILITY MODEL EXAMPLE #3

We will further explore an aspect of serial reliability called hardware reliability. An assembly such as an electronic circuit board can be viewed as the sum of all of its parts, where typically each individual component is required to be functioning in order for the overall circuit board assembly to function properly. Thus the overall circuit board assembly reliability depends on the individual component reliability and can be considered as a serial model of sorts even though in reality there may be some complex path combinations. Hardware reliability allows us to look at the assembly in terms of the sum of its components failure rates.

While it is possible to calculate individual reliability values for each component and then combine them to determine the overall assembly reliability, there is a potential for errors to creep into the calculations. It is actually more prudent to calculate individual component failure rates and then sum up to the overall failure rate for the assembly. After the overall failure rate is determined, it becomes possible to calculate MTBF and reliability. This method of calculation is often called the parts-count method.

This parts-count method can be demonstrated by taking the example of a circuit board that consists of the following individual components and their associated failure rates:

Logic Circuit Board Assembly

Component type	Quantity	Failure rate (ea)	Quantity × Failure Rate
Capacitor (Ceramic)	30	$.00001 \times 10^{-6}$	$.0003 \times 10^{-6}$
Capacitor (Tantalum)	10	$.0003 \times 10^{-6}$	$.003 \times 10^{-6}$
Resistors (Carbon)	30	$.00001 \times 10^{-6}$	$.0003 \times 10^{-6}$
Diodes	10	$.0002 \times 10^{-6}$	$.002 \times 10^{-6}$
Transistors	15	$.0005 \times 10^{-6}$	$.0075 \times 10^{-6}$
IC (Logic)	20	$.001 \times 10^{-6}$	$.020 \times 10^{-6}$

Total failure rate for logic circuit board assembly = $.035800 \times 10^{-6}$

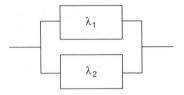

Figure 1-7 Parallel Reliability (Two Components)

The MTBF is = $1/(.035800 \times 10^{-6})$ or 27,932,961 hours.

If we were to examine the performance of this circuit board for a 10,000-hour time period, we would use the serial model of reliability to come up with a reliability value of:

$$R_s = (e^{-.0000000358 \times 10000}) = .99964.$$

The task of calculating hardware or parts-count reliability is a typical task for a reliability program, and there are various methods for calculating individual component failure rates that will be demonstrated in Chapter 3.

1.3.2 Parallel Reliability Model

One of the ways that system reliability is improved is by adding a factor of redundancy to the model. One of the common forms of redundancy is the parallel reliability model where two items are activated up simultaneously with either item being used as a path for system success. This setup is often called active redundancy. In many cases, the failure rate will generally be the same for each item, as in the case of electronic circuitry that uses redundant circuits such as dual power supplies, dual microprocessors, or dual relay switching. The schematic representation of the parallel reliability system model is shown in Figure 1–7.

A way to visualize parallel reliability in terms of probability combinations is where the probability of either event A and event B occurring (or both) is expressed as:

$$P(A) + P(B) - P(A)\,P(B)$$

As the sum of the two probabilities of A and B will include the probability of both occurring together, the product of the two term is subtracted out from the sum of the two terms.

Thus, the reliability of two items hooked up in parallel is represented as:

$$R_s = R_1 + R_2 - R_1 R_2$$
$$R_s = e^{-\lambda_1 t} + e^{-\lambda_2 t} - e^{-(\lambda_1 + \lambda_2)t}$$

R_s can also be expressed in terms of Q or:

$$R_s = 1 - Q_1 Q_2 = 1 - (1 - e^{-\lambda_1 t})(1 - e^{-\lambda_2 t}) = e^{-\lambda_1 t} + e^{-\lambda_2 t} - e^{-(\lambda_1 + \lambda_2)t}$$

which yields the same results.

The system reliability is equal to (1 − probability that both components fail)
The expected value or MTBF is equal to:

$$M = \int_0^\infty R(t)dt = \int_0^\infty e^{-\lambda_1 t}dt + \int_0^\infty \left[e^{-\lambda_2 t}dt - \int_0^\infty [e^{-(\lambda_1+\lambda_2)t}dt\right.$$
$$= 1/\lambda_1 + 1/\lambda_2 - 1/(\lambda_1 + \lambda_2)$$
$$= [(\lambda_1 + \lambda_2) - \lambda_1\lambda_2]/(\lambda_1\lambda_2)$$

If the components have equal failure rate, i.e. $\lambda_1 = \lambda_2 = \lambda$, then:
$$R_s = 1 - Q^2$$
$$R_s = 2e^{-\lambda t} - e^{-2\lambda t}$$

The equivalent MTBF is $1/\lambda + 1/2\lambda = 3/2\lambda$

For three components in parallel with equal failure rate, the schematic of the model is shown in Figure 1–8.

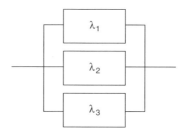

Figure 1-8 Parallel Reliability (Three Components)

The reliability for this model is:
$$R_s = 3R - 3R^2 + R^3 \text{ or } R_s = 1 - Q^3$$

The system reliability is equal to (1 − probability that all three components fails) This then simplifies to:
$$R_s = 3e^{-\lambda t} - 3e^{-2\lambda t} + 3e^{-3\lambda t}$$

This results in an equivalent MTBF = $1/\lambda + 1/2\lambda + 1/3\lambda = 11/6\lambda$.

PARALLEL RELIABILITY MODEL EXAMPLE

Banks have card readers on the outside doors to individual branches for access to the ATM lobby for off-banking hours. In this example, a bank has decided to use two card readers hooked up in parallel redundancy where the customer can use either card reader to gain entry to the lobby. Assume that the percentage of failure for the card reader is 5 percent. Thus,
$$R_s = R_1 + R_2 - R_1 R_2$$
$$= (1 - .05) + (1 - .05) - (1 - .05)(1 - .05) = .9975 \text{ or } 99.75\%$$

This can be solved in terms of the unreliability Q, or:
$$R_s = 1 - Q^2 = 1 - (.05)^2 = .9975$$

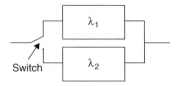

Figure 1-9 Standby Reliability
Note: The switch above is initially connected to the first item (with failure rate λ_1).

Thus, it can be seen that the reliability is improved by adding a redundant card reader from 95 to 99.75 percent.

1.3.3 Standby Reliability Model

In addition to parallel redundancy, another way to add redundancy in a system is when two or more components that perform identical functions are connected together in a standby mode. In the standby reliability model, only one component is activated at a time (as opposed to both components being activated in the parallel redundancy model) and if this component should fail, switching goes to the next component that is hooked up in parallel with the first component. A schematic of this model using identical components with the same failure rates is shown in Figure 1–9.

The overall reliability is calculated as a two-part configuration: the reliability of the first component and the reliability of the second part after the first part fails. Thus, this calculation becomes the unreliability of the first component multiplied by the reliability of the second component if we assume perfect switching. Therefore, we have the system reliability being equal to the probability that no failure will occur plus the probability that one failure will occur.

However, in order to solve this in mathematical terms, it is necessary to refer back to the Poisson reliability distribution and use the identity of:

$$e^{-\lambda t}(1 + \lambda t + (\lambda t^2/2! + \lambda t^3/3! + \ldots) = 1$$

We then extract a portion of this identity, $R = e^{-\lambda t} + e^{-\lambda t}(\lambda t)$, which represents the probability that no failures will occur $[e^{-\lambda t}]$ and exactly one failure $[e^{-\lambda t}(\lambda t)]$ will occur.

If the switching is not perfect, then the reliability of the system is:

$$R = e^{-\lambda t} + R_{sw} e^{-\lambda t}(\lambda t), \text{ where } R_{sw} \text{ is the reliability of the switch.}$$

The expected MTBF with perfect switching is calculated as:

$$\text{MTBF}_0^\infty = \int_0^\infty R_p dt = \int_0^\infty e^{-\lambda t} + \int e^{-\lambda t}(\lambda t)dt$$

$$= -1/\lambda \left[e^{-x} \right]_0^\infty + \int_0^\infty (x)e^{-x}dx \quad [\text{Where we substitute } x = \lambda t]$$

$$= 1/\lambda + 1/\lambda \int_0^\infty (x)e^{-x}dx$$

$$= 1/\lambda + 1/\lambda \left[xe^{-x} \right]_0^\infty - \int_0^\infty e^{-x} dx$$

$$= 1/\lambda + 1/\lambda \left[0 + e^{-x} \right]_0^\infty$$

$$= 1/\lambda + 1/\lambda [0 + 0 - (-1)] = 2/\lambda$$

For three units with identical failure rate λ, the reliability equation for the system is:

$$R_s = e^{-\lambda t}(1 + \lambda t + \lambda^2 t^2/2!)$$

The expected MTBF is:

$$1/\lambda + 1/\lambda + 1/\lambda = 3/\lambda$$

Thus, for n components of identical failure rate λ, the expected MTBF = n/λ

If we have different valued failure rates for components in the standby mode, we now have to keep track of the time segments where t_1 = time to when failure of the first component occurred and t_2 is the balance of the time. Therefore, we get:

$$t_1 + t_2 = t \text{ (where } t \text{ is the total time)}$$

and we will use $t_2 = t - t_1$ in the next equation as follows:

$$f_1(t_1) = \lambda_1 e^{-\lambda_1 t_1}$$
$$f_2(t_2) = \lambda_2 e^{-\lambda_2(t-t_1)}$$
$$R_s(t) = \int_0^\infty f_1(t_1) f_2(t_2) dt$$
$$= \int_0^\infty \lambda_1 \lambda_2 e^{-\lambda_1 t_1} e^{-\lambda_2(t-t_1)} (\lambda t) dt$$
$$= \lambda_1 \lambda_2 \int_{t_1=0}^t e^{-\lambda_1 t} e^{-\lambda_2(t-t_1)} dt$$
$$= \lambda_1 \lambda_2 / (\lambda_2 - \lambda_1)(e^{-\lambda_1 t} - e^{-\lambda_2 t})$$
$$= \lambda_1 \lambda_2 [e^{-\lambda_1 t}/(\lambda_2 - \lambda_1) - e^{-\lambda_2 t}/(\lambda_1 - \lambda_2)]$$
$$R_s(t) = \lambda_2/(\lambda_2 - \lambda_1) - e^{-\lambda_1 t} + \lambda_1/(\lambda_1 - \lambda_2) - e^{-\lambda_2 t}$$

Thus the equation can be quite complex. The expected MTBF = $1/\lambda_1 + 1/\lambda_2$
If the switching device is not 100 percent, we get the following:

$$R_s(t) = e^{-\lambda_1 t} + R_{sw} \lambda_1/\lambda_2 - \lambda_1 (e^{-\lambda_1 t} - e^{-\lambda_2 t})$$

The same type of integration process can be applied to three components with different failure rates with the end result being:

$$\text{Expected MTBF} = 1/\lambda_1 + 1/\lambda_2 + 1/\lambda_3$$

For n components connected in the standby mode, the expected MTBF is:

$$1\lambda_1 + 1/\lambda_2 + 1/\lambda_3 + \ldots + 1/\lambda_n$$

STANDBY RELIABILITY MODEL EXAMPLE

In 1980, the Fairchild Republic Company was conducting a study on a temperature sensor that was located on the compressor inlet of the two engines of the A-10 aircraft that it manufactured. The aircraft had one sensor per engine, but when either one failed, the compressor could stall, which could result in collateral damage to the compressor blades. In addition, when a stall event occurred, the maintenance people had to do a complete engine teardown, which took several hours, and replace all thirty-three blades of the compressor (at thirty dollars per blade) resulting in very high costs. A recommendation came from the engine manufacturer to add a second sensor for each engine that would be hooked up in standby-redundancy.

Thus, Fairchild Republic did the analysis for this addition along with a selector valve that would act as the switch.

- The failure rate of the sensor was based on field data with 35 failures occurring during 130,373 flight hours, yielding a single sensor failure rate of 268.5×10^{-6}.
- In the current setup, the reliability for 100 hours was $R = e^{-(.0002685)(100)} = .9735073$.

The failure rate for the selector valve was taken from field data of a similar valve which yielded a failure rate of 26.8×10^{-6}. Using the equation for standby redundancy with a switching element that has a failure rate, the reliability for 100 hours of service becomes:

$$R_s = e^{-(.0002685)(100)} + e^{-(.0002685)(100)} \times e^{-(.000268)(100)} (.0002685)(100)$$
$$= .9735073 + (.9973236)(.9735073)(.02685) = .999576$$

The approximate equivalent failure rate of the standby-redundant system with the selector valve is calculated using the equation:

$$R_s = e^{-\lambda_s T}$$
$$ln R_s = -\lambda_s T$$
$$\lambda_s = (-ln R_s)/t$$
$$\lambda_s = (-ln\ .999576)/100 = 4.241 \times 10^{-6}$$

We can calculate the expected number of failures against the original 130,373 flight hours using the equivalent failure rate:

$$\text{Expected number of failures} = (4.241 \times 10^{-6})(130,373) = 0.55$$

Thus, by adding a standby-redundant setup, we could expect less than one failure compared to 35 failures for the same amount of flight hours.

An interesting exercise is to show that standby reliability is higher than parallel reliability. A way of showing this is by using the exponential reliability model for two components where we assume $\lambda t < 1$ and by setting up the comparison using $\lambda_1 = \lambda_2 = \lambda$.

Active Reliability		Standby Reliability
$2e^{-\lambda t} - e^{-2\lambda t}$	?	$e^{-\lambda t} + e^{-\lambda t}(\lambda t)$
$e^{-\lambda t} - e^{-2\lambda t}$	?	$e^{-\lambda t}(\lambda t)$
$1 - e^{-\lambda t}$	?	λt

$1 - \lambda t$	?	$e^{-\lambda t}$ [we now expand $e^{-\lambda t}$ into an infinite series]
$1 - \lambda t$	?	$1 - \lambda t + \lambda^2 t^2/2! - \lambda^3 t^3/3! + \ldots$
$1 - \lambda t$	$\leq$	$1 - \lambda t +$ positive terms

(Equality is achieved between active and standby if $\lambda t = 0$.)

Thus, it can be seen that standby reliability has greater reliability than active reliability.

1.3.4 *k* of *n* Configuration Reliability Model

An interesting application of the binomial distribution can be used for certain reliability problems where *k* out of *n* components are required to be working for system success. For such a system, there is a population of *n* items with identical failure rate; there are *f* failures allowed, and *k* non-failures, where *k* is less than *n*.

The basic model for this distribution is based on the binomial expansion of $(R + Q)^n = 1$ and can be expanded as:

$$R^n + nR^{n-1}Q + \ldots \begin{bmatrix} n \\ f \end{bmatrix} R^{n-f} Q^f + \ldots + Q^n = 1$$

We now rearrange the binomial expansion into the *k* of *n* reliability model by moving some of the terms to the right side of the equation as follows:

$$R^n + nR^{n-1}Q + \ldots \begin{bmatrix} n \\ k \end{bmatrix} R^k Q^{n-k} = 1 - \ldots n R Q^{n-1} - Q^n$$

By rearranging the terms as shown above, we can now solve for the case of *k* components required to be working out of a population of *n* components. This can best be illustrated by an example.

k of n RELIABILITY MODEL EXAMPLE #1

What is the reliability of a backlighting panel for a display that has a population of four light bulbs ($n = 4$), and one failure is allowed (where three light bulbs are needed to work, or $k = 3$)?

Using the binomial expansion, we get the following:

$$(R + Q)^4 = R^4 + 4R^3Q + 6R^2Q^2 + 4RQ^3 + Q^4 = 1$$

where:

R^4 = Probability (all four components will survive)
$4R^3Q$ = Probability (exactly three components will survive)
$6R^2Q^2$ = Probability (exactly two components will survive)
$4RQ^3$ = Probability (exactly one component will survive)
Q^4 = Probability (all components will fail)

We can set up the basic equation for three out of four components surviving as follows:

$$R^4 + 4R^3Q = 1 - 6R^2Q^2 - 4RQ^3 - Q^4$$

Thus, if we have a reliability value of $R = .9$ or the light bulb, we can use the equation above to solve for the three out of four components example where:

$$R^4 + 4R^3Q = (.9)^4 + 4(.9)^3(.1) = .9477 \text{ or we can use the alternate equation:}$$
$$1 - 6R^2Q^2 - 4RQ^3 - Q^4 = 1 - 6(.9)^2(.1)^3 - (.1)^4 = .9477$$

There is also a simpler way to calculate the k out of n setup. The average MTBF value can be expressed as follows:

$$M = \int_0^\infty R_S(t)\,dt = \int_0^\infty (R^n + nR^{n-1}Q + \ldots)dt$$

Using the Poisson reliability distribution, this is expanded to:

$$M = \int_0^\infty \left[e^{-n\lambda t} + ne^{-(n-1)\lambda t}(1 - e^{-\lambda t}) + \binom{n}{2} e^{-(n-2)\lambda}(1 - e^{-\lambda t})^2 + \ldots \right] dt$$

Through a series of calculations along with the use of identities from combinatoric theory, we can simplify the above equation through a series of reductions to get a simple formula for the equivalent MTBF:

$$MTBF = \frac{1}{\lambda}\left[\frac{1}{n} + \frac{1}{n-1} + \frac{1}{n-2} + \frac{\ldots 1}{n-f}\right]$$

This is the equivalent MTBF for a population of n items and f failures. This is an interesting exercise to show that by using the standard reliability interval for the binomial distribution, and applying various identities in factorial math, we get the simplified formula for the MTBF shown above.

k out of n SYSTEM RELIABILITY MODEL EXAMPLE #2

What is the equivalent MTBF for a lighting panel that uses 12 light bulbs and only requires 10 bulbs to be working for success (only two failures allowed)? Use a common failure rate of 1 failure per million hours for each light bulb.

Using the simplified formula for MTBF, we get:

$$MTBF = \frac{1}{\lambda}\left[\frac{1}{12} + \frac{1}{(12-1)} + \frac{1}{(12-2)}\right]$$

$$= \frac{1}{\lambda}\left[\frac{1}{12} + \frac{1}{11} + \frac{1}{10}\right]$$

This yields a simplified equivalent MTBF of:

$$\text{MTBF} = 1/\lambda \ (.325)$$

Thus, for a failure rate of 1×10^{-6}, we get a MTBF value of 325,000 for this system reliability setup.

The reliability of the system (where we allow two failures) is:

$$R^{12} + 12R^{11}Q + 66R^{10}Q^2 =$$
$$= (e^{-\lambda t})^{12} + 12 \ (e^{-\lambda t})^{11}(1 - e^{-\lambda t}) + 66(e^{-\lambda t})^{10} \ (1 - e^{-\lambda t})^2$$
$$= (.999999)^{12} + 12 \ (.999999)^{11} \ (.000001) + 66 \ (.999999)^{10} \ (.000001)^2$$
$$= .99999999$$

1.3.5 Combination Reliability Model

Often a system or piece of equipment may be represented by a combination-style reliability model that consists of a number of simpler reliability models that were described in the previous sections. The overall reliability for this type of model can be determined by first calculating the reliability of the smaller or simple individual portions first and then combining each of these portions to determine the overall reliability. We will now present a methodology towards determining the overall reliability calculation of a combination reliability system model.

COMBINED RELIABILITY MODEL EXAMPLE

We are looking at the overall reliability for a LCD display unit that consists of a display, backlighting panel, and a number of circuit boards with the following setup:

1. A LCD panel with a hardware failure rate of λ_1.
2. A backlighting board with 10 bulbs with individual bulb failure rate λ_2 but is still considered good with two bulb failures.
3. Two microprocessor boards A and B hooked up in parallel, each with a total circuit board failure rate of λ_3.
4. Dual power supplies in a standby redundacy, with a failure rate of λ_4 for each power supply (assume perfect switching).
5. EMI board with failure rate of λ_5 hooked up in series with the common input of the power supply A.

The schematic diagram is constructed as shown in Figure 1–10.

There are five basic pieces in this model, and a generalized formula for the reliability of this model can be viewed somewhat in a serial fashion as:

$$R_S = R_1 \times R_2 \times R_3 \times R_4 \times R_5$$

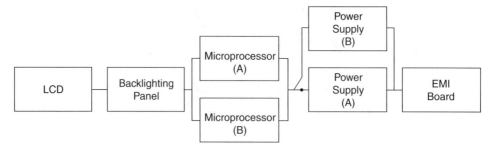

Figure 1-10 Combination Reliability Model Example

The next step is to list the reliability for the individual pieces of the model:

1. $R_1 = e^{-\lambda_1 t}$ (LCD circuit board)
2. $R_2 = (e^{-\lambda_2 t})^{10} + 10(e^{-\lambda_2 t})^9 (1 - e^{-\lambda_2 t}) + 45(e^{-\lambda_2 t})^8 (1 - e^{-\lambda_2 t})^2$ (Backlighting panel) (Microprocessors A and B)
3. $R_3 = 2e^{-\lambda_3 t} - e^{-\lambda_3 t}$ (Power supply A and B)
4. $R_4 = e^{-\lambda_4 t} + e^{-\lambda_4 t} (\lambda_4 t)$ (EMI board)
5. $R_5 = e^{-\lambda_5 t}$

We now multiply each of the terms listed for R_1 through R_5 to get R_s.

Thus $R_s = e^{-\lambda_1 t} \times [(e^{-\lambda_2 t})^{10} + 10(e^{-\lambda_2 t})^9 (1 - e^{-\lambda_2 t}) + 45(e^{-\lambda_2 t})^8 (1 - e^{-\lambda_2 t})^2] \times [2e^{-\lambda_3 t} - e^{-2\lambda_3 t}] \times [e^{-\lambda_4 t} + e^{-\lambda_4 t} (\lambda_4 t)] \, e^{-\lambda_5 t}$

Combined models get quit complicated, as can be seen above, and this can lead to errors when using a personal calculator. Models on this order of complexity can be made manageable through the use of computer programs and computer spreadsheets.

More complex models can arise when there are conditional situations or changing states. For this, Markov-type modeling would be used with the aid of a computer. Markov models will not be covered here as it is beyond the scope of simple reliability analysis and the concept of this text. This subject is adequately covered in detail in other reliability texts such as *Reliability for Technology, Engineering and Management* (Kales, Prentice Hall, 1998).

1.4 SUMMARY

In this chapter, the reader was introduced to important reliability terms, such as failure rate and Mean Time Between Failures (MTBF), that are the staple of reliability analysis. In addition, the reader has been exposed to the various reliability distributions and reliability models. These important reliability terms and concepts will provide the groundwork for understanding some of the more advanced reliability concepts that are presented in the next chapter and for other chapters in this book. These terms are the standard terms that reliability engineers use on a daily basis, so it necessary to be able to understand and "talk the talk" that is used in the profession.

A number of terms are associated with reliability, such as Mean Time to Repair (MTTR), Mean Down Time (MDT), and availability. They have not been discussed in this chapter but instead will be discussed in Chapter 13, Maintainability Concepts, as these terms are more pertinent to the area of maintenance engineering for both military and commercial products.

It is also recognized that adding active redundancy or standby redundancy to an existing serial system will greatly improve the reliability of the overall system. However, adding redundancy will often involve additional costs through the addition of more parts, and this has to be weighed against system requirements along with the needs of the customer.

1.5 EXERCISES

1. Calculate the failure rate and MTBF values for each of the following items that have failed during 1,000 hours of testing. Express the failure rate in scientific notation terms.
   - Microprocessor unit: 10 failures
   - Power supply: 2 failures
   - EMI board: 1 failure
2. What are the individual reliability values for each of the components in Problem 1 if we use the exponential reliability distribution for each component as the model? If these items are hooked up as a serial reliability model, what is the system reliability at the 1,000-hour point and the equivalent MTBF?
3. Using the Poisson reliability distribution, what is the combined probability of incurring one, two, or three failures during a 10,000-hour time period for a component that has a failure rate of 10×10^{-6}?
4. What is the mission reliability for an aircraft that has an overall MTBF of 5 hours during a 3-hour mission?
5. Calculate the parts-count or hardware reliability for a power supply circuit board that uses the components listed in the chart below. First, complete the calculation for each group of components (Quantity $\times$ Failure Rate) and then calculate the total hardware failure rate for the circuit board. Next calculate the MTBF value. Finally, determine the reliability value for this circuit board for a 20,000-hour time period.

Logic Circuit Board Assembly

Component type	Quantity	Failure rate (ea.)	Quantity $\times$ Failure Rate
Transformer	1	$.001 \times 10^{-6}$	?
Capacitor (Ceramic)	30	$.00001 \times 10^{-6}$	?
Capacitor (Tantalum)	10	$.0003 \times 10^{-6}$	?
Resistors (Carbon)	30	$.00001 \times 10^{-6}$	?
Diodes	10	$.0002 \times 10^{-6}$	?
Transistors	15	$.0005 \times 10^{-6}$	?
IC (Logic)	20	$.001 \times 10^{-6}$	?
		Total failure rate =	?

6. What is the reliability, the equivalent MTBF value and equivalent failure rate for the system that has two microprocessor circuit boards that are hooked up in active parallel redundancy with failure rate (λ) of $.21 \times 10^{-6}$ for each circuit board? Assume time (t) = 1 hour.
7. What is the system reliability, the equivalent MTBF value, and equivalent failure rate for the system that has a primary microprocessor circuit board and a secondary microprocessor circuit board that is hooked up in standby redundancy, both with a failure rate (λ) of $.21 \times 10^{-6}$ each? Assume perfect switching (zero failure rate). Assume time (t) = 1 hour.
8. Using the same setup and information used in Problem 6, what is the reliability, the equivalent MTBF value, and equivalent rate if the switch is not perfect and has a failure rate of $.0001 \times 10^{-6}$? Again, assume time (t) = 1 hour.
9. Perform the mathematical expansion for the system reliability for three components that are hooked up in parallel that have different failure rates: λ_1, λ_2, and λ_3.
10. What is the system reliability and the equivalent MTBF for a light panel containing ten lamps, with three failures allowed?

CHAPTER 2

Confidence Limits and Their Use in Reliability Analysis

INTRODUCTION

In the first chapter, the reader was introduced to the basic reliability terms that included, among other things, how failure rate and MTBF calculations are performed. However, the reader must understand that any MTBF calculations obtained from test data or field data are really only best-point estimates of the true MTBF unknown parameter. How reliable are such estimates, and what confidence can we have in them? With this question, we enter into the area of confidence limits and their importance in the area of reliability. This area of statistical analysis is commonly called uncertainty analysis.

The objective of this chapter is to present some statistical theory to the reader in order to lay the groundwork for understanding why we need to use confidence limits and how to apply them. While the theory may initially seem intimidating, it is important to realize that the actual methodology is straightforward in the application of confidence limits against observed failure rate data and calculated MTBF values.

2.1 CONFIDENCE LIMITS

The term *confidence* is a mathematical probability relating the mutual positions of the true value of a parameter and its estimate. Indeed, the larger the sample size, the closer the statistical estimate will be to the true value. Only if there is an infinitely large sample size can we truly obtain 100 percent confidence or certainty that a measured statistical parameter coincides with the true value.

Since we are dealing with a MTBF estimate that is obtained from a reasonably sized sample, it makes more sense that we express this estimate in terms of an interval. We express this interval in the framework of an associated probability of confidence rather than to express the MTBF only as a point estimate.

When we calculate confidence limits of 90 percent, this means that in 90 percent of the cases, the probability is that the true value will lie within the calculated limits, whereas in 10 percent of the cases, it will lie outside these limits. The 90 percent confidence that the true value lies within the calculated limits is called the confidence level. The 90 percent confidence interval, it is expressed as:

$$\text{Prob } (A \leq \text{MTBF} \leq B) = 0.9$$

where the value A (expressed in terms of hours) is the lower confidence limit, and the value B (also expressed in terms of hours) is the upper confidence limit.

With confidence levels, lower and upper limits are defined where the true value lies within.

It is helpful to review basic concepts involving the mean, standard deviation, and variance of a population. The mean estimate of a series of values that is obtained from a sample of n components in the population can be expressed as:

$$X' = 1/n \, (X_1 + X_2 + X_3.. + X_n) = 1/n \sum_{i=1}^{n} X_i$$

The standard deviation is defined as such:

$$S = \left[1/(n-1) \sum_{i=1}^{n} (X_i - X')^2 \right]^{1/2}$$

This calculation is available on most statistical hand calculators, usually in the form of the key, [$\sigma \, n - 1$].

Another term that is used in statistics is the variance. This is the square of the value S or:

$$S^2 = \left[1/(n-1) \sum_{i=1}^{n} (X_i - X')^2 \right]$$

The term $n - 1$ is defined as an unbiased estimate of σ^2 where the symbol σ represents the true standard deviation. Unbiased indicates that the values of S^2 from all of the hypothetically possible samples from the same population will average to σ^2. S^2 is the best estimate of σ^2 in the case of a normal population.

EXAMPLE DEMONSTRATING MEAN, STANDARD DEVIATION, AND VARIANCE

Compute the mean, the standard deviation, and the variance (in times to failure) for a product line that has had six failures during a test with failures occurring at the following times: 45, 56, 67, 69, 83, and 88 hours.

The mean is calculated as an average by taking the sum of the times divided by the number of failures:

$$X' = (45 + 56 + 67 + 69 + 83 + 88)/6 = 68$$

The standard deviation is calculated as:

$$S = [1/5 \times ((45 - X')^2 + (56 - X')^2 + (67 - X')^2 + (69 - X')^2 + (83 - X')^2 + (88 - X)^2)]^{1/2}$$

Substituting $X' = 68$, yields:

$$S = [1/5 \times ((-23)^2 + (-12)^2 + (-1)^2 + (1)^2 + (15)^2 + (20)^2)]^{1/2}$$
$$= (1/5 \times 1300)^{1/2} = 16.125$$

This is also obtained by putting the values into a statistical calculator and then pressing the key, $[\sigma n - 1]$.

The variance is the square of S or:

$$S^2 = 260$$

The concepts are reviewed here in order to build a foundation of basic statistics for the reader.

Now we need to use the concept of standard deviation when calculating confidence limits. The standard deviation for a mean life (M) estimate is defined as:

$$\sigma(M) = \sigma/\sqrt{n}$$

The symbol σ, or sigma, represents the true standard deviation of component life, while the term $\sigma(M)$ of the mean is called standard error.

This allows us to assign confidence limits to an estimated mean obtained from a large sample from a standardized normal curve. By using the standardized normal curve, we can calculate a specific number of sigma or standard deviation from the measured estimate M for various confidence limits. The value k in $k\sigma(M)$ represents the number of standard deviations from the measured estimate M' or expressed as:

$$M' \pm k\sigma(M) = M' \pm k\,\sigma\sqrt{n}$$

The number of standard deviations (k) from the mean is associated with a percentage of the area under the normal distribution curve between the respective interval limits. For example, we get:

$$\pm 1\sigma\,(M) = 68.3 \text{ percent}$$
$$\pm 2\sigma\,(M) = 95.4 \text{ percent}$$
$$\pm 3\sigma\,(M) = 99.7 \text{ percent}$$

The graphical representation of this is depicted in Figure 2–1.

If we were to use the case of 95.4 percent confidence limit, we see that the length of the confidence limit is two times $2\sigma\sqrt{n}$ or $4\,\sigma\sqrt{n}$. We can use this to determine the sample size for this confidence interval. If we use the example of a two-sided confidence interval (length 2) and $\sigma = 8$, we get $2 \times 8/\sqrt{n} = 2$ or $n = 64$.

28 *Chapter 2*

A) Generalized curve

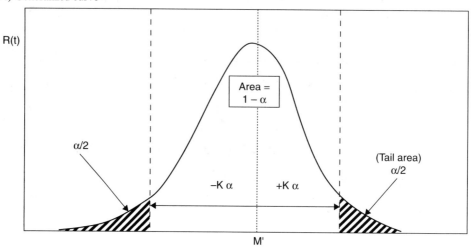

B) 95.4 percent normal curve (k = 2)

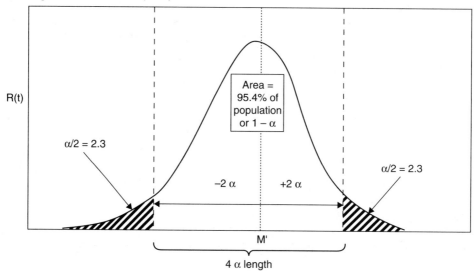

Figure 2-1 Confidence Limits and the Normal Distribution Curve

A two-sided confidence interval at the confidence level of $100(1-\alpha)$ yields two confidence limits, an upper and lower confidence limit which is depicted as:

$$L, U = M' \pm t_{\alpha/2}$$

Thus, for a one-sided lower limit at the same confidence level, we get:

$$L = M' - t_{\alpha}$$

And for a one-sided upper limit, we get:

$$U = M' + t_\alpha.$$

Both the one-sided and two-sided confidence limits are used in many areas of reliability analysis such as risk assessment (including both consumer's risk and producer's risk) and in the development of reliability test plans. Please refer to Figure 2–2 for a graphical representation of the different types of confidence limits. An example of the use of confidence limits in population analysis is provided in the example following Figure 2–2.

A) Two-sided confidence limit

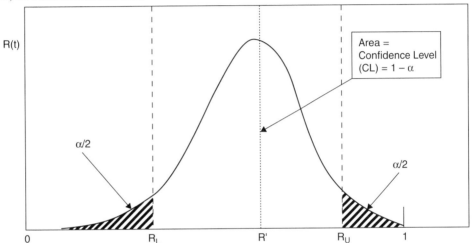

B) One-sided confidence limit (upper)

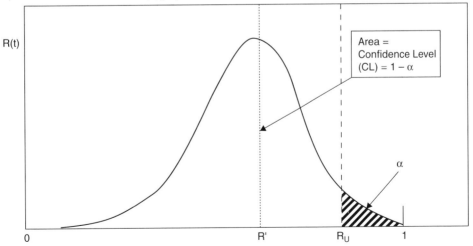

Figure 2-2 Confidence Limits Examples

C) One-sided confidence limit (lower)

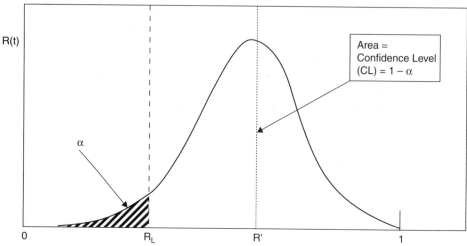

Figure 2-2 Confidence Limits Examples continued

GENERAL EXAMPLE OF CONFIDENCE LIMITS

Item: Aircraft Cockpit Human Factors Design

For many years, the U.S. Air Force had the approach that aircraft cockpit design be ergonomically compatible with regards to human factor requirements. This is accomplished by meeting the needs of the most crew members by ensuring that the size parameters are met at the upper and lower limits of the human population. This is typically accomplished by conducting various measurements of reach and height of both men and women crew members at the 95 percent and 5 percent levels of the population. The measurement parameter that is used for these levels is the height of men and women.

When working for the Fairchild Republic Company, an aircraft manufacturer, the author happened to represent the 95 percentile for men with a height of 6 feet 3 inches, while another member in his reliability group represented the 5 percentile for men with a height of 5 feet 5 inches (average height for men is near 5 feet 11 inches). Thus, when the cockpit design for a new aircraft program was being developed in the company, both the author and the other engineer would be fitted in the cockpit and examined for various size parameters. This would include head clearance, vision through the windshield and canopy, as well as sufficient reach for the controls, in addition to seating comfort. By ensuring that all human factors design requirements were met for these two extremes, it could be reasonably expected that 90 percent of the population would be able to properly fit in the cockpit, as well as being able to use the controls properly. See Figure 2–3 for a graphical representation of this.

Human factors engineering is a discipline that uses confidence limits as a basis for things like cockpit design, automobile console design, and numerous other products that have a human interface. The latter may involve things such as ATMs, automated gas pumps, or other similar devices.

Confidence Limits and Their Use in Reliability Analysis

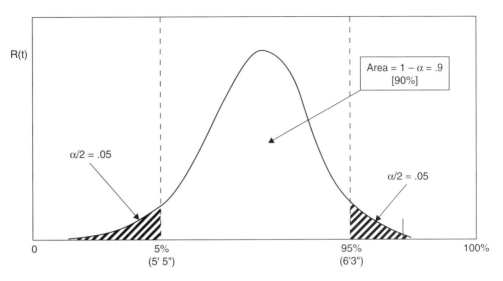

Figure 2-3 Aircraft Cockpit Human Factor Design Example

This generalized example shows a practical situation in which there is a need for the designers of products involving human interface to know the upper and lower confidence limits for the population that the product is being designed for.

2.2 USING THE CHI-SQUARE DISTRIBUTION TO CALCULATE CONFIDENCE LIMITS

The calculation of confidence limits becomes more straightforward in the case of a single parameter exponential distribution, as we can use the chi-square distribution to calculate these limits. This is particularly good when dealing with situations where the parameters M and s can only be obtained from a small sample size, and thus the estimate of $s = \sigma$ cannot be made. Thus, upper-confidence limits have to be used, and the calculation for the confidence limits on standard deviation is based on the chi-square distribution. The variance ratio is:

$$(n - 1) s^2/\sigma^2 = \chi^2,$$

where this chi-square distribution has $(n - 1)$ degrees of freedom.

We now discuss the concept of degrees of freedom that are associated with the area of confidence limits. For example, the sum of n squares of deviations from the sample mean X' is expressed as:

$$\sum_{i=1}^{n} (X_i - X') = (X_1 - X')^2 + (X_2 - X')^2 \ldots\ldots + (X_n - X')^2.$$

This sum is said to have $n - 1$ degrees of freedom. This is because when X' is fixed only $n - 1$ of the X_i values can be chosen independently and the nth value is then determined. The chi-square distribution is one of the many statistics that depend on the number of degrees of freedom.

We obtain the chi-square multiplier factor from the chi-square table of values [listed in Appendix A of this book] by looking up $(n - 1)$ degrees of freedom against a one-sided confidence level of $(1 - \alpha)$. By going through this process, we get a $100(1 - \alpha)$ percent assurance that the true standard deviation of component life does not exceed the upper one-sided confidence limit.

It is often recognized that the sample size of field data selected on a component may have only a handful of n failures against a number of operating hours. Thus with a small size, the chi-square distribution has to be applied. Basically, we are conducting a statistical-based test of sorts, in which we are working with calculations based on the sample size.

There are two variations to the calculation and the application of the chi-square confidence limit. The first case is where the test is stopped at the point when the r failure has occurred. In this case, the number used for the degrees of freedom is $2r$, with this particular test being called a failure-terminated test. Basically, we stop the test (or measurement period) at the point of the rth failure occurring. The MTBF estimate for a confidence level of $(1 - \alpha) \times 100$ percent is:

$$M = 2T/\chi^2_{2r}(\alpha) \; [\text{failure terminated calculation}]$$

In the second case where the test time is at some preselected or predetermined value where a failure did not occur at that point, the number used for the degrees of freedom is $2r + 2$. This is called a time-terminated test and is expressed as:

$$M = 2T/\chi^2_{2r+2}(\alpha) \; [\text{time terminated calculation}]$$

It is seen that the following equation holds true:

$$2T/\chi^2_{2r+2}(\alpha) \leq M \leq 2T/\chi^2_{2r}(\alpha)$$

NOTE: It can be seen that it is still possible to calculate a lower one-sided confidence limit if we have the condition of no failures $(r = 0)$ occurring during the test time period. In this case, we would calculate $2r + 2 = 2(0) + 2 = 2$ degrees of freedom, or:

$$2T/\chi^2{}_2(\alpha)$$

CHI-SQUARE ONE-SIDED CONFIDENCE LIMIT EXAMPLE

Calculate the one-sided 90 percent confidence limit on the MTBF estimate for a piece of medical equipment that monitors a patient's vital signs. Two units of this equipment have accrued a total of five failures during a total of 15,000 hours of field service.

The chi-square value is based on $2r$ degrees of freedom or $2 \times 5 = 10$. The value $\alpha = .10$, corresponding to the 90 percent confidence interval.

Looking up the chi-square table of values in Appendix A for $r = 10$ and $\alpha = .10$, we get a chi-square value of 16.0. We get the following:

$$2T/\chi^2{}_{2r}(\alpha) = 2 \times 15000 // \chi^2{}_{10}(.10) = 30000/16 = 1875 \text{ hours}$$

By comparison, our calculated MTBF estimate is 15000/5 = 3000 hours. When we graph this particular example as shown in Figure 2–4, we see that the chi-square value of 16.0 corresponds to the area of the graph to the right of the 90 percent mark. It is important to realize that each of the values in the chi-square table corresponds to the area of the curve to the right of that value.

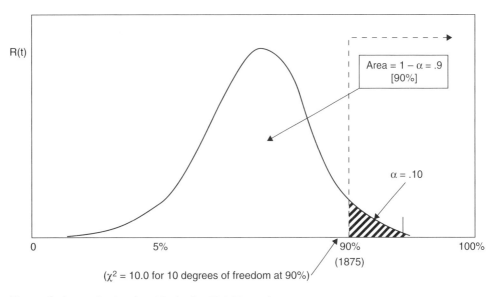

Figure 2–4 Medical Patient Monitoring Unit Example

CHI-SQUARE TWO-SIDED CONFIDENCE LIMIT EXAMPLE

Calculate the two-sided 90 percent confidence limits on the MTBF estimate for a 2-way radio that is in production and has accrued a total of 150,000 hours of field service with a total of 20 failures accrued.

The chi-square value is based on $2 \times 20 = 40$ degrees of freedom. A 90 percent two-sided confidence limit falls in the interval between 5 and 95 percent. Thus $\alpha = .10$ and $\alpha/2 = .05$. After looking up the chi-square table of values for $r = 40$ for percentages at .05 and .95, we get the chi-square values of 55.8 and 26.5, respectively, and then insert them into the confidence limit calculations for MTBF values at these points:

$$2T/\chi^2_{40}(.05) = 2 \times 150000/55.8 = 5376 \text{ hours}$$
$$2T/\chi^2_{40}(.95) = 2 \times 150000/26.5 = 11321 \text{ hours}$$

We see that $P(5{,}376 \leq \text{MTBF} \leq 11{,}321) = 0.9$. The graph is provided in Figure 2–5. Note that the chi-square value for 5 percent (55.8) represents the area under the curve to the right of this point and that the chi-square value for 95 percent (26.5) represents the area under the curve to the right of this point.

By the way, a number of short cuts are used by reliability engineers. For example, if one needs to calculate a 90% one-sided confidence limit with zero failures occurring

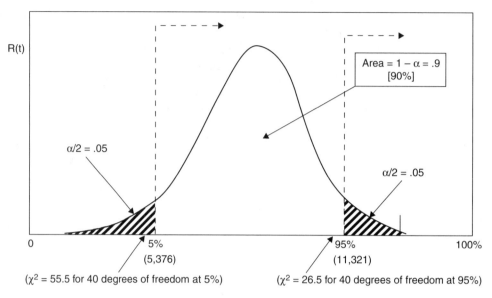

Figure 2-5 Portable 2-Way Radio Example

during time period T, we can use the expression $T/2.3$ to obtain this value. (The complete expression is actually $2T/4.6$). Another shortcut can be used when a table is not available and a rough estimate is needed. This occurs when the 60% one-sided confidence limit is applied; simply divide the time T by 1. (In actuality, the complete expression is $2T/1.83$)

CHI-SQUARE CONFIDENCE LIMIT EXAMPLE—MTBF REQUIREMENT

Determine the amount of flying time in hours that is required to satisfy an MTBF requirement of an aircraft engine such that the MTBF exceeds 300 hours at the lower one-sided 75 percent confidence interval. A further set of clarifications are added in that there are two engines on the aircraft that are installed and, in addition, the engine is operated at a rate of 4 operating hours for every 3 flight hours that are recorded.

The equation is set up as:

$$\theta_L = 300 < (2 \times 4/3 \times T_{min})/\chi^2_{2r+2}(.75).$$ [time-terminated calculation]

The number 2 is used as it implies operation with each engine and 4/3 converts flying hours to operating hours. r is equal to the number of failures and T_{min} is equal to the minimum number of flying hours needed.

Further simplifying the equation we get:

$$T = 112.5 \times \chi^2_{2r+2}(.75).$$

We then set up a table that is based on the number of failures and shows the minimum flight hours required:

K	$\chi^2_{2k+2}(.75)$	T_{min} (in hours)
0	2.77	312
1	4.11	462
2	7.84	882
3	10.2	1150
4	12.5	1412
5	14.8	1671

If we were to word this problem in a different way, in which we wanted to just meet the MTBF value and not exceed it, in other words, where we terminate the test at the point of the rth failure, we get the following equation:

$$T = 112.5 \times \chi^2_{2r}(.75) \quad \text{[failure-terminated calculation]}$$

This will result in the following table which shows the calculated flight hours when the test is terminated when the k_{th} failure occurs:

k	$\chi^2_{2k}(.75)$	T_{min} (in hours)
1	2.77	312
2	4.11	462
3	7.84	882
4	10.2	1150
5	12.5	1412

NOTE: The set of calculations above (time-terminated vs. failure-terminated) is of interest with regard to one of the important military standards used in reliability, MIL-STD-781 (Reliability Test Methods and Environments for Engineering Development, Qualification, and Production). The chi-square table (that is listed in Appendix A of this book) is used to develop a set of multiplier values that are listed in the tables of Section 4.6 of this military standard for various confidence intervals for the two sided-confidence limit for both time-terminated and failure-terminated calculations.

CHI-SQUARE CONFIDENCE LIMIT EXAMPLE—DETERMINATION OF TEST TIME

We can use the chi-square to set a table of values that are based on the number of failures and specified confidence limit values to find the associated minimum number of hours needed to meet that confidence limit value. The following example demonstrates this.

Determine the test time required for demonstrating a minimum MTBF for 60 percent and 80 percent lower confidence levels.

Let T = total test time accumulated in hours. The following equations are then used:
For 0 failures, the MTBF = $2T/\chi^2_2(1-\alpha)$
For k failures (where $k > 0$), the MTBF = $2T/\chi^2_{2r+2}(1-\alpha)$

$$(1 - \alpha) = \text{either .6 or .8}$$

Referring to the chi-square tables with the selection of the appropriate degrees of freedom, we can set up the following tables of values as such:

A. 60 percent lower confidence limit for 0, 1, and 2 failures:

MTBF value to demonstrate	0	1	2
10 hours	9.2	40.4	124.8
20 hours	18.3	80.9	249.6
30 hours	24.5	121.3	374.4

B. 80 percent lower confidence limit for 0, 1, and 2 failures:

MTBF value to demonstrate	0	1	2
10 hours	16.1	60.0	171.6
20 hours	32.2	120.0	343.2
30 hours	48.3	240.0	514.8

We can also solve this type of problem in reverse. Let $T = 400$ hours and $k = 2$ failures; then compute both the 60 percent and 80 percent confidence limits as follows:

Lower 60% confidence limit = $2 \times 400/\chi^2_6(.6) = 800/6.24 = 128.2$ hours,
Lower 80% confidence limit = $2 \times 400/\chi^2_6(.8) = 800/8.58 = 93.2$ hours.

2.3 ONE-SIDED BINOMIAL CONFIDENCE LIMITS

On occasions, the reliability engineer may have to determine the sample size for certain types of tests that are based on the binomial distribution. Testing may be performed on products such as one-shot devices like missiles or explosive devices where the product either works or does not work (there is no in-between or degraded state). How many firings are needed in order to be assured that the product is operating within a certain confidence level?

In the first chapter, the binomial distribution was discussed briefly as it was used in the k out of n reliability model (Section 1.3.4). Recall that:

$$(R + Q)^n = 1$$
$$(R + Q)^n = R^n + n R^{n-1} Q + \ldots + Q^n = 1$$

This can be simplified as:

$$\sum_{j=0}^{n} \binom{n}{j} R^{n-j}(Q)^j =$$

$$\sum_{j=0}^{n} \binom{n}{j} R^{n-j}(1 - R)^j$$

For the one-sided confidence limit, we can substitute $R = (1 - C)$ where C = upper confidence level for the probability, p. We also set the upper limit in the summation to be f (which represents the total number of failures that we will allow). Thus we now get the basic equation for the one-sided binomial confidence limits as:

$$\sum_{j=0}^{f} \begin{bmatrix} n \\ j \end{bmatrix} (1 - C)^{n-j} C^j = 1 - \delta = p$$

Thus if we do not allow any failures ($f = 0$) for the test, we simply get:

$$\begin{bmatrix} n \\ 0 \end{bmatrix} (1 - C)^n C^0 = p$$

or just $(1 - C)^n = p$

ONE-SIDED BINOMIAL CONFIDENCE LIMIT EXAMPLE

An aircraft manufacturer is conducting tests on an ejection seat design supplied by a vendor. The company wants to know how many test trials are required to be conducted in order to prove a test plan of zero failures with a probability of success of 75 percent at the 90 percent one-sided confidence limit.

Using the basic equation for the one-sided binomial confidence limit for zero failures, the result is:

$$(1 - C)^n = p$$

Substituting $C = .25$ and $p = .9$, we get:

$(.75)^n = .10$
$n \ln (.75) = \ln .10$
$n = 8.004$ or 8 trials.

We can turn the problem around by asking what is the probability of sucess of 12 trials ($n = 12$) and no failure at 90 percent one-sided confidence limit?

$(X)^{12} = .10$
$X = .825$ or a probability of 82.5 percent.

ONE-SIDED BINOMIAL CONFIDENCE LIMIT EXAMPLE

How many trials must be conducted on a population of a one-shot explosive device in order to obtain a probability of working of .9999 at the 90 percent one-sided binomial confidence limit?

This is calculated by first determining the value for C as follows:

$P = 1 - .9999 = .0001$ = probability of failure
$C = .0001$ and $p = .90$

For 0 failures, the equation is

$$(1 - C)^n = 1 - .90 = .10$$
$$(.9999)^n = .10,$$
$$n = 23025.$$

For 1 failure, the equation is:

$$(1 - C)^n + nC(1 - C)^{n-1} = .10$$
$$(.9999)^n + n(.0001)(.9999)^{n-1} = .1$$
$$(.9999)^{n-1}[n(.0001) + .9999)] = .1.$$

Through trial and error using an iterative computer program, $n = 38,900$.

One can then construct a test plan at probability of .9999 at 90 percent confidence by setting up a table using the calculated results as follows:

Failures allowed	Sample size needed
0	23,025
1	38,900

2.4 SUMMARY

Here is an important date involving the area of statistical confidence: November 7th, 2000.

On that date, the presidential election between Al Gore and George W. Bush took place—and it kept on going. The television news network attempted to be on the cutting edge of being able to predict the results of each state based on early returns. In order to accomplish this, the news organization used a service group that conducted exit polls of voters, and from this data, statistical methods were employed to predict the results. The data depended on various methods of sampling, such as taking representative areas in a state, even when only using a small sampling size that was typically on the order of one out of every ten voters.

By 8:00 P.M. E.S.T., all of the major networks projected Al Gore as having won the presidential race in Florida and its 25 electoral votes based on preliminary returns. However, by 10:00 P.M. E.S.T., the networks had to embarrassingly withdraw this conclusion as subsequent returns showed a much tighter race. It would turn out that the voting difference for the entire state was less than 500 votes between two candidates who amassed well over 2.9 million votes each. How can an exit poll sample of 10,000 accurately predict the results of an election where the difference between the candidates is significantly less than the sample size? The simple answer is that it cannot. What went wrong with the network's early projections?

First it became apparent that the voters who filed out of the exit polls were not fully representative of all of the people who voted. Also, wrong voting precincts were chosen for the sample, in terms of stable prediction results. Some of the precincts had demographic changes from previous elections in recent years with an influx of new people moving in from the northern United States. Good prediction methods require a stable population for proper statistical applications.

A further issue resulted with the actual election results in Florida, where there were some problems with punch card ballots being properly punched out. It would appear the punch card method was less reliable than the voter's actual intentions. During the lawsuit contesting the election results, a number of Ivy League school mathematicians were brought in to discuss the various possibilities according to statistical methods. When Judge Sander Sauls ruled on the lawsuit, he stated, "In this case, there is no credible statistical evidence and no other competent substantial evidence to establish by a preponderance a reasonable probability that the results of the statewide election in the state of Florida would be different from the result which has been certified by the state elections canvassing commission."

A very valuable lesson can be learned here for the reader regarding the way raw data is interpreted and the statistical methods that are applied to the raw data in order to make predictions or projections. In general, the larger the sample size, the more confidence there is in the data. Thus, in the area of confidence in failure rate data, this would translate into additional amounts of operating hours or testing cycles accrued against a component. By applying the statistical methods shown in this chapter, we can come up with confidence limits that will make a calculated failure rate or MTBF value usable with a known risk factor. The same rule of caution has to be followed with failure rate and MTBF calculations in order to recognize them as estimates in a point of time with the data that is available at hand.

2.5 EXERCISES

1. Compute the mean, the standard deviation, and the variance for several test units of a device that has had nine failures during testing with failures at 57, 68, 79, 91, 95, 102, 117, 122, and 131 hours.
2. If the 80 percent two-sided confidence limit is being used, what are the values for the upper and lower percentages? What is the value for α?
3. A new aircraft design that is undergoing flight testing accrues 8,127 flying hours with 27 failures against it. Calculate the MTBF value, and then calculate the lower one-sided 90 percent confidence limit on the calculated MTBF value.
4. A total of 25 items are operating in a test for 1,000 hours with the stipulation that failed units are replaced by standby items instantly. A total of 8 failures are observed. Calculate the MTBF value, and then calculate the two-sided 90 percent confidence limits on the calculated MTBF value.
5. 25 items are operating at 1,000 hours each with no failures observed. Calculate the lower one-sided 75 percent confidence limit on the MTBF.
6. What is the 90 percent one-sided confidence limit of the MTBF value of an electrical unit that has failed once during 7,500 hours of field service?
7. Construct a 90 percent lower confidence level table for 0 and 1 failures for the following MTBF values: 100 hours, 200 hours, and 300 hours.
8. Calculate how many trials must be conducted without failures on a population of a one-shot explosive device in order to obtain a probability of .99 at 95 percent confidence.

CHAPTER 3

Reliability Program Tasks

INTRODUCTION

During the early design stages of a new product, several reliability tasks will need to be performed as part of the design effort. Many of these reliability program tasks have been required by the customer and are to be submitted as a formal data item to the customer by the manufacturer of the product. In addition, reliability is an important part of the design

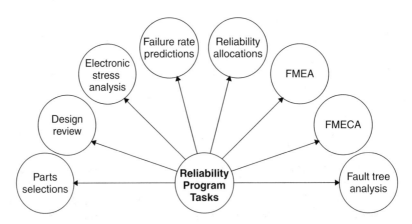

Figure 3-1 Typical Reliability Program Tasks

overview presentation that a company may give to the customer during the Preliminary Design Review (PDR) and Critical Design Review (CDR). Even when no formal data item or presentation to the customer is required, many of these tasks are performed routinely by the manufacturer of the product to provide a level of confidence in the product's ability to perform reliably in the field.

In general, these tasks can be broken into two basic categories: component-level analysis and system-level analysis. Component-level reliability tasks typically include stress analysis and parts failure rate predictions. By performing stress analysis and part failure rate predictions, several of the system-level analysis can be performed. These include failure modes and effects analysis, system level failure rate allocations and fault tree analysis. Figure 3–1 shows the typical reliability program tasks for a product.

In this chapter, basic reliability program tasks will be illustrated by using actual examples of two different systems, one electrical and one mechanical. One is an electronic power supply used in an electronic unit located in the cargo bay area of a commercial aircraft (Figure 3–2), and the other is a mechanical-based emergency oxygen system that is located in the seat of the pilot of a military Navy aircraft (Figure 3–3). The reader will gain more understanding by following how the tasks are performed on actual examples as opposed to providing a series of blank forms and descriptions.

3.1 ELECTRONIC PARTS STRESS ANALYSIS

With electrical components, the first task at hand is to determine the electrical stresses to which each part in the circuit will be subjected. Electronic stress analysis or parts derating is defined as the practice of limiting electrical, thermal, and mechanical stresses on electrical devices to levels below their specified or proven capabilities in order to enhance reliability. The purpose is to have a conservative design approach with realistic derating of parts. Derating increases the margin between the operating stress level and the maximum stress level, thus protecting the system from unexpected design anomalies. The ratio of operating stress over the maximum rated stress level is called the stress ratio.

A number of part derating guidelines were published during the peak of the aerospace era, mainly by Rome Air Development Center. One of these is AFSCP 800-27, which has guidelines for each category of electrical components. For example, in the Derating Guidelines for Linear Microcircuits in this specification, we have the following criteria for stress ratio (Ground Environment is Level III, Flight is Level II and Space Systems is Level I, which is the most conservative):

	Level I	Level II	Level III
Supply Voltage	70	80	80
Input Voltage	60	70	70
Output Current	70	75	80
Max. Junction Temp.	80	95	105

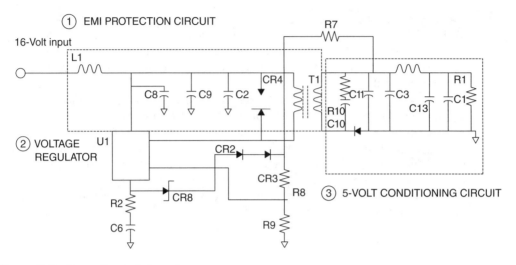

Figure 3-2 Power Supply Schematic

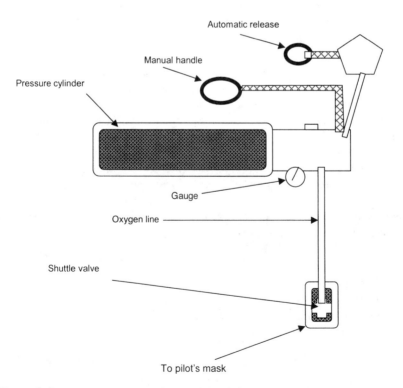

Figure 3-3 Emergency Oxygen System Schematic

The following are general Level II requirements for some of the common electrical components from this specification:

Component	Parameter	Stress
Resistors	Power dissipation	50
Capacitors	DC voltage	60
Diodes	Reverse voltage	70
Transistors	Power dissipation	70

Reliability stress analysis is primarily a task that is performed on electronic components. There are no standard methods for performing reliability stress analysis on mechanical components, although the mechanical engineer may routinely perform mechanical stress analysis on components that have to see certain levels of vibrations during field service. We will be using the example of a power supply for demonstrating the reliability task of electrical stress analysis. Figure 3–2 shows the schematic representation of this power supply circuit for which we will perform stress analysis on some of the components.

Any reliability engineer who performs electrical stress analysis in circuit analysis must have a basic knowledge of Ohm's law as well as the formula for power. This knowledge is needed in order to calculate the voltage and current for components such as resistors, transistors, and integrated components in a circuit in order to obtain the individual stress level for each of these components.

A. Ohm's law

$V = I/R$ (where V is the voltage, I is the current, and R is the resistance)

B. Power dissipation

$$P = I^2 \times R \text{ or } P = V \times I$$

For the first example, we will calculate the stress on the filter capacitor C8 in the power supply circuit. For this example, we will use a voltage rating of 50 volts. The applied voltage is 16 volts. Thus, the applied electrical stress is 16/50 = 32 percent stress, which would be in the guidelines for capacitors as stated above.

For the next example, we will calculate the power dissipation and then the stress levels for each of the .25-watt resistors (.125 watt rating) R7, R8, and R9. Five volts are applied at the beginning of R7 and following this path through R8 and R9, we reach ground. This path is the worst-case situation with 5 volts being dissipated over a total of (3,740 + 301 + 1,240) = 5,281 ohms.

Using Ohm's law of $V = I/R$, we calculate a current of 5 volts/5,281 ohms = .0009 amps. By using this current, we can calculate the individual power dissipation per resistor with the formula $P = I^2 \times R$ or:

$R7 = (.0009)2 \times 3740 = .003$ watts
$R8 = (.0009)2 \times 301 = .0002$ watts
$R9 = (.0009)2 \times 1240 = .001$ watts

It can be seen that the worst-case stress level is for R7 which is = .003/.125 or 2.4 percent, which is certainly well below the recommended 50 percent stress-level guideline. In the case of this circuit, the resistors are grouped together with these specific values in order to set the voltage input to the foldback circuit or pin 3 of the pulse width modulator (PWM) IC. The voltage drop across R7 and R8 is 3.9 volts, leaving 1.1 volt as the input to pin 3 of the IC. Figure 3–4 shows this subset of the power supply in Figure 3–2, along with the associated calculations.

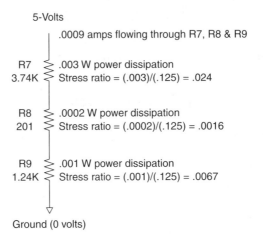

Figure 3-4 Subset of Power Supply Circuit (Figure 3–2)

We have analyzed a few components in this power supply example to give the reader a feel for what type of calculations are involved with electrical stress analysis. Some components require more analysis and more knowledge of electrical circuits in order to perform the parts derating and stress analysis. Components such as transistors and ICs are stressed typically in terms of power dissipation because there are concerns involving thermal rise that have to be considered.

A quick look at the remaining components of the power supply circuit under review shows that all of the resistors and capacitors are reasonably stressed in accordance with the recommended levels. The protection diodes (CR1 and CR2) in this circuit are rated for over 200 volts, so the 16 volts applied in the reverse direction is minimal stress. Other components such as U1 require a little more analysis, but the analysis (not done here) shows that it is not overstressed in this particular circuit.

Understanding electrical schematics is an important skill for a reliability engineer to acquire in order to perform electrical stress analysis, and the engineer will end up developing a basic knowledge as to how an electrical circuit works. The reliability engineer will observe that certain components will fall into the same functional roles between designs such as isolation resistors, bypass capacitors, and blocking diodes.

Stress analysis should be done as soon as possible, usually after preliminary schematics and parts lists have been generated. A preliminary stress analysis has a major bearing on parts selections. Imagine the situation of trying to deal with the issue of overstressed components on a program that is two years into production because no preliminary stress analysis was performed! This has actually happened often because of the lack of a cohesive reliability plan. A true case of this resulted when tubular capacitors in an electrical circuit were overstressed, but there was no room on the densely populated circuit board to upgrade to a higher-rated larger-size capacitor. This problem could have been prevented by this task being done at the beginning of the program so that all size and stress issues are identified right from the start.

The completion of the stress analysis task is the prelude for the next reliability program task of failure rate predictions.

3.2 RELIABILITY PREDICTION TECHNIQUES

With the advent of reliability in the aerospace field, a number of processes were developed by government agencies to develop failure-rate prediction methods. One of the main methods that is used for most electronic parts will be described here. This method was the one used the most during the peak of the aerospace era, as detailed in the MIL-HDBK-217 handbook, *Reliability Predictions for Electronic Parts*. This methodology is a cookbook approach in which different factors are entered into the model to come up with a final failure rate number, and it is described in Section 3.2.1.

We will also discuss the various ways for generating predicted failure rates for non-electronic parts in Section 3.2.2.

3.2.1 Failure Rate Prediction Methods for Electronic Parts

The reliability of each piece of electronic equipment is a product of the individual reliability of each part that makes up the unit. The failure rate of each part adds up to the hardware failure rate of the unit.

Several things may affect a parts reliability performance. These factors include:

- Manufacturing and material quality
- Environment (airborne, ground)
- Operating temperature
- Vibration and miscellaneous mechanical factors
- Electrical stress levels

A number of methods have been developed over the years that attempted to predict the reliability performance of parts based on these factors, with the MIL-HDBK-217 handbook methodology being the most commonly used. The MIL-HDBK-217 handbook was developed by the U.S. Air Force using the past failure-rate data experience of electronic components. The handbook uses a cookbook approach that takes into account the type of device, component quality level (as rated by the manufacturer), and the environment to which the component will be subjected. The latter factor can range from ground environment to aircraft environment to the higher extreme of the space environment. Each of these vehicle environments have a subset: inhabited (such as the controlled temperature environment of an aircraft cockpit) or uninhabited (such as the cargo bay of an aircraft with no temperature control), with the latter being more severe. Each of these factors are developed through the use of tables or formulas presented in the handbook. For most electronic devices, the factors are then multiplied together; other devices, such as ICs, are plugged into a formula with the end result being the predicted failure rate for the component.

FAILURE RATE CALCULATION EXAMPLE

The following is a simple example of how the MIL-HDBK-217 cookbook approach works, in principle, using the example of one of the noise-filtering ceramic capacitors (C1) used in the power supply example from Figure 3–2.

Using the information from Section 10 of MIL-HDBK-217, a predicted failure rate for a CKR-type ceramic capacitor is determined by the formula:

$$\lambda_p = \lambda_b \pi_{cv} \pi_Q \pi_E$$

The base failure rate (λ_b) is determined either by using a complex formula or using a simple table that charts ambient temperature against actual stress. The stress is calculated by the ratio of operating voltage over the rated voltage.

For example, a CKR ceramic capacitor with a value of .1 uf (or 100,000 pf) is used in an electronic unit installed in the cargo bay (uninhabited area) of a transport aircraft that sees an ambient temperature of 20° C. It is under 50 percent stress for voltage, and it is rated at 150° C at a quality level of P. Using MIL-HDBK-217, we get values for the following factors:

$\lambda_b = .0038$	[base failure rate]
$\pi_{cv} = 1.45$	[voltage factor]
$\pi_Q = .30$	[quality factor]
$\pi_E = 8$	[environment factor]

$\lambda_p = .0038 \times 1.45 \times .3 \times 8 = .013 \times 10^{-6}$ failures/hour

However, Appendix A of MIL-HDBK-217 includes a parts count method that uses only two factors (base failure rate λ_b and environment π_E) that yields a failure rate of .009, which is close enough for preliminary predictions and often suitable for final predictions for certain applications.

Several software companies have developed reliability prediction programs that use MIL-HDBK-217 and similar approaches. With these programs, an analyst will enter the various parameters for each component into a menu format provided by the software package. Some of these programs can be quite labor-intensive, requiring the user to enter the individual parameters of the hundreds of electrical parts of a circuit board into a file set up by the software package. The end result usually ends up being a very small failure rate for a group of components on a circuit board. Does the end result really reflect the expected reliability performance of an electronic unit?

In principle, the cookbook approach used by MIL-HDBK-217 and similar-type models would be an ideal way to calculate predicted reliability failure rates. Unfortunately, this approach would only be valid if all components behaved as an empirical model and if all parts were made the same way and with the same quality by all vendors and so on.

But many elements conspire to doom the cookbook approach such that the numbers calculated have no bearing on how the product will perform in the real world. Specifically, the cookbook models have the following flaws:

- It is impossible to keep the failure rates current in keeping with continuously improving processes by part manufacturers.
- Some vendors do a better job with part quality than others, but this difference is not reflected in the base failure rate. It can be reflected somewhat in the quality factor that is chosen, but some experience with the vendor is required in order to make a sound judgment.
- The model does not take into account mechanical factors, such as how the part is mounted or supported. (Mechanical factors will be covered in detail in Chapter 5.)
- No element of software reliability is tracked.
- Aspects of certain manufacturing processes are not taken into account. This is particularly true in the manufacture of multilayered circuit boards and substrate components.

Often, the parts count method that is presented in the appendix of the handbook does suit the needs of most vendors, even in a final design. In fact, there is not a tremendous amount of difference in failure rate (in terms of powers of 10) between the parts count method and the precise cookbook method. Thus, a tremendous increase in man hours can be expended in performing the cookbook method over the parts count method, just to get a small increase in precision. This precision may be meaningless given the factors listed above.

Also, it is unfortunate that predicted numbers can sometimes take a life of their own, in that non-technical managers will quote these predicted numbers as if they were engraved in stone. It is always important to emphasize that these numbers are merely *estimates* and represent a snapshot only!

With the last notice on MIL-HDBK-217, the process of performing failure rate predictions using the handbook has been dropped as a formal requirement on government programs and used as a guide only.

3.2.2 Non-Electronic Parts Failure Rate Prediction Methods

For the example of the emergency oxygen system, no electronic parts are used in the design, hence no failure rate prediction method using MIL-HDBK-217 can be performed. There are some sources for mechanical part failure rates. In the past, various sources provided by the Air Force (66-1 data) and the Navy (3M data) were sufficient sources for using field experiences of similar items. In more recent years, Rome Air Development Center (RADC) has put out a series of books known as *Non-electronic Parts Reliability Data (NPRD),* which contain actual field histories of various mechanical parts such as valves and meters against various operating environments. This is somewhat helpful for predicting the failure rates of some parts, but it is not meant to be a complete compilation of all types of mechanical parts.

In some cases, vendor data may be available for some parts. This data is typically based on historical field experience that the vendor may have accrued on similar components. This same type of philosophy can be done on a larger level, such as assemblies made by a company, and the failure history of these assemblies used on similar products made by the company is often relevant data. For example, if a company made a valve that accrued 10,000 hours of flying time and one unit was returned for failure, the MTBF could be calculated as 10,000 hours and then a 90 percent two-sided confidence limit using the chi-square distribution could be applied to yield the lower limit as:

$$2T/\chi^2 = 2 \times 10,000/9.49 = 2,107 \text{ hours}$$

This would be a safer number to use when there is not enough confidence in the raw MTBF value. How the data is collected and how failure returns are reported or categorized may weaken confidence in the raw number.

The generation of failure rate predictions not only provides an estimate for MTBF values, but it feeds other tasks such as the apportionment analysis, failure modes, effects and criticality analysis (FMECA), and fault tree analysis (FTA). Thus, this initial task of determining equipment failure rate is the building block for several other reliability program tasks.

3.3 RELIABILITY APPORTIONMENT

Reliability apportionment or failure rate allocations is necessary to perform when there is a difference between the predicted or achieved reliability value within the specified field value. Typically, an apportionment may be performed on group of components or subsystems that make up a system that has an overall system reliability requirement. Other cases include allocation of an MTBF value for a piece of equipment at the line replaceable unit (LRU) level down to the individual modules or spares replaceable units (SRU) that make up the equipment.

The process for this task is to calculate a multiplying factor between the two values involved and then use this factor throughout the lower-level predicted values. This multiplying factor is developed by the following relationship:

$$\lambda_{specified} = k\lambda_{predicted}$$

The same factor, k, can be applied to the individual failure rates that comprise the higher-level item, whether it be a piece of equipment or a system. Thus, for the individual component (or subsystem), failure rate predictions make up the end product, and we have the following relationship:

$$\lambda_i = k\lambda_i'$$

Where λ_i' represents the predicted (or achieved field) reliability of each module that comprises the piece of equipment or system.

RELIABILITY APPORTIONMENT EXAMPLE

For this example, we will examine a VHF radio that is used in a military aircraft and has a specified MTBF value of 2,000 hours. The predicted MTBF is calculated to be 8,000 hours, and the overall prediction is based on the individual predicted failure rates of five individual subsystems that make up the end item of the radio as follows:

Module	Predicted Failure Rate
1. Power Supply	30×10^{-6} failures/hour
2. RF Board	40×10^{-6} failures/hour
3. PLL Board	20×10^{-6} failures/hour
4. EMI Board	10×10^{-6} failures/hour
5. Display Unit	25×10^{-6} failures/hour
Radio (total)	125×10^{-6} failures/hour

We then perform a reliability allocation for each of the five subsystems against the specified MTBF value of 2,000 hours.

The factor k is calculated either by taking the ratio between the two MTBF values or the two failure rate values for the radio, $k = 8,000/2,000 = 4$. The allocated failure rates for each module has to be a factor of four times greater than the predicted failure rate, and this factor is applied against each of the individual modules to get:

Module	Predicted Failure Rate	Allocated Failure Rate
1. Power Supply	30×10^{-6} failures/hour	120×10^{-6} failures/hour
2. RF Board	40×10^{-6} failures/hour	160×10^{-6} failures/hour
3. PLL Board	20×10^{-6} failures/hour	80×10^{-6} failures/hour
4. EMI Board	10×10^{-6} failures/hour	40×10^{-6} failures/hour
5. Display Unit	25×10^{-6} failures/hour	100×10^{-6} failures/hour
Radio (total)	125×10^{-6} failures/hour	500×10^{-6} failures/hour

3.4 FAILURE MODES AND EFFECTS ANALYSIS (FMEA)

The failure modes and effects analysis (FMEA) is an analytical tool that is extremely useful to a company or organization during the early design stages of a product. The purpose of the FMEA is to identify the potential failure modes of a product and the possible effects. By performing the process early enough during the design process, there is a possibility of corrective action being implemented by design engineering that would alleviate the effects of the failure mode. In addition, when design engineering is alerted to the effects of failure modes, they may be able to change the impact of the effect of the failure mode by improving detection or by adding redundant circuits to the design. The FMEA is perhaps the most important reliability task that can be performed, other than reliability testing.

The failure modes and effects analysis (FMEA) is set up in table format, and it is a qualitative reliability tool. It presents all of the possible failure scenarios for each major component or assembly (SRU or shop replaceable unit) in a piece or equipment (LRU or line replaceable unit) or an entire system (consisting of several LRUs). Formats vary somewhat, but many are constructed in accordance with the guidelines set up in MIL-STD-1629 for the military field and similar documents in the commercial field.

The first task for constructing the document of the FMEA, as generally required by the customer, is to create a functional block diagram, along with description of the function of the product. For electrical designs, the schematic is generally used in the analysis and is usually provided in the appendix of the FMEA document that will be submitted to the customer.

For the two examples that are used in this chapter, we will construct a simple functional block diagram by using the logical breakdown of the major assembly. By logical breakdown, we mean that components are grouped together by the subassembly or functional level. The reliability engineer can either determine this based on experience or through the assistance of the design engineer. In Figure 3–2, the breakdown has already been described for the power supply example, where it is divided into three main portions or functional blocks: 1) EMI protection circuit, 2) voltage regulator, and 3) + 5-volt conditioning circuit. There is no implied redundancy of the circuit as it is laid out, so it can be constructed as a simple block diagram as shown in Figure 3–5. In fact, this functional block diagram is the same at the reliability block diagram (serial).

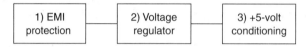

Figure 3-5 Functional Block Diagram for Power Supply
In the diagram above, the functional block diagram for the power supply is broken into the three major areas of the design. It is drawn as a simple serial model as there is no redundancy in this particular design.

For the mechanical system example of the emergency oxygen system that was shown in Figure 3–3, there is an element of redundancy between the manual and automatic emergency oxygen supply and this can be described in the functional block diagram as shown in Figure 3–6.

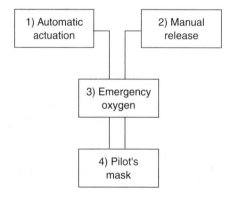

Figure 3-6 Functional Block Diagram for Emergency Oxygen System
The functional block diagram for the emergency oxygen shows an element of redundancy in the area of how the system is activated. It can either be activated manually by the pilot or by automatic means during an ejection sequence in which the pilot is ejected from the aircraft.

From the functional block diagrams presented here, there is a smooth transition in setting up the table format of the FMEA. Each of the blocks in the diagram now become the Item Name in the FMEA table and the functional description of each block is provided in the Function portion of the FMEA. After the functional description, the potential modes of failures are listed, and against each of these modes are the associated effects of the failures from the local effects to the end effects (usually on the pilot or user of the product). Finally, for each listing, the detection methods for the failure are listed along with compensating factors and remarks. Detection may be via visual means or by built-in test (BIT).

The different blocks or functions of a product are analyzed for potential modes of failure. The failure mode analysis of a design is basically a physics of failure approach, and it can be based on either the reliability engineer's or design engineer's experience. Each failure mode listed against each function represents a best guess estimate of the analysis. The overall failure mode for a functional block may be comprised of individual failure modes for a group of components that make up this functional block.

A number of formats can be used for constructing a FMEA. These formats are set to the requirements of the regulating agency that uses the product. For example, the auto industry has SAE requirements for a FMEA, while both the commercial and military aircraft industry generally use MIL-STD-1629 and similar formats.

In this chapter, we will use a format that is based on MIL-STD-1629 requirements, and we will illustrate their meaning of individual columns by using the two examples to help the reader's understanding of this tool. Figure 3–7 shows the power supply FMEA and Figure 3–8 shows the emergency oxygen FMEA. The examples show that the potential failure modes are identified for each item and that in some cases are based on the individual failure modes of smaller components.

The hazard severity that is used in the figures are based on the MIL-STD-1629 format, where the following are the categories:

I *Class 1*—Catastrophic—loss of life or limb to pilot or complete loss of equipment
II *Class 2*—Critical—major damage to equipment and injury to pilot or user
III *Class 3*—Major—malfunction to equipment but no permanent damage or injury
IV *Class 4*—Minor—minor nuisance faults such as power line noise, etc.

Much of the FMEA will be based on the reliability analyst's judgment and experience, along with additional inputs from the design engineer's perspective on what was intended in the design. The FMEA should be performed as early as possible during the early design stage when preliminary schematics and layout drawings have been made. This is because there may be certain failure modes of major concern that are not able to be detected, and there is still a chance for the design engineer to make changes to the design at this stage to improve failure detection. Contrary to some practices by companies, performing the FMEA later in the program does not allow changes to be made, and when customers identify certain conditions, they are likely to be dissatisfied. "The sooner the better" is the best policy for performing this task. Once the customer receives the vendor's FMEA, they are able to absorb this information into their own system-level FMEA and safety analysis, which is usually a requirement for FAA or the military agency running the overall program.

Even though the FMEA, FMECA, and FTA are system-level tasks, they require the failure-mode analysis of individual components, particularly in the case of electrical circuits. There are some variations to these categories, depending what set of requirements are called out. Sometimes, the numbers may actually be reversed in reference to severity, but the above listing reflects the MIL-STD-1629 format, which is generally considered the standard format for industry.

The FMEA is one of the few reliability items that still remains as a standard requirement for a wide range of organizations: military, commercial, and medical.

FAILURE MODES AND EFFECTS ANALYSIS (FMEA)

LRU: Electronic module
SRU: Power supply board

No.	Item name	Function	Failure mode	Failure cause	Mission phase	Haz class	Local failure effects	Next higher effects	End effects	Failure detection method	Compensating factors	Remarks
1.A	EMI protection circuit (primary channel)	Provides regulated 5-volt power output	Loss of 5-volt power supply output	L1 open, C2, C8 or C9 short	All	III	No 5-volt power output	Loss of all logic functions	No data sent to cockpit display	BIT	Select alternate channel	
1.B			Incorrect 5-volt power supply output	L1 short, C2, C8 or C9 open	All	IV	Noise on 5-volt power output	Possible noise	None	None	None	
2.A	Voltage regulator	Provides regulated voltage output to 5-volt conditioning circuit	Loss of all voltage output	U1 high or low, CR1, CR4 short/ open, T1 short open	All	III	Loss of all voltage outputs	Loss of all logic functions	No data sent to cockpit display	BIT	Select alternate channel	
2.B			Loss of or incorrect feedback control	R7, R8 or R9 open	All	III	Loss of or incorrect 5-volt power output feedback signal	Loss of all logic functions	No data sent to cockpit display	BIT	Select alternate channel	
2.C			Loss of or incorrect foldback control	CR2, CR3 or CR6 short or open	All	IV	Loss of foldback control	No short circuit protection	None	None	None	Will be detected on second failure
2.D			Loss of error compensation circuit	CR2 short or open, R2 open	All	III	Loss of error compensation	Loss of all logic functions	No data sent to cockpit display	BIT	Select alternate channel	
3.A	5-volt conditioning circuit	Provides regulated 5-volt power output to logic circuits of unit	Loss of 5-volt output	CR7 open, L2 open, C1, C3, C10, C11 or C13 short, R1 or R10 open	All	III	No 5-volt power output	Loss of all logic functions	No data sent to cockpit display	BIT	Select alternate channel	
3.B			Incorrect 5-volt output	CR7 short, L2 short, C1, C3, C10, C11 or C13 open	All	IV	Noise on 5-volt power output	None	None	None	None	

Figure 3-7 Sample FMEA Worksheet
A) Power Supply Example

FAILURE MODES AND EFFECTS ANALYSIS (FMEA)

LRU: Emergency oxygen
SRU: Actuation system

No.	Item name	Function	Failure mode	Failure cause	Mission phase	Haz class	Local failure effects	Next higher effects	End effects	Failure detection method	Compensating factors	Remarks
1.A	Automatic emergency oxygen actuation	Automatically activates the emergency oxygen system when commanded	Fails to activate	Faulty cable, failed mechanism	In-flight emergency	II	Emergency oxygen is not activated	No emergency oxygen to pilot	Loss of oxygen, possible hypoxia	Visual indication on gauge	Use manual activation of emergency oxygen system	
1.B			Premature activation	Automatic release failure	All	II	Emergency oxygen is depleted	No emergency oxygen to pilot	Loss of oxygen, possible hypoxia	Visual indication on gauge	Descend to below 18,000 feet	
2.A	Manual release for emergency oxygen actuation	Provides a manual method of activating emergency oxygen	Fails to activate	Mechanical failure of manual reset	In-flight emergency	II	Emergency oxygen is not activated	No emergency oxygen to pilot	Loss of oxygen, possible hypoxia	Visual indication on gauge	Use manual activation of emergency oxygen system	
2.B			Premature activation	Accidental pulling of handle	All	II	Emergency oxygen is depleted	No emergency oxygen to pilot	Loss of oxygen, possible hypoxia	Visual indication on gauge	Descend to below 18,000 feet	

Figure 3–8 Sample FMEA Worksheet
B) Emergency Oxygen System Example

3.5 FAILURE MODES, EFFECTS, AND CRITICALITY ANALYSIS (FMECA)

The failure modes, effects, and criticality analysis (FMECA) is also in table format and is similar in layout as the FMEA. It actually replicates information from the first four columns from the FMEA so that items and item numbers can be tracked and found easily, particularly in a large system-level FMEA.

Table 3-1 Electrical Component Failure Modes and Percent of Occurrence

Part Type	Failure Modes	Percentage of Occurrence
Resistor	Open	100%
Potentiometers	Open	5%
	Value Drift	95%
Capacitors (Ceramic)	Open	10%
	Short	90%
Capacitors (Tantalum)	Open	55%
	Short	45%
Diodes (General/Zener)	Open	15%
	Short	85%
Transistors (Bipolar and FET)	Open (C-E or D-S)	4%
	Short (C-E or D-S)	96%
Inductors	Open	25%
	Short	75%
Transformers	Open	20%
	Short	80%
Integrated Circuits	Stuck High (Open)	Output%/N
	Stuck Low (Short)	Output%/N
	Stuck @ Tristate	Output%/N
Voltage Regulators	Not Regulating	50%
	No Output	50%
Relays	Open	50%
	Short	50%
Switches	Stuck Open	50%
	Stuck Short	50%
Thermistors	Open	100%
Filters	Open	50%
	Short	50%
Display Chip (LED)	Open/Short	95%
	Input Control	5%

Notes: 1) Breakdown of the failure modes above are based on MIL-HDBK-338 and other industrial data.
2) N = number of pins for integrated circuits.

The main difference between the FMECA and the FMEA is that the FMECA is a numerical analysis tool that calculates the numerical impact for each failure mode.

In the case where the failure rate is listed against the overall failure mode of the FMEA or in the FMECA, a failure-mode breakdown for electrical components is required. In some cases, the customer may request a full-blown FMECA or an enhanced FMEA that includes failure-rate data. For the latter case, the customer may elect to take the vendor's failure rate and plug into their system-level FMECA that will be used for the fault tree analysis task later on. For either case, component failure-mode ratios will be needed in order to calculate the failure mode for a particular function.

Table 3–1 provides a useful failure mode breakdown that is commonly used in industry. For the example of the power supply, we can see where the failure modes of resistors are 100 percent in the open failure mode. Thus the entire failure rate of each resistor is counted in the line where it appears on the FMECA. In the case of components with more than one failure mode, such as a capacitor (failing short or open), each failure-mode ratio is applied against the component failure rate and placed in the appropriate line of the FMECA. These partial failure rates may be combined with other partial or full failure rates for that line item of the FMECA.

FMEA EXAMPLE

For example, the total failure rate calculation for the failure mode listed in 1A of the FMEA for the power supply example shown in Figure 3–7 is performed as follows:

Des	Component Type	Failure Mode	Failure Rate	Failure Mode Ratio	Revised Failure Rate
L1	Inductor	Open	.00006	.50	.00003
C2	Capacitor (tant)	Short	.00102	.45	.00046
C8	Capacitor (tant)	Short	.00032	.45	.00014
C9.	Capacitor (cer)	Short	.00012	.90	.00011

Total failure rate for this mode is = .00074

It is noted that the component failure modes for the list above actually comprise the failure mode in Item 1B as follows:

Des	Component Type	Failure Mode	Failure Rate	Failure Mode Ratio	Revised Failure Rate
L1	Inductor	Short	.00006	.50	.00003
C2	Capacitor (tant)	Open	.00102	.55	.00056
C8	Capacitor (tant)	Open	.00032	.55	.00018
C9.	Capacitor (cer)	Open	.00012	.10	.00001

Total failure rate for this mode is = .00078

Another task that has to be performed as part of the FMECA is the criticality number calculation. This criticality number will be used by the customer for risk assessment at a system level and for the fault tree analysis. A criticality calculation is performed in a table-type setup and the results will be fed back in the FMECA. This criticality number calculation is based on a number of items, failure rate, failure mode ratio, and loss probability number.

The failure-mode ratio is a straightforward calculation that can be done one of two ways. If we take the FMEA example (Item 1), we can see that the total failure rate for all of the components that make up this item (both 1A and 1B) is the sum of the failure rates for each of the modes in A and B or .00151. This yields a failure-mode ratio of 49 and 51 percent, respectively, for the failure mode of A and B. The other way is to treat the failure rates for each mode as a complete entity (.00074 for 1A and .00077 for 1B) and use a failure mode ratio of 1 for each line item.

Additionally for the FMECA, an assessment has to be made with regard to the effects of the failure on the system as to the likelihood of occurrence. This factor is known as the loss probability and is designated as β and is determined by the analyst's judgment for each of the severity categories. For example, we could have:

Category	Description	β Value
I	Catastrophic	1.0
II	Critical	.5
III	Major	.3
IV	Minor	0

For small failure rates where λ is less than 1, our criticality number is the product of β, failure-mode ratio, failure rate, and at-risk or mission time. For larger values of λ, this calculation should be done using the unreliability formula or:

$$Cm = \beta (1 - e^{-\lambda_m t})$$

where λm is the product of the failure-mode ratio and the overall item failure rate.

FMECA EXAMPLE

A table or spreadsheet can be created that will help in the calculation of the criticality number. In the example of the first item (1A and 1B) for the power supply example, the table would look like the following:

A. Power supply

No.	Item	Failure Mode	Item Failure Rate ($\times 10^{-6}$)	Mode Failure Ratio	At-risk time (hrs)	Loss probability (β)	Criticality Number (Cm) ($\times 10^{-6}$)
1A	EMI Protection Circuit	Loss of 5-volt power supply output	.00151	.49	1	.3	.000222
1B		Incorrect 5-volt power supply output	.00151	.51	1	0	0.0

The table constructed on the previous page can actually become part of the FMECA with the results for this first item and the other items being inserted into the FMECA that is shown in Figure 3–10.

For the emergency oxygen system example, the following table setup for the first item is created:

B. Emergency oxygen system

No.	Item	Failure Mode	Item Failure Rate ($\times 10^{-6}$)	Mode Failure Ratio	At-risk time (hrs)	Loss probability	Criticality Number (Cm) ($\times 10^{-6}$)
1A	Automatic emergency oxygen actuation	Fails to activate	.0000039	.75	1	.5	.0000015
1B		Premature activation	.0000039	.25	1	.5	.0000005

The criticality number that is calculated for 1A and 1B is inserted into the FMECA that is shown in Figure 3–11.

Typically, when a FMECA is called for by the customer, both FMEA worksheets (such as those shown in Figures 3–7 and 3–8) and FMECA worksheets (such as those shown in Figures 3–9 and 3–10) are provided in the submittal. They can either be presented as a separate appendix for each set of worksheets or both sheets can be integrated together for each functional block item that is analyzed. As mentioned before, some customers may request a modified FMEA that includes failure-rate information or criticality number in the last column so that they perform the higher-level analysis themselves. The FMECA format has some flexibility, however, the format is subject to approval by the end customer. The examples presented in the figures are meant to be guides for the reader and not necessarily the definitive format that may be required.

The results of the FMECA can be used to construct a system-level fault tree analysis and submit this as a data item to the customer.

3.6 FAULT TREE ANALYSIS (FTA)

Either the reliability engineer or the safety engineer of the company may perform the fault tree analysis (FTA), with the analysis being based on the results of the FMECA. The FTA will use the failure-mode information and criticality numbers that are provided in the FMECA. The FTA determines in a logical manner which failure modes at one level can produce critical failures at the system level. The FTA provides a concise description of the various combinations of possible occurrences within the system that can result in predetermined critical output events. The FTA helps identify and evaluate critical components, fault paths, and possible human errors. It is both a reliability and safety engineering task, and it is a critical data item that is submitted to the customer for their approval and their use in their higher-level FTA and safety analysis.

LRU: Electronic module
SRU: Power supply board

FAILURE MODES, EFFECTS AND CRITICALITY ANALYSIS (FMECA)

No.	Item name	Function	Failure mode	Failure cause	Mission phase	Haz class	Item failure rate (x10⁻⁶)	Failure mode ratio	At-risk time (hrs)	Loss probability (β)	Criticality number (Cm)
1.A	EMI protection circuit (primary channel)	Provides regulated 5 power output	Loss of 5-volt power supply output	L1 open, C2, C8 or C9 short	All	III	.00151	.49	1	.3	.000222
1.B			Incorrect 5-volt power supply output	L1 short, C2, C8 or C9 open	All	IV	.00151	.51	1	0	.0000000
2.A	Voltage regulator	Provides regulated voltage output to 5-volt conditioning circuit	Loss of all voltage output	U1 high or low, CR1, CR4 short/open, T1 short open	All	III	.224741	.841	1	.3	.056702
2.B			Loss of or incorrect feedback control	R7, R8 or R9 open	All	III	.224741	.001	1	.3	.0000674
2.C			Loss of or incorrect foldback control	CR2, CR2 or CR6 short or open	All	IV	.224741	.003	1	0	.0000000
2.D			Loss of error compensation circuit	CR2 short or open, R2 open	All	IV	.224741	.125	1	.3	.0084278
3.A	5-volt conditioning circuit	Provides regulated 5-volt power output to logic circuits of unit	Loss of 5-volt output	CR7 open, L2 open, C1, C3, C10, C11 or C13 short, R1 or R10 open	All	III	.010182	.58	1	.3	.0015273
3.B			Incorrect 5-volt output	CR7 short, L2 short, C1, C3, C10, C11 or C13 open	All	IV	.010182	.42	1	0	.0000000

Figure 3–9 Sample FMECA Worksheet
A) Power Supply Example

LRU: Emergency oxygen
SRU: Actuation system

FAILURE MODES, EFFECTS AND CRITICALLY ANALYSIS (FMECA)

No.	Item name	Function	Failure mode	Failure cause	Mission phase	Haz class	Item failure rate (x10^{-6})	Failure mode ratio	At-risk time (hrs)	Loss probability (β)	Criticality number (C_m)
1.A	Automatic emergency oxygen actuation	Automatically activates the emergency oxygen system when commanded	Fails to activate	Faulty cable, failed mechanism	In-flight emergency	II	.0000475	.5	1	.5	.0000118
1.B			Premature activation	Automatic release failure	All	II	.0000168	.5	1	.5	.0000042
2.A	Manual release for emergency oxygen actuation	Provides a manual method of activating emergency oxygen	Fails to activate	Mechanical failure of manual reset	In-flight emergency	II	.0000144	.5	1	.5	.0000036
2.B			Premature activation	Accidental pulling of handle	All	II	.0000144	.5	1	.5	.000036

Figure 3-10 Sample FMECA Worksheet
A) Emergency Oxygen Example

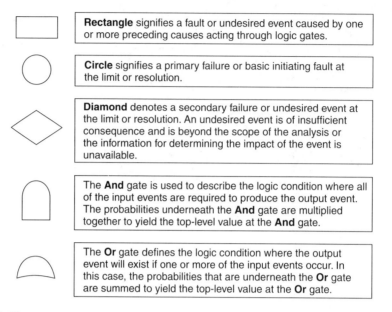

Figure 3-11 Common Flowchart Symbols Used for Fault Tree Analysis

Figure 3–11 shows the flowchart symbols that are used in fault tree analysis in order to aid with the correct reading of the fault tree. These symbols are integrated into each functional block that is identified in the FMECA. See how this is done with the emergency oxygen system FMECA (Figure 3–10) to create a FTA as shown in Figure 3–12.

In the construction of the fault tree shown in Figure 3–12, the AND gate failure rates underneath are multiplied together to yield the number at the top of the gate. In the case of the OR gate, the numbers underneath are summed to yield the number at the top of the gate. Note also the significant use of OR gates in Figure 3–12. This is because in the design, there is a redundant situation between the automatic actuation of the emergency oxygen system and the manual means to activate the emergency oxygen.

Another important exercise can be performed by using the information developed from the FTA, and this is the constructing of a series of event trees in order to form minimal cut sets. The construction of these minimal cut sets help identify the single point failure analysis of a series of events that can result in a specific failure output.

CUT SET EXAMPLE

Using the emergency oxygen system example, the following minimal cut sets can be constructed for A) automatic actuation failure, and B) manual actuation failure.

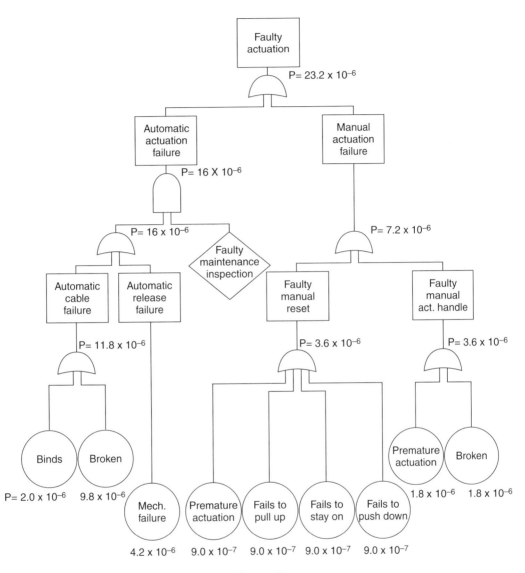

Figure 3-12 Fault Tree Diagram for Emergency Oxygen System

    A. Minimal cut set for automatic actuation failure

No.	Probability	Cut set
1.	2×10^{-6}	Cable binds, automatic cable failure, automatic actuation failure
2.	9.8×10^{-7}	Broken cable, automatic cable failure, automatic actuation failure
3.	4.2×10^{-6}	Mechanical failure, automatic cable failure, automatic actuation failure

The sum of the minimal cut set for A is 16×10^{-6}

B. Minimal cut set for manual actuation failure

No.	Probability	Cut set
1.	9×10^{-7}	Premature actuation, faulty manual reset, manual actuation failure
2.	9×10^{-7}	Fails to pull up, faulty manual reset, manual actuation failure
3.	9×10^{-7}	Fails to stay on, faulty manual reset, manual actuation failure
4.	9×10^{-7}	Fails to push down, faulty manual reset, manual actuation failure
5.	1.8×10^{-6}	Premature action, faulty manual actuation handle, manual actuation failure
6.	1.8×10^{-6}	Handle broken, faulty manual actuation handle, manual actuation failure

The sum of the minimal cut set for B is 7.2×10^{-6}

The overall sum for the minimal cut set for the single point failure mode of faulty actuation is the sum of the minimal cut sets for A and B or 23.3×10^{-6}.

3.7 SUMMARY

The reader has been introduced to several reliability tasks that may be performed on a new program. Not all of these tasks may be required to be performed on a new program as it depends on the nature of the design of the product. Tasks such as electronic stress analysis would be performed on electronic designs, yet tasks such as the failure mode and effects analysis (FMEA) will typically be performed for all types of designs.

It is important for the reader to realize when submitting reliability data items to the customer, that many of them, such as the reliability prediction, have no direct impact on the product. Whether the number is big or small, it is not going to affect the performance of the product in the field and what it actually does. What the number actually has an impact on is the people who use the number for specific tasks. The prediction is an estimate, and other numbers such as maintainability values and spares estimates are based on the reliability MTBF value or failure rate value.

Sometimes reliability numbers take on a life of their own, and managers will quote reliability numbers as if the values were set in stone and cannot be changed. Customers are partly guilty in this process as they will set forth a specified value that is based on their estimates. What is so often forgotten is that there may be a lot of leeway with how the specified value was calculated.

The reliability numbers are very important in certain situations, such as the case of a reliability warranty program (RWP), in which a customer may impose that a field reliability test be conducted. The test is run over a specified period, such as a year, and if the specified value is not met, some sort of penalty may be imposed. This penalty may be in the form of immediate corrective action to be performed by the vendor, free retrofit of repairs, or even free spares. In these cases, reliability values become more important, and it is necessary to develop a sensible methodology when coming up with a specified MTBF value for a product or system.

However, using a cookbook approach like a MIL-HDBK-217 format may not be sufficient to give a level of confidence for the calculation of reliability prediction values. It is not unusual for a company to apply a "safety factor" of 3 or 4 to 1 for cookbook-generated prediction values. This is also a common practice on the system level as customers will routinely degrade reliability values supplied to them by the vendor in order to protect themselves. The reason for the downgrade is that other factors, such as manufacturing, parts quality, or the mechanical aspects of the design can override the overall electronic failure rate. The impact of the latter is discussed in detail in Chapter 5. The important point to remember is that the predicted MTBF of a product is an estimate regardless of what methodology was used.

The reliability task of the FMEA can sometimes develop into a sensitive customer issue. It is important for the reliability engineer to keep in mind certain things as the FMEA is being constructed. It is most important that major failure modes that can occur will have some way that they can be detected, and this should be listed in the FMEA. If a FMEA lists any major failure modes that do not have a means of detection, this is the first thing that the customer will notice and then ask for corrective action. So it is in the best interest of the reliability engineer, as well as the company, to foresee this situation before it happens and look at the design again for improving the means of detection.

3.8 EXERCISES

1. The following .25-watt resistors are hooked up in series in the circuit shown below in Figure 3–13. Using the techniques described in Section 3.1 and Figure 3–4, calculate the stress ratio for each resistor.
   A. Calculate the current that is flowing in the circuit (Hint: sum up the total resistance in the circuit and divide into 12 volts).
   B. Second, calculate the power that is dissipated for each resistor (Hint: remember that power is calculated by $P = I^2 R$).

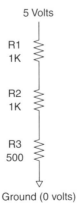

Figure 3–13 *Schematic Used for Exercise 3.8*

C. Finally, calculate the stress ratio for each .25-watt resistor (.250 Watt rating) by actual applied power divided by rated power.
2. What is the part count failure rate prediction for each of the resistors and the total failure rate for this circuit shown in the previous schematic using the information: base failure rate, $\lambda_b = .001$ and environment factor, $\pi_E = .3$?
3. What are the typical failure modes for the following components? (Use Table 3–1 as a guide.)
   A. Resistors
   B. Capacitors
   C. Diodes
   D. Transistors
4. Perform a reliability allocation by completing the chart for the following components of an electronic display unit and allocating to a MTBF requirement of 1,000 hours (failure rate = 100). Use the example in Section 3.3 as a guide for this exercise. Complete the chart by performing the following steps:
   A. First, calculate the predicted failure rate and MTBF of the overall display unit.
   B. Then take the total allocated failure rate and divide by the total predicted failure rate to obtain a K factor. This K factor will be used to multiply the individual predicted failure rate values for each item in order to obtain the allocated failure rate value for each item (the K factor will be the same for all individual items that make up the display unit).

Display Unit Reliability Allocations Chart

No.	ITEM	Predicted failure rate	K factor	Allocated failure rate
1.	Input/output assembly	5.25×10^{-6}		
2.	Power supply	10.50×10^{-6}		
3.	Microprocessor board	20.45×10^{-6}		
4.	Display panel assembly	8.75×10^{-6}		
	TOTAL failure rate			100×10^{-6}
	Corresponding MTBF			1,000 hr. (Goal)

5. Using the data that was presented in problem 4, construct a functional block diagram. The display unit has no internal redundant subassemblies. (Hint: it is identical to the reliability block diagram.)

6. Using the same display unit example from problem 4 and 5, construct a simple FMEA using the following information:

Item	Failure mode	% occur	Cat.	How detected?
1) Input/output assembly	no output	100%	major	visual means and BIT
2) Power supply	A) no output	80%	major	visual means and BIT
	B) noise	20%	minor	none
3) Microprocessor board	no output	100%	major	BIT
4) Display assembly	A) no display	50%	major	visual
	B) missing data	50%	major	visual

7. Using the data presented in problem 6, construct a fault tree analysis using the example that is illustrated in Figure 3–11 as a guide.
8. Using the data present in problem 6 and 7, construct a minimal cut set for the end effect of no output.

CHAPTER 4

The Benefits of Reliability Testing to Product Design

INTRODUCTION

Perhaps the most important task that any reliability engineer can be involved with in a new product design is in the area of reliability development testing. Many products that are to be installed in aircraft (both military and commercial), trains, and cars generally have to undergo some sort of formal reliability test requested by the customer that is conducted during the early stages of the design. Most reliability tests require a significant effort to set up, and completion of the test may be slowed due to various obstacles. Reasons for this include the availability of test units, test facilities and test equipment as well as delays caused by failures. Because of the challenges facing the reliability engineer to perform and complete this test, this chapter could actually be retitled "The Agony of Reliability Testing." However you view this task, the reliability engineer should recognize this as one of the most important tasks to be performed in the overall scheme of developing a reliable product. This is because a properly conducted test can uncover weaknesses in the product design as well as measure the performance of the product under field-like conditions.

The objectives of this chapter will include introducing the reader to what is involved in a reliability development test, how it is conducted along with a description of the different types of reliability tests that are performed during the early design stages of a new program. Much of the emphasis in this chapter will be in the area of electronic or electro-mechanical equipment undergoing reliability testing. Later in the chapter, descriptions of some of the reliability tests that are performed during production will be discussed.

4.1 RELIABILITY DEVELOPMENT TEST OBJECTIVES

The object of performing a reliability test is to subject units of a new design to duration testing simulating field environment in order to do any or all of the following:

- Uncover any weaknesses in components or the packaging of the design and develop appropriate corrective action
- Measure the reliability performance of the product
- Provide a level of confidence for a new product design's reliability performance
- Verify aspects of the functionality of the design
- Determine the breaking point of the product

During the task sequence of a new program, reliability tests are usually performed after qualification testing has been completed on a new design. Qualification testing involves multiple tests performed on the new design in different environments such as humidity, salt air, and rain. Also, vibration and thermal testing may be performed for a unit that is installed in an active environment such as in an aircraft, train, or automobile. A reliability test can only be performed after a thermal survey and vibration survey have been conducted on the unit. Chapter 6 goes into more detail on how a thermal survey is conducted and how a thermal profile is derived.

It is always desirable to start the reliability test prior to the start of full-scale production so that any corrective action that is developed for different failure modes that appear during the test can be implemented into the design prior to production units being manufactured. It is in the company's best interest to reduce potential risks by having timely fixes implemented early in the program before many units are deployed into field service and having the potential of dealing with costly field retrofit actions or product recall issues.

Many of the reliability development tests that will be discussed here are those that are conducted in the laboratory environment or test facility. Those reliability tests that are conducted in the lab will involve two major environmental components: vibration testing and thermal cycling. These components of testing are often referred to as "shake and bake." Sometimes these tests can be combined into one chamber, but more often these two components are conducted separately due to test equipment limitation and availability. A typical reliability test would involve two test units undergoing a total of 3,000 hours of thermal cycling and 100 hours of vibration cycling. A typical reliability test regimen is broken up into a weekly schedule, usually five or six days of thermal cycling and one-half of a day devoted to vibration cycling.

What environmental test actually stresses the component more and brings out failures? It really is a function of the design with how well it is put together along with the overall quality of the components used. Some tests will see both vibration and thermal failures occurring on the test articles while others will see only vibration-related failures. A general rule is to expect more vibration-related failures for a test than those that occur during thermal cycling, but the good engineer will learn to expect anything!

4.2 THERMAL CYCLING

A thermal chamber is needed to conduct temperature cycling on a product for most types of reliability tests. Most thermal chambers require compressor motors to run the chamber at cold temperatures, and for extremely low temperatures ($-40°C$ and lower), liquid

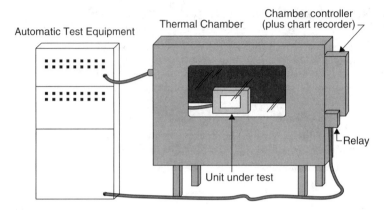

Figure 4-1 Thermal Chamber Setup
During the thermal portion of the reliability test, a setup similar to what is shown above may be used. The unit under test is powered by a cable that goes through a porthole of the thermal chamber wall (that is later plugged) to the automatic test equipment. Power to the test equipment is synchronized and controlled by a relay in the chamber controller. This relay is configured to turn the unit on during certain portions of the thermal profile.

nitrogen bottles may be needed. Thermal chambers have a controller mounted on the chamber that can be programmed with each temperature step of the desired thermal profile. Chart recorders (also mounted on the chamber) are typically used to track the actual chamber temperature profile. Companies that make a number of products for aircraft will often make an investment in obtaining a chamber as opposed to having an outside lab conduct the tests. Figure 4–1 is an example of a thermal chamber setup for a reliability test.

The thermal cycling portion of reliability testing stresses the test unit at a minimum of two different temperature extremes. For example, a thermal profile may involve two temperature extremes such as −40°C and at +55°C. These extremes are typically at the range for the rating of most commercial grade electrical components used in electronic units. Military applications may go as low as −50°C and as high as 70 to 80°C to track with the rating of military-grade components used in electronic units that are located in harsher environments of military products such as fighter aircraft. Some programs may have several temperature plateaus during the thermal cycle such as two at cold, −55°C and −40°C, one at room temperature (27°C) and one at 71°C. Figure 4–2 is one example of a reliability development test thermal profile.

The length of the thermal profile is determined by the results of the thermal survey. The results of the thermal survey are based on a number of factors, including the size of the unit and its individual components, as well as the thermal stabilization factors of individual components. How quickly these components stabilize under temperature determines the length of each temperature plateau of the thermal profile. More details on the thermal survey will be provided in Chapter 6.

The most critical area for electrical component stress in the thermal profile is during the ramp up from the cold extreme to the hot extreme. Just like an aircraft descending sev-

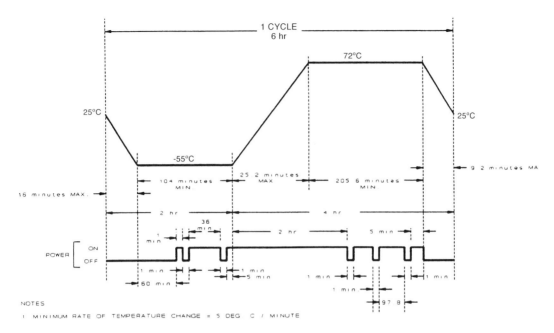

Figure 4-2 Reliability Test Thermal Profile Example
This figure shows an example of a thermal profile used for an electronic unit that is located in the cargo bay of a commercial aircraft. The two temperature extremes are −55°C at cold and at +72°C at hot. The ramp from −55°C to +72°C is accomplished over a 25-minute span. The total duration of the cycle is 6 hours, which means during a single 24-hour period of one day, four cycles can be completed. Note that the unit is powered on and off at critical portions of the thermal cycle.

eral thousand feet over a short amount of time, this ramp up is a particularly stressful time for the electronic components in a box as the rapid thermal changes affects all parts of the components. Ramp rates can cover as much as a 120°C drop in a time period as little as 20 minutes. The changes in thermal expansion during this rapid ramp become a major factor for many electrical components that use ceramic materials with metal pins. As metal and ceramic have different thermal expansion coefficient characteristics, stresses will result during the thermal ramp up. The temperature ramps may cause more stress on the components of electronic units than just staying at one temperature on either the hot or cold end.

It is important to note that for electro-mechanical and mechanical products, the temperature extremes may alter the characteristics of materials (such as rubber) plastic and metal. Chapter 6 will discuss this further.

Typically, time duration for one complete thermal cycle can be two, three, four, six or eight hours, as any of these numbers can divide evenly into a twenty-four-hour day. This is desirable as it aids in the counting of hours and cycles, as well as helping the test technician to maintain a regular schedule for when to perform functional tests at the time when the unit is powered on at temperature extremes.

4.3 VIBRATION CYCLING

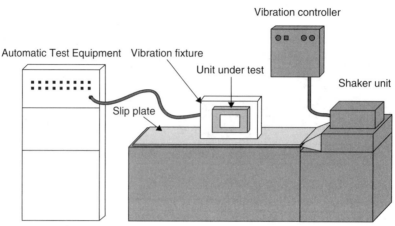

Figure 4-3 Vibration Setup
The figure above shows a typical vibration test setup that may be used for reliability testing. In the setup above, the unit that is being tested is powered by the automatic test equipment for the duration of the test.

In addition to needing a thermal chamber for the reliability test, most reliability tests require a vibration table setup for shaking the unit. The vibration machine consists of a slip plate that rides on a thin layer of oil in a heavy-duty base. The plate is mechanically connected to a control head that can be changed to different orientations for the purposes of doing different axes of vibrations (X, Y, and Z). As with owning a thermal chamber, the vibration table is a major investment for a company to make or to pay a test lab for use. Figure 4–3 shows a typical vibration test setup.

As in thermal cycling, most units that are to be installed into an aircraft or other vehicles will have the requirement of vibration testing to be conducted as part of the reliability test. Vibration testing may be simultaneously combined with thermal cycling for maximum effect, but frequently test facilities limitations may dictate that vibration testing be conducted in separate sessions. When vibration is conducted separately, the unit is usually powered on during the length of the session. Random profiles at the functional vibration level are used to simulate the level of vibration that the unit will typically see during field service. Positioning of electronic units on the vibration fixture with regards to the direction of vibration will depend on the orientation of internal circuit boards or major components. The direction for vibration to be applied that is chosen is usually perpendicular to the internal circuit board orientation of the unit, as this is the worst-case scenario for stressing the components on the circuit board. For some tests, a single vibration session may require that all three axis of vibration, X, Y, and Z, would be performed at a fixed length of time, throughout the duration of the test.

Vibration profiles that are used for the reliability test are designed to reflect the typical field environment that the unit will see and is usually designated as the functional level of vibration. Figure 4–4 shows some examples of vibration profiles used in reliability testing. Profiles can be selected from various sources and tailored to the environment of the pro-

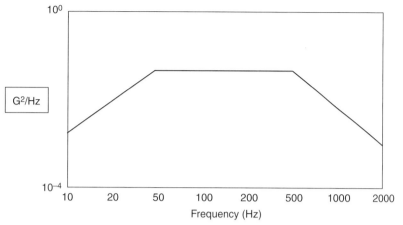

Acceleration = 2.60 GRMS (Root-Mean-Square Acceleration)

A) VIBRATION PROFILE EXAMPLE #1

This is a vibration profile that was used for Accelerated Life Testing of two different electronic units, one that was located in the cockpit and the other that was located in the cargo bay of the MD-90 aircraft. 2.60 GRMS is considered by industry standards to be a moderate functional vibration level.

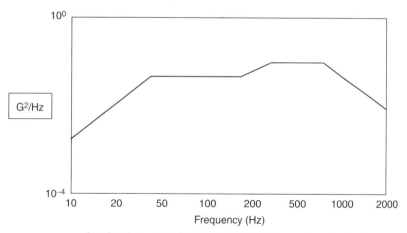

Acceleration = 7.61 GRMS (Root-Mean-Square Acceleration)

B) VIBRATION PROFILE EXAMPLE #2

This is an Accelerated Life Test vibration profile that was used for an electronic unit that was mounted in the wing of a MD-90 aircraft. The higher level of 7.61 GRMS reflects the additional stresses on equipment located in the wing area as there is significant amplitude of motion in aircraft wings.

Figure 4-4 Examples of Reliability Vibration Profiles

gram. Typical government documents for selecting vibration profiles are from MIL-STD-781, *Reliability Tests Exponential Distribution,* or commercial specifications such as DO-160D, *Environmental Conditions and Test Procedures for Airborne Equipment.*

Parts and components that are located in different areas of the aircraft will see different levels of random vibration during flight. Typically, the most benign vibration environment for an aircraft is the cockpit area and the most severe levels of vibration are observed for equipment located inside the wing of the aircraft. Thus vibration levels used during reliability testing are adjusted accordingly. Some of the various reliability test setups are shown in the photos of Figure 4–5.

Either one of two types of vibration testing may be conducted during a reliability test: random or sine. An important aspect of vibration testing is to realize the differences in test results that are obtained when conducting random and sine (dwell) testing, as well the best time to use each. For example, there are certain frequencies for which components or assemblies in a unit will tend to resonate and vibrate in tremendous amplitude shifts. Random vibration covers a wide band of frequencies and thus will sufficiently touch upon the resonant frequencies of components in a unit during reliability testing. By conducting a sine sweep test which begins at 0 Hz and goes up to 2,000 Hz (usually conducted during qualification testing), these specific frequencies can be identified either by the use of a response accelerometer on the component or visually by watching the component vibrate using a strobe light.

One important aspect of vibration testing is to identify the frequencies for which certain components can resonate to the point of failure. At resonance, screws can actually migrate out of a unit, as seen in the real-life example of an amateur radio transceiver that was stowed in a carry-on case, on the floor under a seat in a commercial aircraft during a long flight from New York to Hawaii by the author. It became apparent that a resonant frequency was present in the passenger area of the aircraft as all of the chassis screws migrated out! Thus, it is usually desirable to complete some preliminary vibration testing such as the qualification test prior to the long-term vibration portion of the reliability test so that any obvious vibration resonances are identified early and corrected.

4.4 TEST DOCUMENTATION AND OTHER ITEMS NEEDED FOR RELIABILITY TESTING

In addition to the actual test facility, several other items are needed to run the reliability test. Of course, no reliability test program is finished until all of the paperwork is completed. The documentation of a reliability test is critical and with specific test data items required by the customer as well. The following is the list of the items that most reliability test programs will require.

4.4.1 Test Articles

Obtaining the test articles or units for a reliability test may seem like a trivial matter, but it can actually be a major concern to the reliability engineer. If a new program starts to run behind schedule, company management may elect to route parts and units into production because of customer demands and thus delay delivery of units for the reliability test, even if only one or two test articles are required. Obviously, this delay is a risk with regards to potential design problems not being discovered quickly. Even if test units are assigned, the reliability engineer may encounter difficulty in obtaining replacement parts and circuit boards if allocations are tight because of schedule demands. Case studies presented in Section 4.6 will illustrate how delays in making test units available will cause bigger problems later on in production.

A) The setup shown at the left shows the different units that are undergoing the thermal portion of a reliability test. Note the power cables connecting the units going to the right of the chamber. These cables will subsequently go through an external port in the chamber into automatic test equipment. At a number of points during the thermal profile, power is cycled on and off to the units and functional tests are performed.

B) At the left is an electronic controller unit that is attached to a vibration slip plate, ready to undergo random vibration testing at functional levels. This particular test has vibration sessions that are conducted separately from thermal cycling. Vibration testing that is conducted in this manner is run for a fixed number of hours on the vibration table after each weekly thermal session is completed.

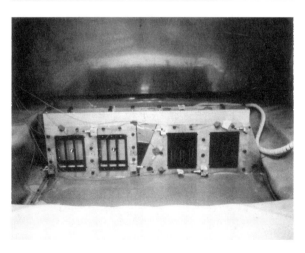

C) The setup shown at the left shows units that are undergoing combined thermal and vibration reliability testing inside a thermal chamber. Vibration testing is applied at only certain points during the thermal cycle. In addition, power is cycled on and off at different points during the thermal profile. Combined testing can save time, but is more difficult to set up and will tie up more test equipment assets. The left side of the vibration fixture has two dummy units installed, and they are used as weights to maintain balance on the slip table.

Figure 4–5 Typical Reliability Test Setups

4.4.2 Test Equipment

The unit under test has to be powered by a piece of test equipment while it is undergoing thermal and vibration cycling. The test equipment will usually have a cable or wire harness that is long enough to connect to the unit while it is in the thermal chamber or set up on the vibration table. There are generally switches on the test equipment that can exercise certain functions on the unit, as well as an automated test sequence that tests the major functions of the unit. One problem that is often encountered is that test equipment is in demand during the early stages of new product development between engineers and production and software developers. Thus, this becomes an issue for the reliability engineer, just as in the process of getting units allocated for the reliability test.

4.4.3 Reliability Test Procedure

The test procedure will provide a description of the ground rules for the test, including the test setup, how many units are to be tested, the duration of thermal cycling, and how many hours of vibration are to be performed. The description of the exact thermal profile and vibration profile to be used during the test are described in the test procedure. Also, the process of failure notification to the customer and failure reports is documented in the procedure.

4.4.4 Test Log Book

A simple log book, usually a binder for listing multiple line entries on a page is generally a necessity even if it is not a requirement. Date and time tracking each portion of the thermal profile that is completed, along with any functional tests completed in the thermal chamber are logged. Each thermal cycle and vibration session that is completed is counted and documented in the log book. Any failures that occur are documented on the appropriate line in the log book when the failure occurred, along with a failure report number. Tests are usually halted at the point of failure. When testing is resumed, the cycle that the failure occurred is usually repeated. Figure 4–6 shows a sample log page.

4.4.5 Failure Reporting

There is almost always a test requirement in which a formal method for the documentation of failures is required by the customer. Along with the date and other relevant test setup informa-

No.	Date	Time	Cum test hours	Environment	Pass	Fail	Cycle no.	Comments	Tech int
126	12/23	5:40	300	+71°C	✓		50		K.E.N.
127	12/23	7:20	301.5	−54°C	✓		51	Visual check	K.E.N.
128	12/23	7:40	302	+54°C	✓		51	Complete 54°C	K.E.N.
129	12/23	8:20	302.5	+71°C	✓		51	Visual check	K.E.N.
130	12/23	11:40	306	+71°C	✓		51	Complete 71°C	K.E.N.
131	12/23	12:10			✓			Post thermal ATP	K.E.N.
132	1/3	8:20	.5	7.61 GRMS		✓	2	Y axis vib–failed F/R#108	K.E.N.
133	1/5				✓			Post-repair ATP	K.E.N.

Figure 4-6 Sample Test Data Log Book Page

tion, an initial analysis is provided. Failure reports are formally closed when the cause of failure has been identified and the corrective action that was incorporated to prevent reoccurrence of the failure in the future. An important aspect of failure reporting is to completely document all environmental conditions when the failure occurred. This recorded information becomes very helpful in the future; for instance, in the situation of a failure not fully diagnosed the first time around, which later appears again, particularly if the failure is intermittent.

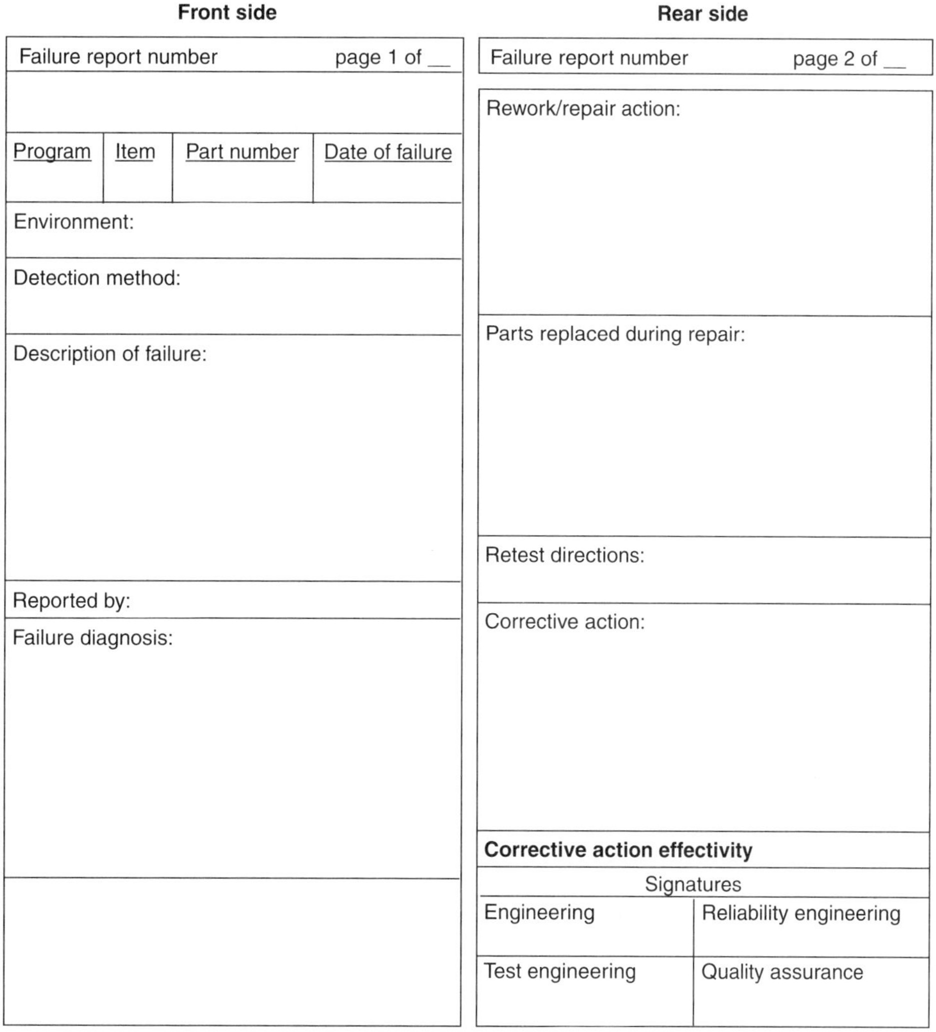
Front side

Failure report number page 1 of __

Program | Item | Part number | Date of failure

Environment:

Detection method:

Description of failure:

Reported by:

Failure diagnosis:

Rear side

Failure report number page 2 of __

Rework/repair action:

Parts replaced during repair:

Retest directions:

Corrective action:

Corrective action effectivity

Signatures

Engineering | Reliability engineering

Test engineering | Quality assurance

Figure 4–7 General Failure Report Form for Test Program
A general form that can be used for failure reports that are generated during a test program is shown above. These forms may vary in the format from company to company as well as reflecting what needs to be reported as per customer requirements.

Figure 4–7 shows a generalized format that may be used for a failure report with the headings of each box in the report showing the type of information that is typically required. All failure reports should be resolved or closed by the end of the reliability test. It is possible that one or two reports may never have a complete answer as to the cause of the failure; in cases like this, the failure may be determined to be random.

4.4.6 Implementing Corrective Action

After a failure occurs during a reliability development test and agreement is reached by all parties involved on what the fix or corrective action should be, the test is allowed to continue with the fix incorporated into the test units in most cases. The general consensus in this process of incorporating and testing with corrective action implemented is that sufficient test time is accrued to prove out the effectiveness of the corrective action.

4.4.7 Final Test Report

After completing a reliability test, a summary of the test results along with failure summary and other documentation is provided. Typically, most customers require full test documentation be collected during a reliability test. This includes items such as log sheets, thermal charts, and vibration profile printouts to be included with the report, typically in the appendix section. Prior to a final test report being written and submitted to the customer as a formal data item, a final failure review is held internally in the company to discuss and close out all failures and action items that occurred during testing.

4.5 THE DIFFERENT TYPES OF RELIABILITY DEVELOPMENT TESTS

As mentioned before, there are many different types of reliability development tests. Reliability development tests that are conducted on new product designs may be called any one of a number of titles and acronyms, depending on the specific test requirements of the customer or agency. Some of the various types of reliability tests are:

- Test Analyze and Fix (TAAF)
- Reliability Development Test (RDT)
- Accelerated Life Test (ALT)
- Durability Investigation Test (DIT)
- Reliability Growth Test (RGT) or Reliability Development Growth Test (RDGT)
- MTBF Field Demonstration Test or Reliability Warranty Program (RWP)
- Reliability Enhancement Testing (RET)
- Highly Accelerated Life Test (HALT)

Many of the tests that were listed are similar with subtle differences. Each of the first four tests listed (TAAF, RDT, ALT, and DIT) will have a baseline or functional vibration level and fixed thermal profile that are used for the entire test period. It is generally not required that the specified MTBF value for the product be measured or verified as a goal of this test, although such data may be collected to show improvement in reliability performance over the duration of the test. Both DIT and RDT have the requirement that if two units are used for the test, no single unit sees more than 60 percent of the total test hours. This is to provide a balance of test hours between the test articles and to verify the manufacturing processes between the units.

Most military agencies and subcontractors require a minimum of two units to undergo reliability testing. A major advantage of having two units undergoing testing is the ability to perform troubleshooting by swapping components from one unit to the other in order to fault-isolate to a bad circuit board. The main reason for using more than one unit is that a significant level of confidence will be attained that the majority of relevant design issues affecting reliability are surfaced during the test.

RGT is similar to the four tests described in the first paragraph with the addition of a MTBF measurement test being imposed. The test is conducted in the lab using the same fixed vibration and thermal profile throughout the test period until completion. RWP measures MTBF in the field environment in lieu of a laboratory environment during a specified time period. RET is a step-stressing test with vibration levels and thermal extremes being expanded during the course of the test. HALT is similar to RET but testing is accomplished during a shorter amount of time using special equipment to simulate and accelerate the levels. Each of these tests will be explained in the case studies provided in Section 4.6 and the HALT is explained in Section 4.7.

Reliability tests can be fairly costly, particularly if the manufacturer does not have test equipment in-house. A number of vendors provide reliability test services with specialized shops that can perform tests such as the HALT. Whatever test is chosen, it is highly recommended that the manufacturer implement some form of a long-term test performed in order to simulate field usage. The benefits are always better to perform a test of this sort as opposed to not performing any kind of reliability test at all!

Reliability testing on a new design is probably the most important reliability task that can be done by a company. It is the final aspect of a reliability test program that supplements the paperwork analysis such as FMEA and reliability predictions. By testing the new product in a field-simulated mode, problems that are discovered during testing can be corrected before full-blown production begins.

A company may have the attitude, "Why perform a reliability test when qualification tests have already been performed?" The answer is that qualification tests usually involves short term testing of a particular environmental parameter (shock, endurance-level vibration), but it does not simulate long-term field environment usage. Reliability testing is, therefore, of major importance to the customer who wants to have a reliable product in service and not have to worry about major design problem surprises.

4.6 RELIABILITY TEST CASE STUDIES

The following case studies involve true examples of reliability tests, and some of the significant design problems that were uncovered. In all of the cases, qualification testing was completed on the design, but several design problems were not uncovered at that time and they were discovered during the long-term test environment of a reliability test.

These case studies are provided in order to give the reader a general overview of how tests are conducted and the difficulties that can arise in getting these tests started and completed.

Some information is provided as to the failures that occurred but most of the details are discussed with regards to specific failure modes such as mechanical or intermittent failures, in other chapters of the books. Each reliability test performed on a specific unit takes on a life of its own, and there is always a basic theme or lesson that is learned from each test.

4.6.1 Reliability Test Case Study #1

Test Performed: Reliability Development Test (RDT)
Equipment: Engine Nozzle Position Indicator for U.S. Navy fighter aircraft
Test Setup: Two units undergoing 3,500 hours of thermal cycling combined with short periods of vibration
Lesson Learned: There is no such thing as an easy reliability test, even on simple units

A Long Island aerospace company manufactured a cockpit indicator for a navy aircraft that measured the position of the engine nozzle inside the engines of the aircraft. The indicator was a small dial face that was one inch in diameter and six inches long. The indicator consisted of an external connector, two circuit boards with a handful of basic electronic components with no microprocessing involved, along with a meter movement consisting of a pointer and dial face that was manufactured by another vendor located in Ohio.

As the unit was simple in design, along with the fact that it completed qualification testing with a minimum of problems, the RDT was considered by the manufacturer to be a matter of a simple formality. It was expected that the two units would breeze through the 3,500-hour test that consisted of combined thermal cycling and vibration cycling in one chamber and the data item requirement would be completed to the satisfaction of the aircraft manufacturer.

Due to immediate customer production need for units, the RDT was delayed until two units were finally allocated. Testing commenced smoothly and readings of two test values were recorded at both high (+70°C) and low (−55°C). However, after a week of testing, the test technician noted on one of the units that the meter movement of the pointer was sticking at different values. At this point, the aircraft manufacturer's test representative agreed with the technician that the unit had failed, and testing was halted while failure analysis was conducted.

Microscopic investigation of the meter movement was focused on the area of the pivot, which consisted of a metal point that was centered on top of a jeweled cup. Examination under the scope revealed a significant amount of metal particles that had accumulated on the pivot and in the jewel cup, apparently as the result of vibration-related damage. At this point, the meter movement vendor was consulted, and the meter movement was sent to the vendor for further failure analysis.

The vendor confirmed the initial findings and determined that the metal material was actually the high metal coating from the pivot shaft. The vendor had recently changed to a pivot design where instead of one uniform material being used throughout the pivot, a hard metal coating was applied on the outside of the pivot. This normally would be satisfactory for most products but is apparently not for the case of this test which simulated the fighter aircraft environment where high levels of vibration are an inherent condition in the cockpit. See Figure 4–8.

After some experimentation with other materials that subsequently failed, it was decided to go with one uniform material for the pivot, carbon-based steel that would work or burnish itself against the jewel cup. In addition, the product specification was reviewed and was able to be changed to allow for a wider tolerance for units that had seen periods of test-

A) FRONT VIEW OF FACE

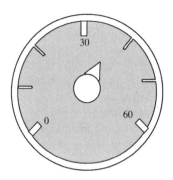

B) SHAFT AND PIVOT ASSEMBLY (SIDE VIEW)

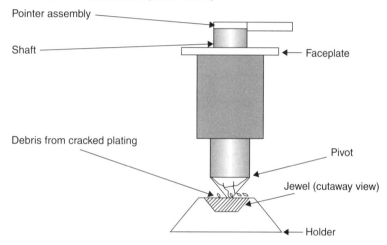

Figure 4-8 Description of Meter Movement Failure

ing. Although this was a simple unit, the test took over one year to complete with the occurrence of a half dozen failures, all pertaining to the pivot design.

The point of this example is to illustrate that there is no sure thing as far as an easy test, even for the simplest designs such as a meter movement. It also points out the dichotomy of qualification testing and reliability testing where some areas of the designs are not tested in qualification testing but significantly tested in a long term reliability test.

4.6.2 Reliability Test Case Study #2

Test Performed: Reliability Development Test (RDT)
Equipment: Engine parameter display unit for U.S. Navy fighter aircraft
Test Setup: Two units undergoing 3,500 hours of thermal cycling combined with short periods of vibration
Lesson Learned: Things get complicated when starting a reliability test well after full-scale production begins

This program resulted out of a cockpit upgrade where the original aircraft that came out in the early 1970s used cockpit indicators that were primarily analog tape units. By the 1980s, LCD technology came out and an upgrade of many of the cockpit instruments to this technology resulted for the current model of this aircraft. One instrument that was upgraded to the new technology was the engine parameter display unit, which contained three distinct LCD displays. As part of the data item package, a new qualification test and reliability test program had to be conducted on the new upgraded unit by the vendor.

The reliability test for the display unit ran into a major scheduling problem. This was because there were various program delays in completing the design of the display unit, and the production of the aircraft was moving ahead before the development testing of the display unit was completed. As a result, demands for this display unit for production aircraft were being made despite the fact that qualification testing was still ongoing and that reliability testing had not begun. In fact, it was not until after fifty units were built that two units could be assigned for the vendor to perform RDT. As can be expected, many reliability problems were encountered on the field that probably would be discovered during a long-term test such as RDT and corrected.

Several meetings occurred between the engineers of the aircraft manufacturer and the vendor, in order to address the causes for a number of field failures of the unit. A number of failures that surfaced were clearly design issues, and it became apparent that because the RDT did not start, the field environment was becoming the actual testing ground. The customer was not very happy about the situation, particularly since there were limited spares to put on the aircraft in place of failed units.

Despite the need for RDT to start, units were still being allocated to meet the production schedule of aircraft rolling off of the line. Thus, when some failures started to appear in the field, an investigation had to be conducted to discover the cause of failure using means similar to what was used in a reliability test. For example, some limited environmental testing was initiated in the meantime to investigate a mysterious problem with the power supply board where traces in the middle of the board broke along with broken leads on larger components such as the transformer and a large capacitor. See Figure 4–9. This ad hoc testing included vibration testing at sine sweep frequencies under the strobe light that revealed a resonant frequency where the circuit board took off and had flexing action up to 0.25 inches in each direction. The complete failure description along with the solution to this particular problem is described in detail in the next chapter on mechanical reliability of this book.

After 50 production units were delivered, two units were allocated for reliability testing. What seemed odd is that the 50 units had already accrued more than 5,000 flight hours of field service which seems to dwarf the 3,500-hour RDT test. When RDT finally commenced, it took two years to complete and additional design problems involving both the unit's software and the motherboard circuit board flex assembly were surfaced. Thus the test was still very much relevant to perform, despite problems having been solved previously from the field experience.

The theme for the lesson learned in this particular story, is that reliability tests have to be completed as soon as possible in order to preclude field failures and the potential wrath of the customer. The basic reason for the number of field failures that occurred on this display unit was that no reliability test of any sort was completed. Because of various program delays and the need by the aircraft manufacturer for units, two units could not be assigned

Figure 4-9 Power Supply Circuit Board Failure
A power supply board that experienced numerous vibration-related failures during field service and reliability testing. Due to the insufficient securing of the heat sink design of the board, pivoting action was introduced that appeared at certain frequencies that occur on the aircraft. When this type of situation occurs, larger components such as the square-shaped transformer and the large metallic capacitor located in the middle of the board would experience broken leads. Note that the board is severely overpopulated resulting in very little room for properly securing the heat sink on the left side of the circuit board.

for the Reliability Development Test (RDT). It would be over 50 production units before units could be assigned for reliability testing. There is no escaping the fact that design problems in a unit will surface at some point. The preferable choice for when this should occur is during reliability testing and not during field service. The delay in starting the reliability test in this example was certainly not an acceptable situation and as could be seen in this case, serious problems resulted.

4.6.3 Reliability Test Case Study #3

Type of Test: Durability Investigation Test (DIT)
Equipment: Suppression gas controller unit for U.S. Air Force transport aircraft
Test Setup: Two units undergoing 3,000 hours of thermal cycling, with separate sessions of vibration totaling 100 hours
Lessons Learned: While reliability tests as a rule do not run smoothly, some tests may be worse than others because of unresolved design issues

Some reliability tests may run longer than anticipated when many failures result from a lack of quality put into the design concept. Under these circumstances, it would not be unusual for a reliability test to take two or three years to complete. The case study described here is one story of a very difficult reliability test, a durability investigation test (DIT) that was conducted by the reliability engineering department of an electronics company that manufactured a control unit for a U.S. Air Force transport aircraft. This unit controlled and monitored the amount of suppression gas that was pumped into the full cells of aircraft to prevent fire during combat service. The unit had an LED display that allowed the aircraft crew and ground crew to view system faults for this suppression gas system. The unit consisted of eleven large circuit boards that were populated with parts to the hilt resulting in a weight of almost thirty pounds.

The first reliability engineer assigned to conducting the test was kept busy with the many failures that occurred during testing for two calendar years. Only 500 hours of the 3,000-hour test was completed! This engineer then went to another program, and a second reliability engineer was brought into running this reliability test; this took two more calendar years to complete. Both reliability engineers were pulled into being active participants in the troubleshooting of the test units. There were so many things wrong with the two test articles that it was impossible to complete more than one or two thermal cycles at a time. The test rules were such that if a failure occurred with a unit while testing, the test would stop and troubleshooting would be performed. Testing would only be allowed to continue if a corrective action was implemented to preclude the reoccurrence of that failure. However, during testing of the two test articles, multiple types of intermittent failures occurred that could not be reproduced during troubleshooting on the test bench. Thus it worked out that for every hour of test that was completed, over 20 to 30 hours of troubleshooting were performed.

A typical routine with this DIT test was that the unit would be going through a temperature change from cold ($-40°C$) to hot ($+55°C$), and failures would show up on the LED display of the unit. At that point, the test had to be stopped and the unit forwarded to the bench for troubleshooting. But the failure would no longer be present, and then the customer would have to be contacted about resuming the test again. This happened on a weekly basis to the point that only 500 hours of a 3,000-hour test was completed after two years using two test units.

Some serious design and quality problems were starting to be surfaced by reliability testing of the controller unit. Yet, management of the company was not grasping the seriousness of the problems that were surfacing, as well as the changes that would be needed to correct the design. The program as a whole was running late for the company, and it was becoming the last outstanding data items that needed to be completed to close out the first stage of the contract. Because of the destructive nature of some of the failures that were occurring during vibration testing and thermal cycling, a major investment in building several replacement multilayer circuit boards at over $1,000 apiece was needed.

The manufacturer of the circuit boards had used an inferior process during the development of the layout design of these complex multilayer circuit boards, and the result was that the traces of the circuit board would separate from the solder barrel holes during thermal cycling. For a period of time, jumper wires were added to the circuit board to bypass the failed traces. However, the boards were starting to look like spaghetti as more and more jumper wires were added. At a certain point, it became obvious that the four sets of ten-layered circuit boards from each test unit would have to be scrapped and replaced.

The second reliability test engineer assigned to the test was able to convince his customer counterpart, the reliability engineer from the aircraft company, to allow a little leeway where the test could continue to run even after certain intermittent failures occurred.

The reasoning was that eventually an intermittent failure would stay "hard" or not disappear. By doing the test in this fashion, all of the failures would eventually be discovered, and the proper corrective action could eventually be implemented. In addition, the engineer decided on a hunch to solder a wire to a key voltage point on the power supply to the outside of the chamber to a volt meter while the thermal profile was being run. It was hoped that it would be possible to capture an intermittent failure that occurred during the cold to hot ramp-up in the thermal cycle. This paid off as the voltage was found to vary by one volt and was traced to a flaky potentiometer on the power supply board that changed value during temperature changes. One by one, the causes for the intermittent failures were determined and test hours were starting to get accrued.

A total of thirty-eight failure reports were written during the test, with a number of failure reports documenting multiple failures. If this test were actually a formal MTBF measurement test, the actual MTBF would have been less than 100 hours, well below its specified value of 700 hours MTBF. The "unit from hell" has provided a number of good examples that were used in other sections of this book. Figure 4–10 shows a general overview or roadmap of where the failures occurred on the unit. It is interesting to note that the production units were able to perform fairly well in the field after many of the problems were fixed during durability testing. This shows that there is value in doing this test, even at the expense of stressing out a few reliability engineers!

One will always learn more from a test in which many failures occur. If efforts are made to understand why these failures occur and to use this understanding for future design effort, then the test will have been a success for that reason. Any engineer involved in reliability testing may occasionally have a "unit from hell" test experience such as the one described above. But this is a good thing in terms of what lessons will be learned. One learns more from failures than from success.

A number of very important lessons are illustrated here in this case. Intermittent failures are the hardest to find and to troubleshoot, particularly those that show up for a brief period during the ramp-up from cold to hot temperatures. By contrast, an intermittent failure that occurs during vibration is generally suspected to be a broken component, usually a broken lead that is intermittently making and breaking contact when shaking. What is even harder than an intermittent failure that occurs during thermal cycling is the situation where any one of several different types of intermittent failures occur during thermal cycling. Troubleshooting intermittent failures is akin to chasing ghosts so reproducing the same conditions during troubleshooting are essential to getting to the root of the failure.

Another lesson learned for this particular test was the fact that cooperation was secured from the customer in order to allow flexibility in the ground rules of the test. Without this flexibility in allowing the test to continue despite failure, the test could have never been able to move forward and reach a successful conclusion.

More importantly, the original design engineer had a responsibility to do a better job in making a more robust design than the product that actually resulted. Designing a kluge-type unit that will fall apart as it undergoes significant levels of test serves no purpose for anyone, particularly the company. All design-related issues have to be resolved in order for the unit to complete long term testing. The engineers on this program did not do their homework with regards to anticipating whether the original design and packaging was capable of enduring the levels of reliability testing and qualification testing.

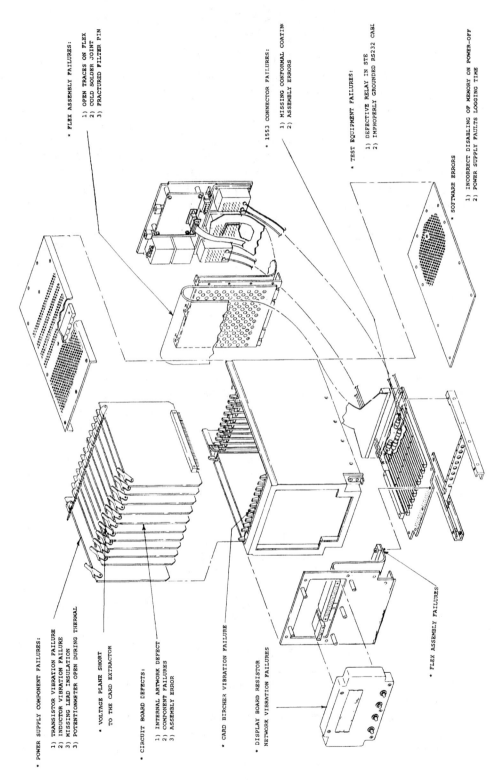

Figure 4-10 Location of Durability Test Failures on Test Units

4.6.4 Reliability Test Case Study #4

Type of Test: Accelerated Life Test (ALT)
Equipment: Commercial Aircraft Fuel Gauging components (three different LRUs)
Test Setup: One unit of each type undergoing 1,500 thermal hours with 100 hours of vibration cycling for two LRUs and 50 hours for the other LRU
Lesson Learned: Up front design work makes a big difference later on for reliability tests running smoothly

This program initially evolved as a redesign of existing fuel quantity gauging components that were originally made by an electronic company for a commercial aircraft until the aircraft manufacturer launched a newer version of this aircraft. Engineers from the aircraft manufacturer were very much surprised when the vendor told them at the preliminary design review that they were planning to do a reliability test—all of this prior to any request being made by the aircraft manufacturer. The manufacturer then requested that the reliability test be formalized under their data item for Accelerated Life Testing.

Engineers from different disciplines in the company (reliability, mechanical and electrical) worked together right from the beginning towards addressing issues from the previous fuel gauging system used on the original aircraft. In particular, various failure modes that occurred previously were addressed and corrected in the new design. Also, the basic generic type failures encountered by the reliability engineer and the project engineer from previous reliability tests and qualification tests on other programs were also addressed.

The ALT went smoothly for all units, for the most part, with the longest test taking six months of duration and the shortest test taking just three months and no failures. Test time was scheduled to be optimal such that when a weekly thermal session was completed, the unit could go right on to the vibration table for five hours. Truly, the high amount of preparation work performed during the design stage paid off with excellent performance seen during testing and eventually during field service.

Results of the ALT

LRU	Failures	Calendar time to complete test
Electronic Module Unit	0	90 days
Cockpit Display Unit	7	184 days
Ground Refueling Unit	10	187 days

Testing on the first two units was completed prior to the second production unit being shipped. In the case of the ground refueling unit, the test article was delivered late for testing, and thus the test was completed around the time of the fifth production unit being delivered. The ground refueling unit underwent the highest level of vibration because it was located in an area of the plane's wing that would typically see high vibration level during flight, even though the unit was powered off at that time. Most of the ten failures that occurred on the ground refueling unit occurred during vibration where either wires or component leads fractured. Corrective action resulting in engineering changes was developed for these failures and the balance of the seventeen failures uncovered during testing of the cockpit display unit and the ground refueling unit.

The fuel gauging system field performance clearly demonstrated the positive impact of completing an effective reliability test in a timely manner before many production units were delivered. After 86,000 flight hours were logged by this new aircraft during its first four years of service, the following performance was achieved in terms of flight hours for the Mean Time Between Unscheduled Removals (MTBUR) and compared to the predicted MTBUR as calculated by the MIL-HDBK-217 methodology:

	Field MTBUR	Predicted MTBUR
Electronic Module Box	21,600	13,500
Cockpit Display Unit	28,800	7,500
Ground Refueling Unit	12,350	6,000

It can be seen that as opposed to the normal practice of having to derate the reliability predictions by a factor of three or four times, the actual field performance was 1.5 times to 4 times better than the reliability prediction values! It would seem to imply that reliability predictions have no real bearing to field performance. However, the bigger lesson that was learned here was the importance of up-front work by the reliability engineer and the engineering team on the program to address as many issues as possible before the design was finalized. In addition, full support was given by the engineering team while reliability testing was going through so that when failures occurred, speedy and effective corrective action was developed and implemented!

4.6.5 Reliability Test Case Study #5

Type of Test: Reliability Development Tests (RDT)
Equipment: Various electronic equipment for the T-46A U.S. Air Force Trainer
Test Setup: No test completed
Lesson Learned: Unrealistic requirements placed on vendors may result in no RDT completion!

The T-46 A Jet Trainer contract was won by Fairchild Republic in 1984, and it was terminated by the Air Force on Friday the 13th in March of 1987 after only two production planes were built. A number of factors led to this termination, one of them being difficulty in getting vendors to comply with data items and testing programs.

The T-46A program was hamstrung by the fact that it was a fixed price program and that any cost overruns were to be borne by Fairchild Republic and not the Air Force. In order to win the contract, Fairchild had bid several million dollars less then the next competitor. As there was a fixed amount of monies available, the company looked to the vendors for cost savings.

What made this difficult was that a number of requirements could not be changed and reliability testing was one of them where all new equipment, or equipment that was drastically modified had to have a reliability development test (RDT) performed. In all, over twenty-one different pieces of aircraft equipment had to have RDT performed. As if this requirement was not enough, the Air Force mandated that test time be equal to twice the specified MTBF value for a piece of equipment by the vendor. This condition punished vendors who had units of high MTBF reliability values as they would have to test with more units in order to meet the test time requirement. For example, a unit that a had a specified MTBF of 10,000 hours would have to accumulate 20,000 hours of test time, far more than the standard 2,000 or 3,000 hours for most reliability tests.

Despite efforts by Fairchild Republic, the "two times MTBF" requirement would not be relaxed by the Air Force personnel monitoring the T-46A program. Some vendors reluctantly started reliability testing, while many of the others held out. Many vendors during the original bidding process had not been thoroughly aware of what reliability testing would involve and then balked when they realized the true costs involved. As a result, this was one of many things that undermined the ability of Fairchild Republic Company to effectively manage the T-46A program. The final straw came in February of 1985 when the first rollout of the T-46A aircraft was a plane that had many of its equipment missing from the rollout aircraft. The company closed in 1987 after the program was subsequently terminated.

Perhaps the biggest lesson to come out of a failed program is the need for all subcontractors and vendors to thoroughly understand the requirements. The unrealistic testing criteria using the strict interpretation by the Air Force of the MIL-STD was a major factor towards vendors not cooperating with Fairchild Republic and was one of the many things that ultimately led to program cancellation. Experience shows that a 2,000- or 3,000-hour test would have been more than sufficient to achieve the necessary results of identifying failure modes that will require corrective action prior to field deployment. After the collapse of this program in 1987, along with the Berlin Wall in 1989, defense spending was drastically reduced. Along with this situation, funds were reduced for developing "gold-plated" requirements, one of them being "twice the MTBF" test time. Testing is good but only if it has a chance of being completed in a timely manner!

4.6.6 Reliability Test Case Study #6

Type of Test: MTBF Field Demonstration Test
Equipment: Display Unit for U.S. Air Force aircraft
Test Setup: None (conducted on field units)
Lesson Learned: Some risks involved on conducting MTBF demonstration test

On occasion, a customer or contracting agency may request that a vendor forego a formal reliability test that is conducted in a lab and instead submit to a MTBF field demonstration test. Sometimes this test may be the result of mutual agreement between the vendor and the customer because there is no test facility readily available or because the program may be behind in schedule. In cases like this, it makes better sense to track the unit on the field as opposed to running a formal test in a lab. The test involves the tracking of failures against operating hours along with the calculation of MTBF on a monthly basis.

In this particular example, this test was imposed on the vendor, because no formal reliability test was conducted during the development stages of the program. In addition, the aircraft manufacturer wanted to impose a reliability warranty program (RWP) as part of this test. The ground rules were that the test would go on for one year and all failures and operating hours would be tracked and a specified MTBF value of 800 hours had to be met. If the vendor did not meet this value, some sort of spares arrangement at no charge to the customer would take place. In addition, it was specified that the test be run in accordance with the math tables in MIL-STD-781C, in which confidence intervals of 80 percent were imposed. Thus, a confidence factor or multiplier applied against the achieved field MTBF value. This table is created through the use of the chi-square statistical distribution (as discussed in Chapter 2) where the MTBF multiplier is equal to:

$$\frac{2r}{\chi^2[(1-c)/2, 2r]} \quad \text{for the lower limits}$$

$$\frac{2r}{\chi^2[(1-c)/2, 2r+2]} \quad \text{for the upper limits}$$

A number of test plans are provided in MIL-STD-781C, some at different confidence limits and some with different parameters such as discrimination factors built into the plans. A discrimination factor of 2 means that that the ratio of $\theta_0/\theta_1 = 2$ where θ_0 represents the lower test MTBF that is unacceptable (which will result in a high probability of equipment rejections using the test plan) and θ_1 represents the upper test MTBF that is acceptable (which will result in a high probability of equipment acceptance using the test plan). Terms such as consumer risks and producer risks that were discussed in Chapter 2 are also parameters that are used to identify test plans from this military standard.

Testing began in the fall of 1991 and ended a year later. A total of 116 failures accrued against 46,800 flight hours which using a operating hour to flight hour factor of 1.3 to 1, equated to 60,853 operating hours. This yield an achieved operating hour-based MTBF of 525 hours and with the applying of the 90 percent lower limit multiplier of .92 against this value, a revised MTBF of 490 hours was the result. This was below the specified value of 800 hours. A total of 96 failures of the 116 total failure were directly attributed to the problem of voids or bubbles in the LCD. (This particular problem is discussed in depth in Chapter 9.) However, the customer realized that the vendor was hamstrung by production problems that the LCD manufacturer was going through with the production of LCDs. Thus, a reasonable agreement was reached with regards to spares where the LCD panel assembly would be the free spares that the vendor would provide as soon as the production problems with the LCD manufacturer were resolved.

Note that at least there is a defined end or conclusion to the RWP test shown here. The same vendor had another program in house, in which it built an indicator for the cockpit of another fighter aircraft and had the requirement of Reliability Growth Testing (RGT) imposed. The indicator had a specified MTBF value of 5,000 hours and this number had to be demonstrated in the lab. Ten units were initially allocated for this test, with the hopes that each unit would accrue about 500 hours each with no failures occurring to demonstrate the MTBF. Unfortunately, this did not happen as failures occurred and this would now reset the clock back to zero and one MTBF or 5,000 hours had to be demonstrated after the fix for the failure. At some point, units were starting to experience wear out failures that further prolonged testing and new replacement units were needed, further prolonging the test where it lasted for more than three years.

The point here in comparing the RGT and the RWP is that sometimes demonstrating a high MTBF value in a test as the conclusion of the test may be a difficult, if not impossible goal to achieve. The RGT can eventually start sucking in additional units as well as test facility and engineering hours. Thus RGT is a case involving the politics of a MTBF value.

4.6.7 Reliability Test Case Study #7

Type of Test: Reliability Enhancement Test (RET)
Equipment: Display Unit for U.S. Air Force fighter aircraft
Test Setup: Two units undergoing thermal and vibration cycling in incremented steps
Lesson Learned: The Breaking point is defined

The aircraft manufacturer of a U.S. Air Force fighter aircraft requested that the vendor of a display unit perform a reliability development test that incorporated the concept of step-stressing for thermal and vibration tests in lieu of conducting a conventional reliability test that used fixed thermal and vibration profiles. This test was called reliability enhancement test (RET). The objective was to accelerate the reliability growth by subjecting production hardware to environmental stresses that was tailored to stimulate failures which might be encountered over the service life of the equipment. The RET employed the following test sequence:

Step	Vibration level (GRMS)	Temperature range
1	3.5	$-54°C$ to $71°C$
2	4.2	$-60°C$ to $75°C$
3	4.9	$-65°C$ to $80°C$
4	5.6	$-65°C$ to $85°C$
5	6.3	$-65°C$ to $90°C$

Vibration was conducted for one hour in each axis. Each thermal cycle was conducted for 100 hours (fifty cycles of a four-hour thermal profile). An acceptance test procedure was conducted after each thermal and vibration session.

Two failures involving the flex circuit rubbing against the chassis and causing an electrical short during the vibration sessions that were conducted during the latter part of the test, step 4 and step 5. Corrective action involved additional reinforcement of the flex in the area of the failure. The test was completed in less than six months.

4.7 HIGHLY ACCELERATED LIFE TEST (HALT)

Highly accelerated life test (HALT) is one of the newer trends for reliability tests to arrive on the scene. The philosophy of this test is to conduct stress testing on the component in order to quickly uncover weak links in the design and fabrication processes of a product. The main features of this test are:

- Omni-axial vibration (six degrees of freedom vibration)
- Rapid temperature transitions (High/low extremes)
- Voltage margining

The goal of this test is to do this within a two to three week time period at labs that have the special equipment to perform this type of testing. Usually, in addition to the lab cost, the company will send an engineer and a repair technician along with a quantity of spare parts, so that troubleshooting of any failures can be done quickly and repair implemented quickly.

Usually, the test is conducted in a special chamber that can be programmed to produce a high rate of change in temperature while exposing the test items (typically two or three units) to six degrees of simultaneous random vibration. The six degrees of freedom vibration include:

1. Motion in the X-axis
2. Rotation about the X-axis
3. Motion in the Y-axis
4. Rotation about the Y-axis
5. Motion in the Z-axis
6. Rotation in the Z-axis

Quick temperature changes coupled with the vibration can bring out any weakness in the soldering process or even problems with component quality.

The results can be very good with this type of test, as the weak links of a design are discovered quickly. After completion of this test, the manufacturer can have confidence that the design will be able to meet the stresses of the field environment. In order to conduct this test, company management has to be able to make the commitment in terms of money to be allocated in a short period of time. Sometimes HALT is not able to be accomplished and a close second to this test is RET, where step-stressing is performed by expanding each thermal and vibration level in consecutive steps.

4.8 PRODUCTION RELIABILITY TESTS

After reliability tests are performed on new designs, some sort of testing is generally required on production units to ensure the quality and reliability of these units. Most units undergo what is known as environment stress screening (ESS) tests, commonly known as burn-in testing. In various cases, some units will undergo production reliability acceptance testing (PRAT) where failures are tracked against a customer requirement.

Independent of reliability development testing, there are production reliability acceptance tests that are performed on production units going on military aircraft. These tests typically involve a fixed amount of vibration time and thermal cycling, independent of the burn-in. If units fail during this test, repairs must be made and typically two thermal cycles are performed for verification of the repair. The purpose of PRAT is to catch the situation of a bad lot of components as well as verifying a manufacturer's assembly process of the unit during production. If a production line does well during PRAT, testing can be relaxed at the discretion of the customer. Unfortunately, if a unit has developed a problem, it is conceivable that excessive amounts of vibration can be accrued on the unit, making it a fatigued unit prior to entering the field! The overall consensus of this type of testing is that they are basically useless if a rigorous qualification testing and reliability testing has been performed in the program along with continued burn-in testing. The trend has been to go away from this type of testing and concentrate on the design and the burn-in.

4.8.1 Environmental Stress Screening (ESS) Tests or Burn-In

Environmental Stress Screening (ESS) testing (also known as burn-in) will vary depending on the type of equipment and what the field environment will be. ESS testing for most aircraft equipment involves a two-part process: vibration and thermal vibration. A typical ESS test might involve 10 minutes of total vibration with the axis of vibration being applied perpendicular to the printed circuit board. Where electronic components in the equipment are oriented in more than one plane, the equipment should be shaken in each of the three orthogonal axes (X, Y and Z), typically at 5 minutes apiece. Standard burn-in vibration levels used in industry is chosen at a level around 6.06 GRMS or 0.04 G^2/Hz level (U.S. Navy Spec. NAVMAT P-9492 profile) as shown in Figure 4–11. After the unit successfully completes an acceptance test procedure after vibration, it will then be installed in a thermal chamber for thermal profiles, typically two days worth of eight to ten complete cycles. At different points in the thermal cycle (hot and cold points), the unit will be functionally tested. After successful completion of thermal testing, a final Acceptance Test Procedure (ATP) is performed.

ESS testing for stationary-type equipment such as computers, TV, or radios that will be installed on a desktop may involve a simple power on test for an hour of operation. As this environment is generally benign in that little vibration will be seen and that the equipment will be in room temperature for the most part, there is usually little need for the manufacturer to conduct thermal or vibration testing as in aircraft equipment. However, for consumer electronics, the biggest vibration problem is seen during shipping, so particular attention to packaging is needed. Some manufacturers run vibration tests on packaged units to simulate shipping, including random vibration and shock testing. Shipping companies such as UPS have defined the vibration profile and levels that will correctly simulate actual shipping environments.

ESS failures should be of concern to the reliability engineer as potential reliability problems may be uncovered, particularly if the vendor did not yet start a formal reliability test program. A case study of this particular scenario is provided in Chapter 5 involving connectors.

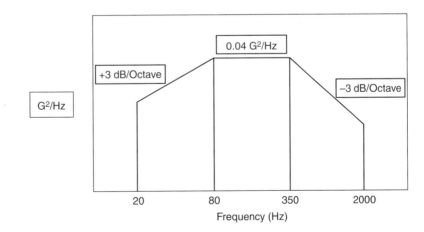

Figure 4–11 Typical ESS Random Vibration Profile Used for Aircraft Equipment

4.8.2 Production Reliability Acceptance (PRAT) Tests

Some military programs may require additional reliability tests to be performed once ESS has been completed. These tests are designed to give the customer a level of confidence that all design-related problems have been found and that parts quality and manufacturing issues are discovered quickly during the production process.

These tests tend to be conducted on military aircraft products and involve a test sequence similar to ESS, where a specified amount of vibration is performed (usually 10 minutes), eight to ten thermal cycles performed and ATPs performed after each test. The vibration that is chosen is usually the functional level of vibration specified for the equipment and the thermal profile can either be the one used during ESS or in some cases, reflect the same profile that was used in reliability development testing.

There is a threshold level for the total number of failures that can occur during cumulative PRAT testing, typically at two failures. If this point is reached during cumulative testing, the equipment manufacturer has to enter into a corrective action cycle and testing is allowed to continue with the cumulative failure counter set back to zero.

Unfortunately, experience has shown that PRAT can sometimes prematurely age a unit too much beyond its life if too many failures occur and verification thermal and vibration cycles have to be run. The accumulation of vibration time is a big concern towards aging a unit before it is allowed to enter into field service. When a situation like this occurs, it is indicative of a parts quality problem in the unit or even a design problem that has not been resolved. When the latter is the case, it may be that the reliability test that would be conducted during the design stage was not performed yet, and thus PRAT is taking over this particular function!

Because of the possibility of premature wear-out and the amount of expense that this additional testing can add to a program, PRAT testing has become less of a requirement. In most cases, ESS has served the needs of both the equipment manufacturer and the customer and thus there is generally no need for additional production testing.

4.9 TROUBLESHOOTING TIPS FOR RELIABILITY TEST FAILURES

There are some basic patterns that failures follow throughout the different types of reliability tests that will help in the area of troubleshooting. Some of these basic patterns are presented as follows.

4.9.1 Thermal Testing Failures

- The ramp-up from cold to hot produces the most severe stress to electrical components because of the different materials used—ceramic, plastic, and metal.
- During the ramp-up from cold to hot, condensation is generally present with the buildup of frost inside the thermal chamber now melting. This is an additional "bonus" or stress for the electrical components in an electronic unit, particularly in the area of analog circuits and exposed voltage points on a circuit board.

4.9.2 Vibration Testing Failures

- Vibration testing is particularly harsh to unsupported wires and to any components that are located in areas of the circuit board that may not be well-supported. Many of these things can be spotted prior to any reliability test beginning and should be addressed during the early stages of the design.
- Power supplies with large components have significant weight and are particularly prone to breakage if there are any weaknesses in the way the circuit board is supported. The use of switching power supply designs can reduce weight and the use of too many large components.
- Extra-long components such as ICs with many pins or resistor networks are prone to pin breakage during vibration.

4.10 GUIDELINES FOR THE TIMELY COMPLETION OF RELIABILITY TESTS

It is really a desirable goal for the reliability engineer to complete the reliability test in a timely fashion. Tests that take too long can drain the company's resources, and this could create a loss of interest in the program. A few things can be done to ensure that the test can move forward at a decent pace. The following represents a list of general guidelines for engineers trying to complete a reliability test in a timely manner:

- Do as much up-front work during the design stage to preclude the occurrence of certain failure modes. Work with the different engineering disciplines on the program to accomplish this. This will save much grief to the reliability engineer later on when reliability testing begins!
- The way that an electronic unit is put together represents a sort of history of a company and its processes. Therefore, it would be impossible to escape testing without failures if there are flaws in the company's processes or if there are quality problems in the parts used. Identify these problem areas during a design review or physics of failure approach as early as possible in the design stage.
- A reliability engineer needs to keep in direct regular contact with his counterpart on the customer side and develop a good working relationship during the testing. This counterpart may be more supportive than the managers and engineers in the reliability engineer's company! This engineer will work with you if you are truthful and are cooperative with information.
- Reliability testing may sometimes be viewed by managers as an unnecessary burden to the company. If a test uncovers problems, it may often be viewed as a bad thing by management because of the corrective action that is needed to be developed along with the other logistical tasks (retrofit, documentation, etc.). Yet it should be looked at by the company as a process towards making the product better able to endure the field environment. Unfortunately, many managers in a company are not able to make this connection and the reliability engineer may often be caught in a no-win situation when failures occur during testing.

4.11 SUMMARY

This chapter has illustrated the different types of reliability tests that can be performed on a new design. As can be seen from the various examples provided in this chapter, the task of conducting a reliability test can often be very difficult to run, not only for logistical reasons but also if many failures should occur. However, the engineer that conducts the test must not lose sight of the end goal of discovering any major design issues and must keep the test going forward in a timely manner.

When a reliability test is properly conducted, it is very helpful to the design process. A manufacturer of a product has to buy into the fact that reliability development testing for new designs and ESS testing for production units are both beneficial to the long-term reliability of a product and to the company's way of doing things. Both of these tests can identify design or manufacturing processes that need to be improved and this will benefit future product lines by the company. Any company that avoids performing either test strictly because of cost and schedule reasons may end up falling into difficulties during field service of the product. When major failures are discovered during field service, there is a significant cost impact to the company when retrofit action is required.

It may be a very difficult task for the reliability engineer to convince the company to perform reliability testing when the customer does not make it a formal requirement. Too often, companies may be short-sighted in looking at the costs and schedule impact of conducting such a test rather than recognizing the benefits of conducting such a test. Any type of reliability test that can be performed before production will be beneficial, whether it be a formal environmental test as described in this chapter or even a pilot test where a unit is installed in a test environment and monitored carefully.

A case that illustrates this is the situation of a company that was able to manipulate itself out of an informal agreement with the customer that some sort of reliability testing be performed on fuel system components that the company made for a business jet aircraft program. The company used the argument that it made no sense to give up a unit for testing when it could be used for production. Also, the company argued that reliability testing was not a formal requirement set forth by the customer. This thinking proved to be very short-sighted, as all units from the first production lot of 20 units had to go through a major retrofit where they ended up being returned to the company for extensive redesign and repair caused by major design issues. The costs in running this test proved to be prohibitive as well as the reputation of the company being somewhat tainted. Certainly, a reliability test would have been a viable and less costly option in discovering these failures early!

4.12 EXERCISES

1. Describe some of the reliability tests that were discussed in this chapter and how they differ from each other? Which ones are identical but have a different designation?
2. What reliability development tests are conducted in order to measure the actual MTBF performance against a specified MTBF for a product? What reliability tests are not set at a fixed length in duration?

3. What tests have the highest risk or expense to the company running the test with regards to the number of units that will be required for the test as well as the amount of test time needed? What part of the thermal test profile applies the most stress on electrical circuits?
4. Based on what has been discussed in this chapter, what types of components and failure modes are seen during thermal testing?
5. What is a possible thermal profile length for a unit that would meet the requirement of at least twenty cycles completed during a five-day period?
6. What axis of vibrations applies the most stress to electrical circuits and components?

CHAPTER 5

Mechanical Design Impact on Electronic Component Reliability

INTRODUCTION

The mechanical or structural aspect of a design is of major importance in how well the product performs in the field. Even in the area of electrical products, a weak mechanical design can undermine the reliability and performance of the product.

The discipline known as mechanical engineering can cover anything in the range of small mechanical assemblies up to the level of an entire aircraft. Fatigue and stress analysis are some of the tasks that are performed by mechanical engineers on both small and large components. Volumes of books cover how to perform these types of tasks.

There is, however, an area that may not get the same level of attention that the structural parts of an aircraft may get in the area of mechanical analysis, and this is the mechanical aspects of electrical circuits. The way that circuit boards are put together or the way that components are mounted on circuit boards are issues of concern for a new design and can adversely affect reliability. It is necessary for the reliability engineer to identify these issues, particularly since they are rarely covered in textbooks and the knowledge typically is gained through experience.

As mentioned in previous chapters, the mechanical design aspect has more of an impact on reliability of electrical equipment than the electrical design aspect itself. In fact, the general experience from both testing and field service of electrical equipment has shown that only about 10 percent of failures are caused by inherent electrical part failures while the rest are caused by quality issues, manufacturing process errors, and mechanical related problems.

Electrical engineers generally do not look at the mechanical packaging of the overall design, even though poor mounting or packaging can adversely affect the reliability of the electrical components in a unit. It is beneficial to the program that the reliability engineer or the project engineer look at all of the mechanical-related factors at the beginning of the program and minimize the impact of negative conditions. It is important for these issues to be identified and addressed right away during the development of the design.

In this chapter, we will examine mechanical factors that affect electrical equipment designs, both on the component level and in the overall unit assembly. Remedies for these factors will be discussed as well.

5.1 BASIC CONCEPTS OF STRUCTURAL RELIABILITY

Both mechanical and electrical components have a physical presence and, as a result, will have a structural reliability. Any structural component is subjected to a series of cyclic stress throughout its life. The fatigue life can be defined as the total number of stress cycles required to initiate a dominant fatigue crack, plus the effects of additional cycles required to propagate the crack to a final failure state.

Failure of a structural component is based on the strength or the resistance of a structural component versus the load applied to the component. Both the strength and load are typically functions of time (or the cumulative number of stress cycles). Also, when dealing with metal structures, degradation factors such as corrosion are also based on time. Thus the probability of failure is a function of time and is called time-dependent reliability.

Various methodologies for time-dependent reliability have been developed for larger structures such as bridges and ships. This concept can also be brought down to the level of smaller structural components such as valve seats in water faucets, in which each cycle is based on the force of both the water and the application of the washer. The concept of stress cycles can be applicable to smaller structural electronic components that are installed on circuit boards, particularly in the area of the leads and the packaging, as the component will see stress cycling during both testing and field service.

5.2 MECHANICAL FAILURE MODES OF ELECTRICAL COMPONENTS

During the design stage, an electrical component may be viewed strictly in terms of its electrical characteristics. However, it is also important to view an electrical component as a structural device that is subjected to the forces of mechanical stresses. In a benign environment, limited amounts of shock and ambient vibration may occur resulting in little stress on the component. However, in an environment where the device is situated inside a moving vehicle, there will be some applied mechanical stresses that require reliability evaluation.

Virtually all electrical components will have some external leads extending from the package of the device. Other components will have some internal wiring such as the filaments of lamps. Any pivoting action that is applied to a section of the lead, the filament, or wiring can be considered as mechanical stress cycling. Sometimes the cycling may be occasional or random, whereas other times the cycling may be applied at a regular frequency because the component is in a moving vehicle. The effects of mechanical stress cycling could actually be amplified if the natural frequency of the component or parts of the component is present in the environment causing resonance. If the situation of resonance is reached, amplification allows for large amplitude or displacement of the movement or cycle.

As a very rough rule of thumb, the larger the electrical component in terms of size and weight, the more concerns there are with regards to mechanical reliability. Another version of this rule is that the more area and dimension that a component has, the more concerns there are with regards to structural reliability. Any electrical component that is electromechanical in nature, where there are moving parts such as potentiometers, relays, fans, and motors, is always a concern with regards to reliability.

Components that should be on the list for a reliability engineer to investigate regarding mechanical issues in new electrical equipment design are:

- Wiring and wiring harnesses
- Large transformers, inductors and capacitors (as used in power supplies designs)
- Terminals
- Printed circuit board quality, component population density, and mounting inside the unit
- Lamps
- Connectors

Certain components are more susceptible to mechanical failure because of their larger size or because of unique mechanical issues associated with their design. These will be examined in detail in the following sections.

5.2.1 Potentiometers

Potentiomenters (or pots as they are commonly called) are always a major reliability concern when used in an electrical circuit. Pots are electromechanical devices that consist of a wiper blade and circular plate on which this blade rides on, giving it the ability to provide resistance from zero ohms to the maximum rated resistance of the pots. Unfortunately, there is a tendency for film or corrosion to develop on the plate, causing inconsistent resistance reading when the wiper is in contact with it. For example, one would notice this when using a volume control on a musical instrument or amplifier where static is heard as the knob of the pot is turned.

A number of manufacturers have used miniature pots in power supply design in order to allow them flexibility to adjust the various voltage settings. After the final voltage adjustment is made, a drop of colored epoxy is placed over the turning shaft in order to hold the position in place. However, reliability issues with pots range from open circuit to resistance changes. Please refer to Figure 5–1.

As illustrated in the figures, the potentionmeter is primarily a mechanical device, and there are a number of potential areas for failure, particularly since the slider unit is only making a mechanical connection with the metal plate and carbon element. If there are any defects with the slider unit or the plate, the resistance could change. Also, if the unit is not sealed properly and if any moisture is captured inside the unit itself, it could lead to problems during thermal cycling.

A) Rectangle model

   1) Side view (cut away)

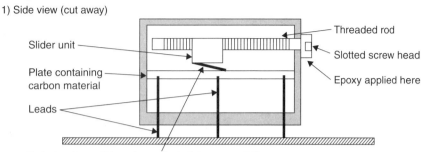

   2) Top view of plate and slider (cut away)

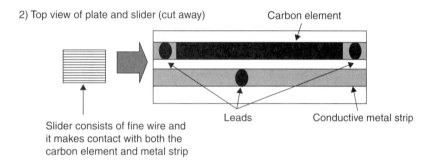

B) Circular model (Cut away view)

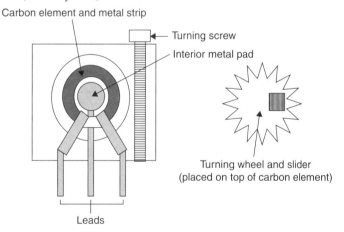

Figure 5-1   Cutaway View of Different Types of Potentiometers

For the circular model potentiometer shown above, the turning screw turns the wheel and slider, with the latter making contact with both the interior metal pad and the carbon element.

Figure 5-2 Pots used in Interface Circuit for Field Adjustments
This is a photo of an aircraft fuel interface unit that uses potentiometers. On the right side of the photo is a movable panel above a series of pots in the unit that are adjusted in accordance with fuel readings from the fuel probes in the tanks.

Another issue with the use of pots is that once they are set to a certain spot, such as in the focus adjustment of a CRT, they have to be held in place by an adhesive. Otherwise the pot will move during normal shipping and installation.

New power supply circuit designs depend less and less on potentiometers. The goal would be to use precision resistors instead and depend less on a variable resistor and final voltage adjustments to be made. However, pots are still used for certain circuit adjustments that are performed on the field. One specific example of this application is in the area where aircraft fuel probes capacitance values feeding into an interface unit require in some cases the use of pots to adjust the correct sensing of each input line by the interface unit. The photo in Figure 5–2 illustrates an example of this particular case. Again, newer circuit designs using the principle of feedback control are reducing the need for additional field adjustments of pots in situation like this.

5.2.2 Resistor Networks

Resistor networks come in a number of package sizes, from four pins in series to ten pins in series, and they are commonly used as isolation resistors, pull-up resistors and whenever there is a repetitive use of the same resistor value in an electrical circuit. The longer-style resistor network packages are especially prone to vibration problems. Typically, the failure mode is that one of the end pins will crack and the two halves are still pressing together, yet will open up during some application of vibration (see Figure 5–3). As the long package size of the resistor network is generally left unsupported on most circuit boards, it is espe-

A) Last pin of resistor network is usually affected

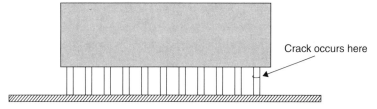

B) Rocking motion is induced by vibration perpendicular to body

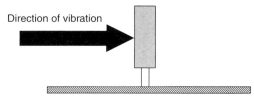

Figure 5-3 Resistor Network Mechanical Failures

In a fairly high-vibration-level environment, it is generally a good idea to add some kind of filleting or filler material such as silicon or epoxy around the pins of the resistor network in order to reduce the effects of rocking.

cially prone to rocking motion that causes this cracking. The natural frequency for vibration resonance for the larger package size is in the range of 200 to 500 Hz, well in the range of what most electrical equipment will see when installed on aircraft or even in automobile applications.

5.2.3 Tubular Capacitors

The larger value capacitors are usually in the form of tubular or axial-style capacitors, such as the tantalum and the electrolytic capacitors. These capacitors are physically larger than the ceramic material capacitors, and because of this larger size they may have mechanical issues. Specifically, the leads of these type capacitors have a tendency to break on occasion during assembly and often during vibration, whether in the field or during testing, as shown in Figure 5-4.

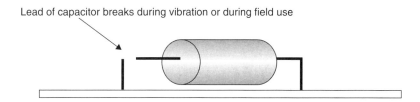

Figure 5-4 Tubular Capacitor Mechanical Failures

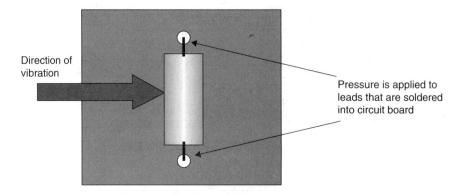

Figure 5-5 Direction of Worst Case Stresses on Tubular Capacitor

In recent years, the availability of some tantalum capacitor values in surface mount configuration has reduced mechanical issues. However, for the larger values and high voltage ratings, this is not always possible, and mechanical fixes have to be implemented to reduce the impact of vibration and shock.

The failure is usually induced during vibration testing or field service and is worst when the vibration is applied in a perpendicular direction into the body of the capacitor (this can either be the X- or the Y-axis, depending on how the circuit board is mounted). This particular vector causes a rolling action that puts more pressure on the leads as shown in Figure 5–5.

There are some fixes that can be implemented to reduce the effects of vibration and shock on the leads of tubular capacitors. These are illustrated in Figure 5–6.

5.2.4 Lamps

Incandescent lamps are a continual reliability problem for electronic displays that use them for the backlight circuit. Because of the internal structure of the lamp, where a fragile filament wire is used, certain types of lamps may be susceptible to the high-vibration levels that would be seen in aircraft and automobile applications. The worst direction of vibration is when it is applied perpendicular to the filament as shown in Figure 5–7.

A detailed case study of lamp failures is provided in Section 5.4. It is basically impossible to avoid incandescent lamp failures when they are used in high-vibration environments, even when all attempts to physically isolate the lamp from direct vibration. The need for lamps being used in backlighting is becoming less as higher-intensity LEDs are being used. The LED has the advantage of being a more compact and more durable package, along with being able to take higher levels of vibration than lamps.

5.2.5 Transformers and Large Inductors

Many types of electrical equipment that are used in our world will have a built-in power supply inside the unit. The function of the power supply is to take external power and convert the

A) Add service loops to the leads

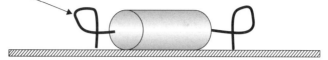

B) Cut the leads and solder flexible wire from the lead to circuit board

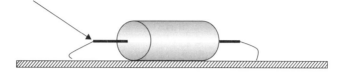

C) Add RTV or silicon potting compound to hold capacitor to circuit board

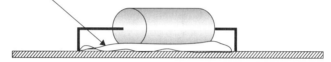

D) Add clips to hold capacitor to circuit board

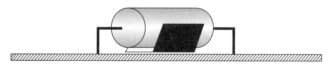

Figure 5-6   Various Fixes for Capacitor Mechanical Failures

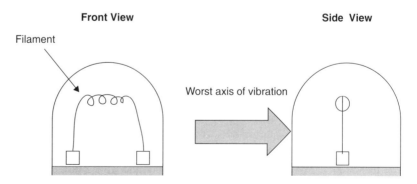

Figure 5-7   Incandescent Lamp Mechanical Failures

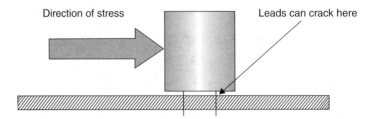

Figure 5-8 Transformers and Inductors Mechanical Failures

necessary voltages to the voltages required for its internal circuitry. For example, aircraft equipment usually works off of a 28-VDC voltage bus, so there has to be a step-down power supply that converts this external voltage that is coming in into 5- and 12- volt circuits for its internal digital and analog circuits. Likewise, household appliances have an internal power supply that converts 110-VAC wall voltage to a step-down DC voltage.

Many of these internal power supplies used in electrical units will have large components such as transformers and large inductors, as well as large tubular capacitors. And like the tubular capacitor described in the previous section, transformers and inductors have similar situations involving mechanical stressing and breakage of the leads. If the body of the transformer or the inductor is to be used in a high-vibration-level environment, there is concern about whether the leads coming from the inductor will be able endure the induced mechanical stresses as shown in Figure 5–8. It is hard to believe, but some manufacturers may actually bring the coil wire out of the body of the device to act as leads, and these leads may break under certain vibration conditions.

As in the tubular capacitor, epoxy or silicon may be needed to act as a fillet and add support to the body of the inductor for high-vibration-field applications.

5.2.6 Wiring

Wiring concerns fall into two areas: internal and external wiring. Wires that are situated inside of units have two major mechanical issues. One concern is how they are connected on either end, whether to a circuit board or to a connector. Another concern is whether there is any pivot action along the routing of the wire. This pivot action could be on a regular basis, as in the example of wiring going to a magnetic head on a credit card reading device. Inadvertent pivot action could occur when the unit is taken apart or serviced on a regular basis and the wiring is moved.

Internal wiring requires adequate securing, routing, and support in order to prevent both pivoting and pulling actions that are induced by the ambient environment. When they are not, then potential for breakage exists based on the function of the field environment. A number of these cases are presented in the second case study in the later part of this chapter (Section 5.4). Generally, the need for internal wiring in a device has been reduced in product designs over the years with more of point A to point B connection being incorporated inside multilayered circuit boards.

External wiring is still a necessary evil when connecting multiple units in a system, such as in an aircraft or an automobile, through the use of wiring harnesses. General areas

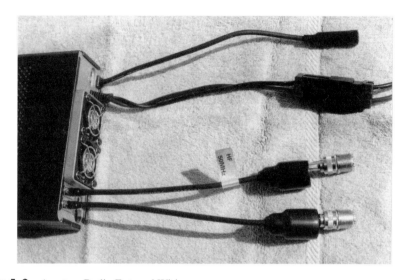

Figure 5-9 Amateur Radio External Wiring
The two cables coming out of the unit at the bottom of the photo are to be connected to antenna feedlines. This can be a concern regarding mechanical stresses when the feedlines are a bigger diameter cable than the cable coming from the radio.

of concern include how well the ends of the wires are connected to the connector pins and whether they should be crimped or soldered. Also, assuming that the wiring is made correctly, there should always be concern with how the wiring is routed in a system, so that no pinching or pivoting action should occur to stress the wire. Also, with external wiring, a concern is maintenance, where it helps to have the wiring bundle move with ease so that it does not catch on other structures and components during maintenance action. Further discussion on the maintenance aspect of wiring is discussed in Chapter 14.

An example where external wiring is a concern is shown in Figure 5–9, which shows the rear of an amateur radio transceiver that, in addition to the power cable, has introduced two external cables to coax connectors for the purpose of hooking up to an antenna. Most radios have these connectors built on the rear panel of the radio, but because this is a very small radio with a lot of functionality, the only way for the manufacturer to provide these connectors was via the external cable method. This would normally not be too bad if the feedlines to antennas were more or less permanently attached; however, as this radio is primarily designed for portable and mobile use, the cycling of antenna connection removal and replacement will cause breakage in this cable. This case is made worse by the fact that the coax of most feedlines can be as thick as .5 inch in diameter, which when connected to this .25-inch cable will induce additional stresses.

5.2.7 Printed Circuit Boards

Printed circuit boards have become more and more complex over recent years. Early printed circuit board designs consisted of two sides only; currently, circuit boards can be

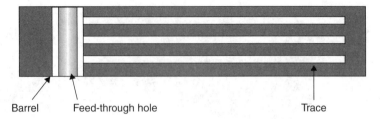

Figure 5-10 Cross-Section View of Multilayer Circuit Board

made with as many as ten distinct internal layers. Because of this increased functionality of circuit boards, significant amounts of internal wiring can be eliminated. Printed circuit boards can be made commercially by a number of manufacturers, and even though they may be thought of as passive-type devices, a number of reliability issues must be addressed during the process of constructing them. Circuit board reliability involves several items such as the manufacturing process in the construction of the plated-through holes and the number of layers.

The plated-through hole is used in double-sided and multilayer printer circuit boards to connect component leads to board circuitry. The plated-through hole can often be the largest contributor to circuit board failure for these types of boards. This is because there are differences between the thermal expansion of the epoxy glass base material and the copper plating. The epoxy glass and the copper expand and contract at different rates during thermal cycling. This results in axial-direction strains on the plated-through hole barrel wall; it also weakens the mechanical properties of the copper plating and eventually leads to open circuits through the effects of thermal or vibration cycling. If there are any inherent weaknesses in the processes or layout by the vendor, the problem is made much worse. Poor drilling or excessive acid etching during the plated-through hole cleaning process can lead to imperfections in the barrel wall and can lead to open circuits as well.

Multilayer boards have additional layers of circuit separated by epoxy glass laminations. While this leads to higher packaging density, it creates additional problems involving the plated-through holes. The same type of failure modes are experienced here as in the double-sided boards.

The manufacturing process used to make multilayered epoxy glass circuit boards is a major concern. As more and more functionality is incorporated into circuit boards while not increasing the overall size or footprint, more layers are added. It is not uncommon for a complex board like a microprocessor circuit board to consist of ten distinct layers. Figure 5-10 shows the basic structure of a multilayered circuit board.

A second major issue concerning circuit boards is the thickness of the boards. The decision on the appropriate thickness to use for a circuit board should be made during the early stages of the program, right after it has been determined how many internal layers will be needed to support the artwork. In addition, the mechanical engineer has to determine the appropriate circuit board thickness that will be able to sustain the levels of vibration that the unit will see during field service, as well as being able to endure normal maintenance and repair tasks.

Companies may fall into the trap of using the thinnest board possible for weight savings, however this can sometimes lead to some major problems. Look at the case with computer monitors in which it is typical to have a very large circuit board that has heavy components such as transformers installed. The board tends to be thin and subsequently endure damage when shipped or when repaired. Figure 5–11 shows an example of this scenario.

Another major concern with printed circuit boards is rework, in which components have to be removed and replaced from the board. As the hand soldering operation can introduce excessive amounts of heat when removing and replacing components, a high probability of damage being introduced to the through holes and traces of the circuit board can occur. Any rework that occurs on a circuit board has to be monitored carefully and kept to a minimum. The integrity of the circuit board can start to be questioned when there are a number of wire jumpers added to replace failed traces.

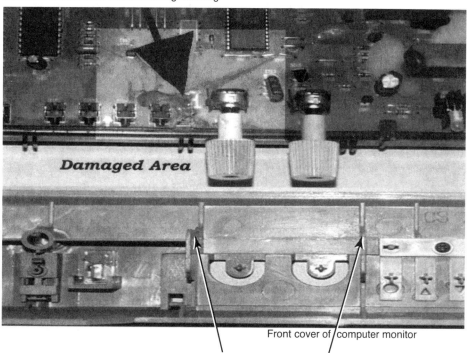

Figure 5-11 Computer Monitor Main Circuit Board Failures

Photo A shows the type of damage that can be inflicted by a design that does not take into account handling problems.

B) Mechanical failure of circuit board in area of the power supply transformer

Damaged Areas

Figure 5-11 Computer Monitor Main Circuit Board Failures *(continued)*
Photos in B show the amount of damage that can be inflicted on to very thin circuit boards during field use. The circuit board is only .06-inch thick and has some massive components mounted on it, as seen by the side view of the power supply transformer shown in the lower photo. The weight of the transformer has actually ripped the board apart as shown in the top photo, even though the board has a structural frame underneath for support.

Finally, the artwork or layout of a printed circuit board is also critical to the reliability of the fully assembled circuit board. The traces must be routed so that there is no conflict between other traces where inadvertent electrical shorts can occur. In addition, the layout of the board must be such that components are spaced evenly as much as possible to prevent mechanical issues regarding overall weight balance of the circuit board. This is typically the responsibility of the mechanical engineer, but the reliability engineer needs to monitor this carefully as defects in the circuit board construction, layout, and population of components will definitely impact reliability during testing and field service.

5.2.8 Flexible Circuit Boards

Flexible circuit boards or flex circuits are used frequently in electrical design in areas where it is difficult to use standard-size wiring. As opposed to a wire bundle or ribbon cable, flex circuits can be made very thin such that they can fold into the mechanical structure of the design. The hand-soldering of connectors to the ends of the flex circuit board is a major source of concern. Because flex circuits generally cannot be wave-soldered in the area where the connectors are attached, hand-soldering has to be done in this area. The amount of heat from the skill of the solder operator will determine whether any weaknesses are introduced into the flexible circuit. Hand-soldering may also be used to attach a flex circuit to individual pins that are used for mating with connectors and because of the number of pins involved, concern may be warranted. Please see Figure 5–12 for example of this type of design.

As with glass *epoxy* material printer circuit boards, the integrity of the manufacturing process of the flex circuit is also critical, particularly with the trace connection to feed-through holes for multilayer circuit boards. Another concern is the way that a flex circuit is folded into a unit. It is important to prevent excessive cycling or bending in certain areas of the flex because of the way it is installed or when it is removed.

Figure 5-12 Flexible Circuit Board with Connector Pins Installed
The photo shows the end of a flexible circuit board that has three separate sections with connector pins installed. Each of these pins must be carefully hand-soldered into the flexible circuit board. If heat is applied too long during the soldering, there is a possibility of weakening the circuit board and introducing a potential latent condition that may fail in the future.

5.2.9 Connector and Sockets

Electrical connectors are used in some form in just about every electronic product made in the world, and they are a major reliability concern. For instance, connectors are heavily used in products such as automobiles and aircraft. Connectors are a necessary evil, and the integrity of the connection becomes a major concern when bringing the electrical signals and voltages from one circuit board to another circuit board of a unit, as well as external connectors to the outside world. The subject of electrical connector reliability can fill a book by itself concerning the different failure modes of different types of connectors.

Electrical connectors provide an electrical junction from one circuit to another, with a male half and a female half comprising the connector assembly. Connectors conduct electricity, but they are mechanical devices with the main failure mode being the loss of continuity. The loss of continuity may be caused by factors that are mechanical in nature, such as dimensional problems, or environmental in nature (such as corrosion, humidity, or temperature changes). Also, there is a concern with multiple-pin connectors where individual pins can be pushed in and not make appropriate contact.

The two halves of the connector can be positively locked in many ways, but because of its mechanical nature, failure is always possible. A manufacturer may use a locking tab to hold the halves together in a conventional connector, as shown in Figure 5–13. If the connector is a circular type, locking action may be accomplished through the use of threads, as in a coax connector, or through locking in place, as in a bayonet connector.

The reliability engineer may be involved when connectors fail during field service because of temperature factors, vibration factors, and humidity factors. Any reliability engineer involved in reliability testing or field failure analysis will need to develop a practical working knowledge of connector failure modes and various corrective actions. Part of the reliability engineer's general charter is to look for ways to reduce the number of connectors in an electrical design, and when they are needed, to ensure that they do not break apart because of environmental factors. Another area of concern is with quick disconnect or bayonet-type connectors, where there is the element of wear or damage caused by regular disconnect during maintenance or disassembly. An electronic unit that has a high failure rate on its internal circuitry may also experience secondary connector failure if the unit is frequently being removed and replaced, as well as being disassembled frequently.

Some mechanical test setups can be constructed in order to simulate the cycling action of connecting or disconnecting a connector pair. The results of these tests can be used to help set up a predicted reliability value, in which the connector pairs can be rated by the number of mate/unmate cycles it can endure during the life of the product in the same way that relays and switches are rated. The value used to measure this is mean cycle between failures (MCBF), and as each connect and disconnect cycle represents an event, this aspect of reliability can be viewed as a discrete function. Unfortunately, as cycle field data is not always available, the number of cycles may have to be estimated, or as is often the case, the connector reliability may be measured in terms of time (MTBF) for the purpose of expediency.

A major area of concern in the assembly of connectors, is how the wires themselves are connected to each connector half. Crimp-type connections and solder-type connections

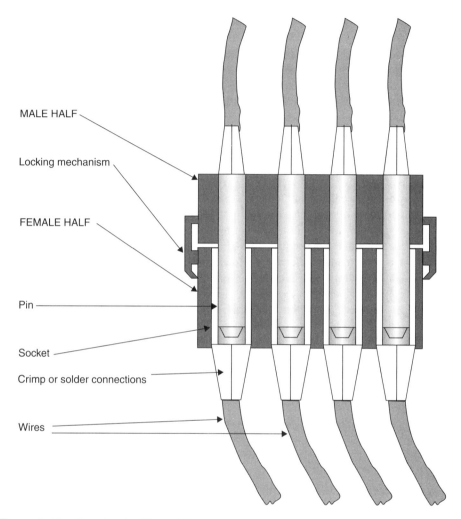

Figure 5-13 Cross Section View of Connector Pin and Socket

bring their own set of potential failure modes, particularly those involving workmanship quality. Weaknesses in the crimping or soldering process may be uncovered during long-term reliability testing, production burn-in tests or field service.

Another component that is similar to connectors is the socket that is used for integrated circuit (IC) chips. In this case, the pins of the IC play the part of the male connector half and the IC socket is the female connector half. Reliability issues are concerned with the IC not firmly installed into the socket or in the area of film developing on the pins of the IC that will not make good connection with the socket. Certain chips such as EEPROMS may be updated frequently with new programs and thus a socket is necessary. However, widespread use of sockets for standard logic chips should be discouraged because of reliability concerns.

Thermal factors can adversely affect connectors, particularly during rapid thermal changes. Indeed, just like the different materials that may make up a circuit board (fiberglass, metal, and ceramics) connectors may be constructed using two or more different types of materials (metal, plastic, rubber) that have different thermal coefficients that may result in failures if the design does not allow sufficient tolerance. Thermal coefficients are covered in some more detail in Chapter 6, but as a generalized case they can be described here to show the nature of the problem.

This case involves an electronic unit that was experiencing unusual intermittent failures during production testing. After much investigation, these failures were isolated to the multiple pin circuit board connectors that were losing contact because of factors of thermal expansion where the lack of stress relief in the design for the different material expansions would cause the circuit to break on an intermittent basis. Thermal cycling is one way to expose weakness in connections; vibration cycling is another way.

In the case of vibration cycling, the applied vibration may cause movement where a break occurs in the contact between connectors. A case that describes this is where a manufacturer made an electronic module box that was to be used in a business jet. The unit consisted of two circuit boards that were connected to each other by a long multiple-pin connector located in the middle of the circuit boards. The two boards were supported by standoffs in each of the four corners. Each unit had to go through vibration as part of ESS burn-in at a level of 6.06 GRMS for about ten minutes.

The first production units went through ESS (no reliability test was performed on this unit), and failures occurred shortly after random vibration began. The failure was particularly serious in that the unit's main microprocessor IC, an ASIC (application specific IC) component which was expensive and made to special requirements, was being blown beyond repair. It took a while to get to the true root cause of the failure, and special vibration tests had to be conducted as no units could pass ESS. Eventually the failure cause was found by a lab support technician when he conducted a strobe light test during a sine vibration sweep on the unit with a hole cut out of the chassis in order to view the circuit boards. At a specific frequency, it could be seen that the connectors were actually pulling apart from each other. Some of the connections that were pulling apart were critical voltage and signal lines. When this condition occurred when power was on, any interruption in voltages to one part of the ASIC while voltage was still on to other parts of the ASIC would permanently lock up or fry the internal lines of the IC in one position (see Figure 5–14).

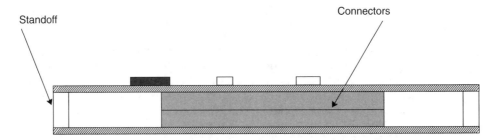

Figure 5-14 Circuit Board Connector Vibration Failure (applied vibration caused separation of the connectors)

The fix was to add another standoff in the area as opposed to using the connector as the main supports in this area of the circuit board sandwich. The problem of inadequate support on circuit boards is described in detail in Section 5.4.

5.2.10 Relays and Switches

Relays and switches are electromechanical devices, so there is a real concern that mechanical failures can occur that can cause electrical failures, either shorts or opens. Failure modes in the relays can occur with the weakening of the relay spring or coil, as well as arcing effects that may damage the contact areas. Switches also have similar mechanical failure modes as relays in the area of worn contacts or loss of spring action that will eventually result in a short or open. Both switches and relays are often rated by the number of cycles (on/off) that they can endure during their lifetime.

One way that design engineers have been able to work around the problem of relay failures is by using Field Effect Transistors (FETs) in lieu of relays, particularly in circuits in which a great deal of cycling takes place on an hourly basis. Cycling is done electrically instead of mechanically with the FET.

A case study that illustrates this vividly is the failure of relays involved in a display unit that is used during the ground refueling operation of a commercial airliner. The relays are cycled a lot during the refueling operation, and eventually, they would wear out. When this unit was redesigned, FETs were used in lieu of the relay, and the reliability performance achieved on the field was many times better.

Another case study involves a piece of automatic test equipment used to power up a controller unit used for a military transport aircraft, while it was being thermal cycled for either ESS testing or reliability testing. After 600 hours of reliability thermal testing that used an eight-hour thermal profile, a fault light appeared during thermal cycling on one of the discrete signal lines that the unit under test had to monitor. After troubleshooting, it was revealed that the problem was with the test equipment where a relay was stuck in the "on" position. This relay would be cycled a minimum of twenty times during the eight-hour thermal profile. A decision was to allow the test to continue with the relay being replaced.

However, another 1,000 hours of thermal cycling had passed when the same signal line failed again. It was again traced to the same relay in the circuit. Because this signal line was cycled often during the course of one thermal cycle, it became apparent that the relay was not going to cut it, and the designer of the test equipment redesigned the circuit to use an FET in lieu of a relay.

An important point to consider regarding failure rates for relays and switches is the type of measurement criteria that is used. Using the example of the test equipment relay that failed, one could calculate the MTBF in terms of test hours or:

$$MTBF = 1600 \text{ hours}/2 \text{ failures} = 800 \text{ hours}$$

However, a truer indication of the reliability performance would be more in terms of the number of times that the relay is cycled. Using the fact that the relay is cycled 20 times per 8-hour cycle and that 200 cycles were conducted (1,600 hours/8 hour/cycle), the Mean cycle between failures is calculated as:

$$\text{MCBF} = 20 \times 200/2 = 2000 \text{ cycles}$$

Typically product specifications for relays and switches will rate them in terms of total cycles that the device can meet during a lifetime. The example shows the dichotomy between measuring the reliability of this type of device using the time domain (continuous function) and in terms of cycles (discrete function).

5.2.11 Fans, Motors, and Meters

All three of these items are used often in electrical equipment, and their failure modes are of major concern in any design. Lubrication of the bearings for cooling fans and motors can be an important issue, both during the initial assembly and during future preventive maintenance actions. Because of the location and air flow motion involved with fans, it may be necessary to implement a regular preventive maintenance schedule to clean out dust that is sucked in.

Motors are a combination of internal windings, magnets, bearings, metal housings, and external wiring. Any of these areas can be of concern with regard to reliability. In addition to the common failure modes of wire breakage and bearing wear, there is concern with electrical noise that can be generated by defective motors. When this noise is present as the result of internal defects of the motor, it can actually affect the adjacent components in a subsystem. For example, a motor with defective brushes or winding that generates voltage spikes can actually affect the performance of a nearby sensor circuit. Quality is a major concern when evaluating motors.

Meters are electromechanical devices that use pivot action for moving the meter arm. Like a motor, windings and magnets are involved, however failure of the pivot movement can be a major issue. A close look at a detailed example of a meter failure that failed during reliability testing is provided in Chapter 4. In this particular case, the mechanical pivot was the failure mode but was not anticipated by the electrical engineers involved with the original design.

5.3 THE RELIABILITY ENGINEER'S ROLE IN MECHANICAL DESIGN

The reliability engineer has to work with the design engineer to reduce the negative impact of various mechanical issues. This may include better tie-down of wiring, extra stiffening of circuit boards, and extra supports for larger components. The reliability engineer has to make a mental or a written checklist of his or her various concerns involving the mechanical features of the design. This is a physics of failure approach that is based on the reliability engineer's experiences. In addition to the individual components covered in Section 5.2, the overall assembly has to be reviewed for:

- Interference between components or assemblies
- Potential structural weak points in the chassis or support
- Any wiring or cabling that is rubbing against other components that could result in chaffing

With all of these items, the reliability engineer has to be involved directly with the engineer overseeing the mechanical design. The following case histories cover many of these items, and they are worth reading to see how certain problems were discovered and ultimately fixed.

5.4 MECHANICAL RELIABILITY DESIGN CASE STUDIES

The following case studies are mechanical issues that came during testing.

5.4.1 Case Study #1

Item: Engine Instrument Display for Navy Fighter Aircraft

An engine instrument display was manufactured by an electronic display vendor for a navy fighter aircraft as a replacement for the older analog tape drive unit in the previous version of the aircraft. The new display unit used three separate LCD displays and consisted of about ten three-inch by four-inch circuit boards. Because of transition of personnel in the company, there was never a consistent group of mechanical, reliability, and project engineers that was present throughout the early part of the program, which spanned about three years. As a result, some inconsistency was present in the design, and not enough attention was paid to the mechanical aspect of the design.

A number of failures were found during field service of the display unit during early production prior to qualification testing and reliability testing being completed. In addition, some long-term failures were discovered on the field related to the cumulative effects of vibration. The following is a shopping list of failures that occurred and is of interest to engineers of other designs using the same type of component or technology.

Power supply circuit board failure

During early field service, a number of failures related to the power supply card, were discovered by the aircraft's manufacturer. The power supply card was a three-inch by four-inch card that was inserted into a slot in the chassis, as were the other ten circuit boards that comprised the unit. One failure on the power supply board was traced to a broken lead of a large tantalum capacitor located in the middle of the circuit board, while another failure was traced to a broken lead of a large transformer, also located near the center of the circuit board. A third failure was more puzzling, where a trace opened up in the center area of the circuit board. In the investigation of the last failure, the aircraft's manufacturer provided laboratory facilities where the circuit board was sectioned to see where the open was located in the trace. Unfortunately, no firm conclusions could be drawn.

Meetings between the aircraft manufacturer and the vendor determined that a vibration survey should be conducted on the unit. This would involve a sine sweep vibration profile that began at 1 Hz and climbed up to 2,000 Hz. Response accelerometers were placed on various spots of the circuit board and monitored during the sweep.

The sweep was conducted at a 2-G sine input level and at around 200 Hz, the output recorded by the accelerometer was in excess of 100 Gs, a factor of over 50 times the input level! This result was verified with subsequent sine sweep tests. Then a strobe light test was performed with the top cover of the unit removed so that the circuit boards that were situated in the slots of the chassis could be observed during the sine sweep. The strobe light is matched to the frequency so that any unusual movement could be detected. At 200 Hz, the power supply circuit board hit resonance and began to vibrate wildly in large amplitudes exceeding over a .25-inch of displacement.

Further analysis revealed that the reason for the power supply circuit board movement was the way that the circuit board was constructed and inserted into the slot of the chassis. All of the circuit boards had connectors at the bottom of the circuit board that were inserted into the motherboard of the unit. In addition, each of the circuit boards was held in place inside the machined grooves in the chassis by use of wedge-lock fasteners, which were attached to two sides of the circuit board. All of the wedge-lock fasteners were attached directly to the board, with the board material actually behind the wedge-lock fastener. The one exception was that one side of the power supply board did not have a wedge-lock fastener behind the board. This was because the designer needed a heat sink attachment that had to make electrical and thermal contact with the chassis, and thus a modified attachment arrangement was used as shown in Figure 5–15A. Unfortunately, this created a serious vibration condition that caused all of the failures noted above to occur. Please refer to Figure 5–15B.

A limited amount of corrective action could be implemented at this late stage after all of the circuit boards were designed and fully populated with components. However, there was a small amount of space at the top of each circuit board that had a small hole in this area for inserting the extraction tool to pull out the circuit board from the chassis. It was decided to put a retaining strip made of soft plastic into the top cover that could clamp onto this space at the top of the circuit board (see Figure 5–15C). This vibration snubber reduced vibration enough that the amplification factor of over 100 at 200 Hz was significantly reduced.

The vendor had also incorporated the idea of a retaining strip into other units that also had similar vibration problems involving circuit boards resonating at different frequencies. At high levels of vibration, it appears that supporting circuit boards on three of the four sides may not be enough.

Cracks in the chassis

After two years' worth of field service had been accrued on the F-14D, the navy and Grumman noticed that cracks were developing in different areas of the chassis of the display unit, particularly in the corners of the opening where the cover was mounted. This failure was traced to the way that the unit was mounted in the cockpit of the F-14. Because the unit was about nine inches long, it had a cantilever effect where significant stress was placed in certain areas of the chassis when installed in the cockpit panel of the fighter aircraft (refer to Figure 5–16).

As a result of the field failures, the vendor had to review the mechanical design and had to shore up the corners of the chassis with extra material and support.

Lamp filament failures

During various pre-production testing that involved vibration testing, it was occasionally noticed that one or two lightbulbs that made up the back-light of the LCD display were not illuminated. Yet a short time later the lab technician would see that the previously failed light had become illuminated again. The technician then thought that it might have been an anomaly in the test equipment that provided voltage to the backlight assembly and thought nothing of it.

Later in the program, after customers had made numerous complaints of lamp failures, a microscopic investigation while the unit was under vibration testing (as shown in Figure 5–17) was performed by the vendor's reliability engineer, and it revealed a very interesting phenomenon. A comparison of new lamps with lamps on returned units revealed that sev-

A) Front View of Power Supply Circuit Board

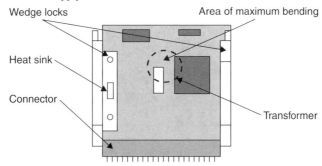

B) Top View of Chassis and Circuit Board Mounting

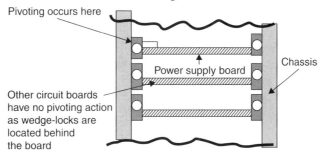

C) Vibration Snubber Added to Cover (Bottom View)

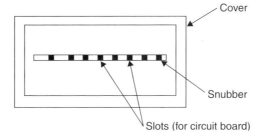

Figure 5-15 Power Supply Circuit Board Mechanical Failure Analysis and Corrective Action

eral of the lamps had experienced broken filament sections but had re-welded themselves to the other sections, apparently when the unit was vibrated. This was possible because a broken filament section is still very hot and swings freely until it comes in contact with another filament section as shown in Figure 5–18.

This unusual phenomenon of self-healing filaments was confirmed while low-level vibration was applied to the lamp board assembly as seen in the photo in Figure 5–17. While the aspect of a self-healing filament was possibly desirable, a decision of reducing the vibration levels to the circuit board by the use of rubber isolators on the standoff hardware was the final fix implemented. This reduced the occurrence of broken lamp filament.

The lesson learned in this particular situation is that engineers have to believe their eyes (when the lamps originally failed during testing) and that a microscopic investigation

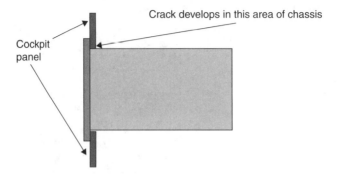

Figure 5-16 Cross-section View of Unit Installation

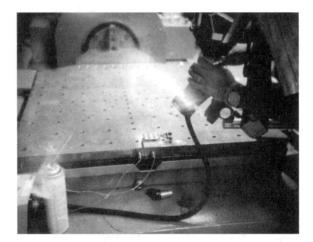

Figure 5-17 Microscopic Investigation of Lamps During Vibration Testing
In the photo, the reliability engineer is focusing the microscope on the individual lamp filaments while the lighting panel is being subjected to sine sweep vibration on the vibration shaker table. The filaments experience tremendous displacement at certain frequencies where resonance occurs. Although it is difficult to do, it is desired to isolate the lamp panel from the direct effects of vibration in the design.

should be performed as soon as there is a pause in the test. The engineer almost has to expect the unexpected during any test program!

5.4.2 Case Study #2

Item: Ground Refueling Unit for Commercial Aircraft

The ground refueling unit was manufactured by an electronic display vendor for a commercial aircraft built by Douglas Aircraft. It was mounted in the wing and used only during ground refueling where the fuel quantities could be viewed by the ground crew while

Mechanical Design Impact on Electronic Component Reliability 119

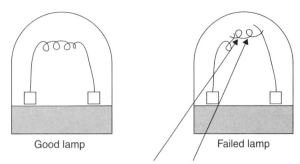

Figure 5-18 Lamp Filament Failures

adding fuel into the aircraft's fuel tanks. The unit consisted of a number of LED displays and pushbutton switches. It was about four inches high, one foot wide, and six inches deep. The unusual shape was needed in order to fit in the wing area. The unit consisted of a number of switches and associated wiring and two long and narrow circuit boards.

Because the unit was mounted in the wing, it would be subject to high levels of vibration because of the wide amplitude displacement of the wing. Even though the unit would be powered off, it was specified that the unit should be able to work properly under this environment when it was used by the ground crew after and before flight.

A number of failures were observed during accelerated life testing, which is a formal reliability test that was performed prior to full-scale production (please refer to Chapter 4 for more details on this type of test). Most of the failures occurred during vibration testing, which was conducted at 7.6 GRMS, a level conducive to what the unit would see while mounted in the wing of the aircraft during field service.

Insufficient circuit board support

During accelerated life testing, the power transistor in the power supply section of the top circuit board had suffered fractured leads at two different points during the test. A review of the mechanical design revealed the obvious reason for this failure: an apparent lack of support under the circuit board in the area of the transformer as shown in Figure 5-19.

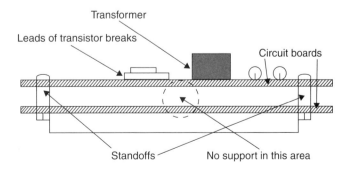

Figure 5-19 Circuit Board Failure Due to Lack of Adequate Support (Side View)

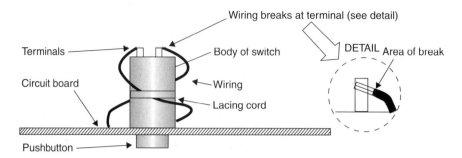

Figure 5-20 Switch Wiring Mechanical Failure

As there was no room to put a conventional standoff in this area with screws to hold it to the circuit board, a plastic post was made that would be glued to one side of the circuit board and fit snug against the other circuit board when assembled. This post would provide damping action in the presence of any ambient vibration.

Switch wiring failures

Again as in the last failure, this failure occurred during accelerated life testing. After the vibration test session was completed, it was noted that some switches for the control panel were not functioning. Opening up the unit revealed that the wiring going to the switch terminal was broken at the point where the leads were soldered to the terminal post. What was interesting was the fact that there was lacing around the body of the switch to help secure the wiring, but apparently this was not enough, as shown in Figure 5–20.

Because of the high vibration applied to the unit and even though the wiring was not free-swinging and secured to the body of the switch, there was still enough movement in the area where the wire connected to the terminal post. This movement or pivoting motion was enough to break the wire in the area where the lead with solder met with the rest of the wire (which was not stripped).

As a result of this discovery, it was decided that an RTV potting compound would also have to be added to the area of the terminal posts and where the wiring connected to them. This fix proved to be effective in the subsequent vibration session. The lesson that is learned here is that even with the best of intentions in securing wiring, it may require more securing.

Terminal post failures

Another wire-related failure also appeared during one vibration session of the accelerated life test. It occurred in the area where terminal posts were installed in circuit boards for which larger-gauge wire could be attached and soldered. Unfortunately, in certain axes of vibration, the terminal post would rock back and forth. The problem was made worse by the fact that the terminal post had no plated-through-hole in which it could be soldered to in addition to the solder spot on the circuit board (refer to Figure 5–21).

There were two potential fixes: one was to make the hole a plated-through hole to aid in the support of the terminal. The other was to eliminate the terminal post altogether and wire directly into the circuit board. The vendor elected to do the latter.

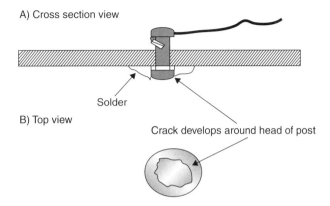

Figure 5-21 Terminal Post Installation into Printed Circuit Board
Note in this drawing that the terminal post is installed in a hole in the circuit board material that does not have a solder barrel for solder to wick down.

5.5 SUMMARY

The different mechanical failure modes discussed in the two case studies were not the type of failures that could be predicted by a standard reliability prediction-type process such as the MIL-HDBK-217 document. Structural items are not covered in the handbook and neither are mechanical failure modes such as fractures of circuit traces and wires covered. The mechanical aspect of the design cannot generally be quantified with the conventional methods of a reliability prediction handbook. Yet, these failure modes are still valid reliability issues that must be recognized and addressed by a reliability engineer during the early stages of new product development. One can not ignore that these factors exist, even if they cannot be numerically predicted!

The reliability engineer has to play the role of a safety net for the mechanical engineer for new products. Good mechanical engineers will be aware of many of these issues but some may not, because they may not have had the benefit of much reliability test experience or field failure investigation experience. The reliability engineer can address these potential failure modes by using a physics of failure approach that could be put in checklist form during the early part of a new program. The checklist can be used as the mechanical aspect of the design evolves.

5.6 EXERCISES

1. The relay that is used in an electronic device has accrued three failures during 100,000 hours of field service. It is estimated that the relay has cycled 50,000 hours. Compute both the mean time between failures and the mean cycles between failures for this relay. Which parameter would be more meaningful in describing the reliability performance of the relay? Give reasons why.

2. Find any electronic unit or electrical appliance that is either in your house or at your job that is about to be thrown out. The unit should be unplugged as you take off the outside cover when you perform the following mechanical reliability design review:
   A. Locate the power supply of the unit. Identify the large components in the power supply as discussed in Section 5.2. Are they adequately supported? List all concerns on a piece of paper. What steps would you take to improve support of power supply components?
   B. With a scale or micrometer, measure the thickness of the power supply circuit board. Is the thickness sufficient for the environment that it will be used in? Is the board amply supported?
   C. Look for the circuit board(s) that contain(s) the microcircuits and logic circuits. Are there any mechanical issues that can be identified? Are the larger chips of concern?
   D. Examine the switches and associated wiring of the unit. Identify and write down any issues. What steps would you take to resolve issues in this area?
   E. Examine the wiring and connections between circuit boards. Again, are there any issues that can be identified?

(Save this junk appliance or unit and leave it disassembled. It will be used for another field exercise in Chapter 6.)

CHAPTER 6

Thermal Factors and Reliability

INTRODUCTION

The last chapter discussed how issues involving mechanical design can adversely affect electrical components and assemblies. In addition to vibration-related problems affecting the mechanical reliability of components, thermal factors are another area which can adversely affect the performance of both electronic and mechanical components. This chapter will discuss some of the physical mechanisms involved with thermal failures.

There is a tendency to think that thermal failures are the result of the design not being able to handle low temperature and hot temperature extremes, but many failures can actually occur because of the effects of rapid temperature variations. Understanding thermal factors, along with a general knowledge of thermal rise, thermal expansion, and other terms can be very helpful when troubleshooting failures. By taking a physics of failure approach, one can get to the root cause of a thermally induced failure. As many electrical components are constructed from a variety of different materials such as metal, ceramics, and plastics, rapid temperature change becomes an area of concern with the thermal-mechanical behavior of these materials.

The reliability engineer is most likely to encounter thermal-related problems either during reliability testing or during investigation of failures that occur during field service. Those failures that occur during testing can usually be isolated to either a thermal problem or to another issue such as humidity. Those failures that are caused by a thermal issue that occurred in the field will take more effort to diagnose since the unit is returned after the failure has occurred. For these cases, some sort of in-house thermal testing may be helpful in isolating the problem and will be touched upon in the case studies presented later in this chapter.

6.1 THE THERMAL ENVIRONMENT

Equipment that are located on board aircraft and space vehicles have to endure very severe thermal environments. For example, equipment on an aircraft will see both cold and hot temperature extremes while the plane is parked on the tarmac of airfields depending on the location. Rapid thermal rises will be experienced when the plane takes off from the ground ambient temperature to generally cooler temperatures of higher altitude. In addition, the effects of cold temperature extremes will be felt during flight at higher altitudes on much of the equipment located throughout the aircraft.

Electrical components have temperature ratings based on manufacturer's testing. The three main temperature grades along with associated ranges that are used for most equipment are:

Commercial	0°C to +70°C
Industrial	−40°C to +85°C
Military	−55°C to +125°C

Vendors certify that their components meet these ranges through an in-house screening process. When components operate at temperatures outside these ranges, there is a greater potential for failure occurring on these components.

6.2 THERMAL ANALYSIS

One of the duties that a mechanical engineer will perform on a new product that uses active electronics is a thermal analysis. Components such as resistors, diodes, transistors, microcircuits, and transformers will generate heat as a result of power being dissipated. The cumulative effects of each individual component's power dissipation is calculated and measured against the effects of any cooling factors employed in the design. These factors include the location of components on circuit boards, venting in the chassis, cooling fans, and heat sinks that are attached either to the chassis or individual components.

The reliability engineer will use the thermal analysis results for a number or reliability tasks. This includes part stress de-rating as well as determining the nominal ambient temperature for the unit in order to perform failure rate predictions for electrical components. For example, if we have a 54 HC00 logic device, the thermal rise is 12 mW/°C or .083° C/mW. If we were to reach the maximum power dissipation rating of 500 mW, this would result in an internal component temperature rise of:

$$500 \times .083 = 41.5°C$$

There has to be a concern with the collective thermal rise for the overall product that is based on the cumulative thermal rise effects of each individual component that makes up the product. It is particularly important in the case where the ambient temperature may be high for the environment that the product is installed in. In some cases, there may be a customer requirement on what the maximum temperature for a product can be. In cases like this, additional external cooling may be required to keep the temperature below the maximum required value.

6.3 THERMAL SURVEY

While the collective thermal effects of individual electronic components of a product as the overall product temperature increase can be determined by analytical means, this is only an estimate. The real proof in the pudding is to conduct an actual thermal survey on an actual unit in order to get the real-world thermal rise. It is typically performed on a prototype or first production unit, and it has to be one of the first engineering tasks performed. This test is a task that may be conducted by either the mechanical engineer or the reliability engineer. The results of the thermal survey dictate how long the thermal profile has to be for ESS burn-in as well as for any reliability and qualification thermal testing. Also, any abnormally high thermal condition of a component may be identified during the thermal survey.

During a thermal survey, the test temperature sensors (or thermocouples) are placed on the following areas of the prototype unit of the product:

- The chassis
- The internal component with the largest thermal mass (usually a power transformer)
- The internal ambient of the unit (not connected to any components)
- Any component that may generate significant heat because of its power dissipation. This is usually any component that has a heat sink attached to it. Examples of this include: microprocessors, power supply regulators, power supply transistors, power rectifiers, and power transformers

The temperature sensors are hooked up to a cable that feeds to a data tracker that records values at a regular time interval, typically every minute. The data that is collected can be entered into a computer, and thermal graphs can be generated from the results.

6.3.1 Purpose of Conducting a Thermal Survey

The purpose of conducting a thermal survey on electronic equipment is two-fold:

1. Determine the stabilization time (non-operating temperature dwell) for all components in the unit to reach the target temperature. Because of the thermal mass of the unit, and even though the exterior unit temperature may match the thermal chamber temperature, internal components lag behind the outside temperature. The amount of thermal lag is difficult to calculate because of heat transfer factors that are done via conduction, convection and radiation. However, this thermal lag is easy to measure during a thermal survey. By knowing the stabilization times, the thermal profile for ESS testing and reliability tests can be constructed using the minimal overall test time.
2. Verify the electrical/thermal component design so that individual component junction temperature ratings are not exceeded. These ratings are developed by the component manufacturers.

6.3.2 Thermal Survey Ground Rules and Setup

The basic ground rules for the thermal survey are based on the physical weight of units. Units that are less than two pounds in weight and made with a metal chassis and covers generally require a stabilization time of less than thirty minutes for each temperature plateau. Units that are greater than fifteen pounds in weight with a metal chassis and covers, along with many components, may need a stabilization time of more than two hours.

6.3.3 Thermal Survey Example

TASK: Conduct a thermal survey for a unit that is rated for operation from $-40°C$ to $+70°C$ that weighs 1.7 pounds.

1. With the unit placed in the thermal chamber with power off and all temperature sensors in place, start recording all temperatures.
2. Bring the thermal chamber temperature to $+70°C$. Measure the time from when the thermal chamber temperature first reaches $+70°C$ until the internal component with the largest thermal mass reaches a constant temperature (no change in temperature reading for five minutes). This temperature will be at or very near $+70°C$. This time is the stabilization time for the hot temperature extreme.
3. Bring the thermal chamber temperature down to $-40°C$. Measure the time from when the chamber temperature first reaches $-40°C$ until the internal component with the largest thermal mass reaches a constant temperature.
4. Turn on unit and bring the chamber temperature to $+70°C$.
5. Monitor all temperate sensors, and continue the test until all monitored components reach a constant temperature.
6. Review the data:
   - The stabilization time at high and low temperatures should be close.
   - The maximum case temperature of monitored components should be less than their ratings. If not, there is a design problem. Some solutions include: larger heat sinks and improved ventilation (adding a fan, added perforations on covers).

The procedure varies slightly for heavier units. See Figure 6–1 for an example of a thermal plot of a five-pound unit for a military aircraft that went through a thermal survey that covers the temperature range of $-55°C$ to $+71°C$. In this particular survey, units are powered off during the first ramp down to $-55°C$ ambient chamber temperature (ambient chamber temperature is designated by thermocouple #5 in this figure). The unit is switched on during the ramp up to $+71°C$ ambient chamber temperature. After the unit is stabilized at $+71°C$, the unit is then powered off during a second ramp down to $-55°C$. After fifty minutes at this temperature, the unit is then powered on.

In this unit, the largest single component in terms of mass is the transformer (as designated by T1 with thermocouple #12). Also, it has the highest amount of thermal rise when powered on at $+71°C$ and at $-55°C$.

The reliability engineer should be an integral part of the team that conducts the thermal survey, along with the mechanical engineer assigned to the program. It is important for the reliability engineer to be aware of those components with the largest thermal rise during power up as they are candidates for potential temperature overstress conditions and should be watched during subsequent testing.

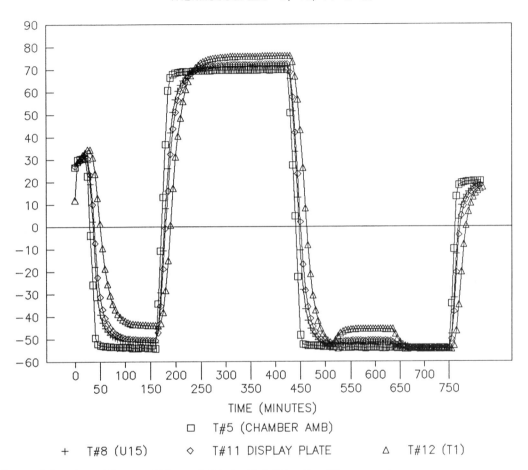

Figure 6-1 Thermal Survey of Electronic Display Unit Using Four Thermocouples

6.4 COMPONENTS AFFECTED BY COLD TEMPERATURES

A number of components, both electrical and mechanical, fall into the category where they fail to operate properly when operating at cold temperatures. Cold temperatures that affect these components do not necessarily mean subfreezing temperatures like −40°C but cold enough to affect the physical properties of the components in the equipment being used.

6.4.1 Liquid Crystal Displays (LCDs)

At very cold temperatures, from 0°C down to as cold as −40°C, LCD displays can be significantly affected. Displays tend to have ghosting effects and can appear to be very sluggish when numerical values are changed. This is a function of the liquid in the LCD being affected by the cold temperature. It becomes an issue with a number of LCD displays that are used occasionally in cold environments. This includes:

- Aircraft that may either be situated at cold locations or when very high altitudes are reached
- Portable amateur and CB radios used in cars
- Portable test equipment used in the field—antenna SWR meters, gauging devices
- A LCD-based GPS display
- Gas pump displays
- Fuel oil truck transaction units
- Outdoor signs using LCDs

6.4.2 O-Rings

As o-rings are made of rubber and similar synthetic materials, there is a major concern when using them in cold temperatures as this material contracts and becomes less pliable. In the space shuttle *Challenger* disaster in January of 1987, an o-ring was one of the items that failed because of the cold temperatures on the day of the launch. Air temperature at launch was 36°F, which was 15 degrees colder than any previous flight.

A number of things contributed to the failure. The cold was certainly one of them, but o-ring blow-by had been observed on nine previous flights of the space shuttle, and a total of twelve flights had o-ring anomalies of some description. The cold did, seemingly, contract the o-ring further, allowing yet more blow-by. But the largest contributing factor (besides a basically poor design) was some severe upper-level winds that had increased the torque in the stack more than usual. These tended to separate the joint sealed by the already-contracted o-ring, allowing yet more hot gases through, which eroded the o-ring further, allowing more gases, etc.

It was not out of specification per se, only because there was no specification on temperature, but the situation was certainly outside of the experience base. A detailed analysis of this temperature-related failure is provided in the book, *Space Shuttle: The History of the National Space Transportation System* (Dennis R. Jenkins, Specialty Press).

Also note that other rubber parts and plastic are affected by cold temperature extremes. For example, the plastic insulation of wires gets stiff and less flexible at low temperatures.

6.4.3 Plastic Parts

Plastic parts are used in a lot of designs to save weight and cost. When used as material for gears, there is a big concern when operating in cold temperatures as plastic tends to become brittle and breakage can result. An example of this is provided in the case study later in this chapter.

Another item to be aware of is the use of lubricating grease for these plastic gears. The grease can become harder (or have less viscosity) and can possibly impede or bind gear movement. Both the gears and grease should be reviewed against the field environment that they will see.

6.5 COMPONENTS AFFECTED BY HOT EXTREMES

Several components have degraded performance when they are running hot. Many of these components are electrical components.

6.5.1 Transistors and Integrated Circuits

Power transistors dissipate a lot of power. If there is no path for the heat to move anywhere, the device temperature can rise above its specified value, resulting in failure. This is where heat sinks come into play and will be discussed in Section 6.7.

Certain integrated circuits like the microprocessors used in computers draw high current and dissipate power. As some of these devices are significantly larger than transistors, a fan may be used in addition to heat sinks to remedy the situation. These devices run very hot as they are being driven at high speed which ultimately results in high-power dissipation. The device will fail in a catastrophic manner if there is insufficient cooling; thus it is important to remove as much heat as possible through the use of heat sinks, fans, and thermal grease.

6.6 COMPONENTS AFFECTED BY THERMAL VARIATION

All components may be affected by temperature variations. This causes concern when there is an environment of rapid temperature change, such as the situation that would be experienced by an aircraft going from a hot runway environment into a cold temperature environment at high altitudes in a matter of minutes. Some of the temperature variation may be of smaller thermal range, yet over the period of years, it can cause failures. This becomes a major issue when dealing with different materials at once.

As each material has a thermal coefficient of expansion where the material expands at a certain rate as temperature rises, the thermal coefficient of expansion becomes important when there are different materials involved in an assembly. This is why thermal cycling is important in both production and reliability tests as these conditions are uncovered.

6.6.1 Circuit Boards and Solder Connections

Circuit board assemblies have many different materials with different thermal coefficients of expansion. For example, solder joints in a circuit board assembly are always of concern because different materials are involved: the fiberglass circuit board, the plated feedthrough holes, and the leads of the component itself as shown in Figure 6–2.

Each of the materials in the printed circuit board has different thermal characteristics as temperature changes with contractions and expansion. Table 6–1 shows the difference between ceramic and glass epoxy, the common materials used in the construction of most printed circuit boards.

The following example demonstrates the amount of expansion that is experienced in a typical circuit board.

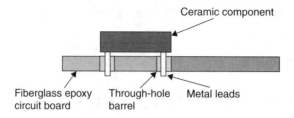

Figure 6-2 Circuit Board Thermal Expansion Conditions

Table 6-1 Approximate Coefficient of Thermal Expansion

Material	Coefficient of Thermal Expansion (Parts per million/°C)
Glass epoxy fiberglass in the length and width of a circuit board (X-Y Plane)	15
Ceramic (aluminum oxide)	6

THERMAL EXPANSION EXAMPLE:

A glass epoxy circuit board that is five inches wide by seven inches long experiences temperature changes from 85°C to −40°C. What is the approximate change in length and width?

1. The temperature change is $85 - (-40)$ or 125°C.
2. The coefficient of thermal expansion is 15 parts per million per degree C, or $15/1,000,000 = 1.5 \times 10^{-5}/°C$.
3. The change in length is 7 in. $\times$ 1.5×10^{-5} or 1.05×10^{-4} in./°C, the change in width is 5 in. $\times$ 1.5×10^{-5} or 7.5×10^{-5} in./°C.
4. For a 125-degree change, the change in length is $125 \times 1.05 \times 10^{-4}$ or 0.013 in.
5. For a 125-degree change, the change in width is $125 \times 7.5 \times 10^{-5}$ or 0.009 in.

It can be seen that this amount of expansion by the glass epoxy glass is significant in comparison to the smaller amount of expansion of the ceramic material such that potential problems could result.

6.7 REMEDIES FOR THERMAL RELIEF

There are a number of ways to reduce the effects of high and low temperature extremes as well as ways to reduce the effects of thermal changes. The following is a discussion on the remedies.

Components with leads allow any thermal coefficient mismatch between the component and the circuit board to be adjusted for by the leads. This built-in "expansion joint" reduces the high stress in the connection. The move toward leadless surface-mount components requires knowledge of the product's intended environment and the thermal characteristics of the components used.

The amount of expansion or contraction for a given temperature change is proportional to length. Larger components exhibit more problems than smaller components (SMD resistors and capacitors). Ceramic leadless integrated circuits (ICs) mounted on a glass epoxy PCB work well at room temperatures, but if exposed to temperature extremes, the very low coefficient of thermal expansion for ceramic and the much higher coefficient or expansion for glass epoxy put tremendous stress on the solder joint. Over many temperature cycles, the joint will fracture, leading to intermittent failures and, ultimately, unhappy users of the equipment.

Some solutions include using polyamide PCB material instead of glass epoxy, and use of ICs with gull wing or J leads.

6.7.1 Heat Sinks

Heat sinks are metal structures that are attached either to individual components (such as transistors and integrated circuits) or attached to larger-type structures such as the chassis of electrical equipment. The basic principle of the heat sink is to provide a thermal path for the high temperatures generated by individual components into the ambient air environment. Basically, this is a thermal conductive path that is being transferred to a convection-type situation.

Because of this process, heat sinks are designed to provide the maximum amount of surface area that can be exposed to the ambient air environment. This accounts for the somewhat unusual shapes for heat sinks that one might see on components or on the chassis of electrical equipment. Heat sinks for individual components will take on numerous shapes that fan out into the ambient air environment.

Heat sinks that are employed on a unit such as those attached to the back of radios (see Figure 6–3), are designed to remove the heat that is generated by power transistors and integrated circuits in the output stages. A tremendous amount of power is generated in certain amateur radio modes, typically on the order of 100 watts of output power. The heat that is generated by this amount of power dissipation has to be taken out of the internal air environment of the radio as much as possible or else the internal temperature could be higher than the thermal rating of the components.

The beauty about the use of heat sinks for dissipating heat is that it is a passive device with no reliability issues. Most heat sinks are painted or anodized so there are not issues involving corrosion. Figure 6–4 shows an example of how much square area of surface that a heat sink can give compared to a flat surface of both an electronic component and the overall equipment packaging. In these examples, it can be seen that for an electronic component, the heat sink provides sixteen times the surface area and for a piece of equipment, the area of the heat sink is almost eight times the area of a plain flat surface.

Even if heat sinks are used, there may be limitations on how much surface area can be created. In addition, there may be too much power dissipation than the structure can handle properly. Thus, additional devices such as fans may be required.

6.7.2 Fans

Fans are deigned to move air mass of various temperatures. Common applications are for cooling, sometimes to supplement existing heat sinks. In this scenario, fans either bring

Figure 6-3 Heat Sink Design

Amateur radio transceiver designs use large heat sinks for dissipating the heat from the output stage transistors. The heat sink designs are maximized for the largest amount of surface area to allow for as much heat to be dissipated into the open air. Heat sink designs on radios are typically located at the back of the unit as shown in the photo.

A) Structure attached to power transistor

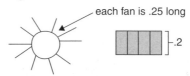

Surface area of heat sink
1) area of fan elements = 10 x .2 x .25 = .5 in.² plus
2) area of remaining surface = .16 x .2 − (.02 x 10 x .2) = .028 in.²

TOTAL SQUARE AREA = .528

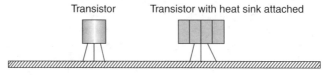

Surface area of power transistor (can style) =
1) area of top cylinder (πr^2) = .002
2) area of the surface on the cylinder ($2\pi r \times H$) = .031

TOTAL SQUARE AREA = .033

The heat sink surface is 16 times the area of the surface of the transistor.

Figure 6-4 Examples of Different Types of Heat Sink

B) Structure attached to back of electronic unit

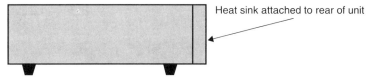
Heat sink attached to rear of unit

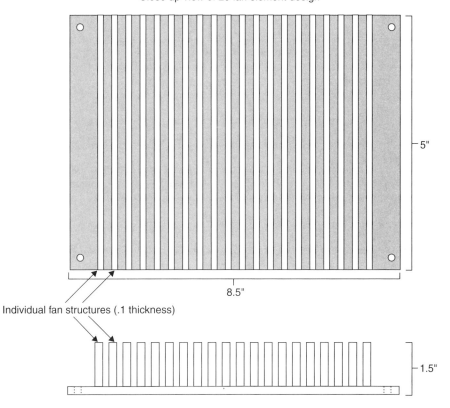

A) Surface area of heat sink
  1) Area of fan structures = (20 x 2 x 1.5in. x 5in.) + (20 x .1in. x 5in.) + (20 x 2 x .1in. x 1.5in.) = 316 in.^{2}
  2) Area of remaining surface = (8.5in. x 5in.) − (20 x .1in. x 5in.) = 32.5 in.^{2}

Total surface area of heat sink = 316 + 32.5 = 348.5 in.^{2}

Compared to the surface area of a plain square surface (8.5 in. x 5 in. = 42.5 in.^{2}), this is over 8 times the area!

Figure 6-4 Examples of Different Types of Heat Sinks *(continued)*

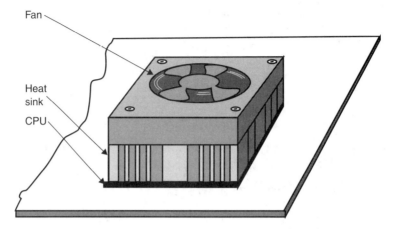

Figure 6-5 Computer Microprocessor Unit Heat Relief
The figure shows the tremendous amount of thermal relief needed for the amount of power that is dissipated by the CPU (computer processor unit) used in many personal computers. The CPU IC itself may be in the area of 4 to 6 in.^{2} size at a height of 1/8-in. However, in order to cool the IC, the heat sink unit with all of the surface areas created by the extrusions represents a cooling area of a few hundred times the size of the area of the CPU. The heat sink alone is not sufficient for cooling the CPU, so the heat sink is constructed to accept the mounting of a cooling fan on top of it.

cooling air in or are used in the exhaust mode to move hot air out of an internal environment into the larger outside environment. A roof fan in the attic of a house is an example of this.

Fans can be small enough to be placed on individual components. Many personal computers have small fans on the microprocessor device as part of the heat sink structure as well (see Figure 6–5).

6.7.3 Heaters

Portable heater units can be used for large devices such as ATMs that may have to be placed in the outdoor environment. By using these heaters inside the unit, the internal temperature will be high enough to keep the plastic parts from becoming brittle.

6.8 THERMAL FACTORS CASE STUDIES

A number of examples that follow show how the different aspects of thermal factors can cause failures.

6.8.1 Thermal Factors Case Study #1

ITEM: Air Force transport aircraft cockpit display unit

Two cockpit display units were undergoing a formal reliability test known as durability investigation testing where a total of 3,500 hours of thermal cycling had to be completed along with over 100 hours of random vibration. During thermal cycling, one of the units experienced a hard failure where it failed the self-test at the high temperature extreme (70°C) after completing only eight thermal cycles and sixty-eight hours total of thermal cycling.

Examination of the failed unit revealed that on the discrete circuit board, a number of the solder pads had lifted up off of the glass epoxy circuit board. These solder pads were the larger solder pads on the entire board, and it became apparent that all of the electrical components were hand-soldered in place instead of the normal machine wave soldering operation that the company employed. Apparently, this board was a sort of prototype, one of the first ones built. It was also apparent that while thermal cycling had aggravated this failure condition, the original cause of the failure was the excessive amount of heat that was applied to the large solder pads during the soldering operation.

This was not the only problem. The layout of the circuit board also had an inherent design problem. The large pads that failed were determined to be much too large for what was really needed. The large pad sizes would typically be used for components with thick leads or for large components requiring firm mounting. In this situation, neither case applied as the components were moderate size with moderate-size leads. There was no need for the large pad size. With a large pad size, the soldering iron has to be kept on the pad longer in order to properly melt the solder to cover the pad. This results in a lot of heat being placed on the pad and causing it to start to separate from the glass epoxy circuit board material.

It is probable that this problem would not have occurred if the board had been wave-soldered by machine instead of hand-soldering. However, this was not a sure thing, and the manufacturing engineer put the board layout in for a revision change where a smaller solder pad size would be used.

This example illustrates how two aspects of thermal factors can be involved—the effects of thermal variation that induced the failure during testing, and the damage to the circuit board that was caused by application of high amounts of heat during soldering.

6.8.2 Thermal Factors Case Study #2

ITEM: Outdoor customer service units: Vending machines, ATMs

A number of customer units are used in the outside environment. This includes vending machines and ATMs. These machines are not entirely electronic, as there are a number of mechanical components that are needed to dispense the money to the customer. These mechanical components have issues involving cold temperature, as seen in this particular case study.

During a particular cold January of single-digit temperatures in the New York area, a service company experienced several service calls with vending machines used at rest stops in the New York area along the interstate.

Field engineers dispatched to fix these vending machines found that in several of the machines, the plastic gears in the dispensing mechanisms had sheared and fallen off. Most of the damage was in the area of the shaft of the gear where stress fractures had occurred.

Plastic becomes brittle as the temperature gets colder and when used in a machine environment in which stress is applied, there is a potential for the plastic to break. Figure 6–6 shows the area where most of the breaks had occurred on the plastic gears.

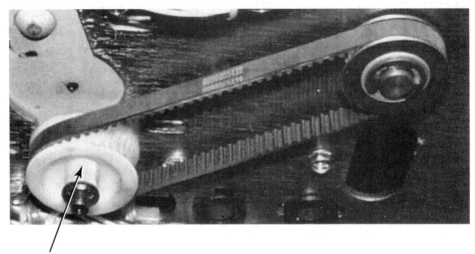

Stress fracture occurs on shaft part of plastic gear

Figure 6-6 Damage to Plastic Gears by Cold Temperatures
In the photo, the plastic gear shown is one of several that is at risk for breakage when the cash dispenser is used during cold temperature environments, such as outdoor customer service units. The plastic gear becomes brittle, as well as contracts, causing added stress in the area of the shaft. When the machine is activated, the gear can actually fracture in the area of the shaft and fall off.

6.8.3 Thermal Factors Case Study #3

ITEM: Amateur radio transceiver

A number of excellent amateur radio transceiver designs have been developed in recent years. Amateur transervers include both a receiver and transmitter in one package. These designs include many "bells and whistles" that have been made available because of advances in software and improved components. As design features are made available, certain failure modes that have not been seen before have raised their ugly heads. This particular case study involves a failure mode on a particular transceiver caused by an aspect of water that the engineering designers in the company had no inkling of being a factor in the design.

The transceiver featured a LED-based display and microprocessor-controlled digital phase lock loop (PLL) circuit that controls the frequency through a single crystal oscillator with high accuracy and stable frequency control. This being said, a major failure mode appeared about four or five years after the radio was introduced on most units in field service. When the failure mode appeared, the audio on the radio would become severely distorted and the display would blink. This was the result of the PLL unlocking because of the frequency varying.

The author experienced this failure mode with his radio, fifteen years after he had bought it new. The radio was used in the radio room in the basement of his house and the failure appeared during a particularly hot and humid summer.

The failure mode was analyzed by the company and was traced to one of the voltage control oscillators (VCO). This VCO is actually a circuit, primarily consisting of transis-

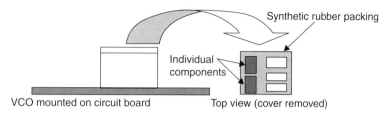

Figure 6-7 Amateur Radio Voltage Control Oscillator Failure

tors, that is contained in a can-type structure (see Figure 6–7). The failure condition was caused by the fact that the components of the VCO were packed in a synthetic rubber that was incorporated for mechanically stabilizing the VCO during mobile operation. Over a period of several years, the rubber's insulation properties had broken down as a result of repeated temperature variations. In addition, the presence of humidity on the brittle rubber was an additional aggravating factor. This caused the VCO to vary in frequency, which in turn causes the PLL to unlock as well as causing the audio to warble.

The synthetic rubber that was used to pack the components in the VCO would actually deteriorate over time because of the effects of temperature variations from cold to hot. In addition, humidity factors would accelerate this deterioration process further. The fix was simply to remove the synthetic rubber and in more severe cases, replace the individual components of the VCO. The company alerted the owners of this particular problem and the appropriate repair action to be taken by the issuance of a service bulletin.

6.9 SUMMARY

This chapter has illustrated how temperature factors can be a major reliability concern for a product when deployed into field service. Any issue involving potential thermal factors on a new product design has to be a major item that the reliability engineer has to review. While many aspects can be determined via analytical means, the real proof in the pudding comes with actual testing, both by the conducting of a thermal survey and by conducting thermal cycling during reliability and production testing.

6.10 EXERCISES

1. What types of components are affected by extreme cold temperatures?
2. What are the three temperature grades and their associated temperature ranges?
3. Describe some of the remedies that can be used to provide heat relief to electrical components.
4. Calculate the thermal rise for a 54 HC02 logic IC that is rated at 12 mW/°C and is dissipating at 100 mW of power. If the internal ambient was initially at 25°C, what is the new internal ambient temperature?

5. What is the amount of thermal expansion for both the length and width of a glass epoxy circuit board that is three inches wide and four inches long during thermal change from −50°C to 70°C? There are two ceramic chips that are .5-in. wide and 1-in. long on this board, what is the thermal expansion for these chips over this same temperature range? (Refer to Table 6–1 for thermal coefficient values.)
6. Examine any LCD equipment such as a radio, clock, watch, or calculator and operate it outside during cold weather or inside the freezer. Do you witness any degradation to the display?
7. With the junk electronic appliance that was used for the lab exercise in Chapter 5, conduct a thermal analysis investigation as follows:
    A. Locate the power supply. What components will probably have the largest thermal lag?
    B. Locate any component such as a microcircuit or transistor that has a heat sink attached to it. Are there are any heat sink structures built into the chassis?

CHAPTER 7

The Impact of Water on Product Reliability

INTRODUCTION

The adverse effects of water on the performance of both electrical and mechanical designs are things of which the reliability engineer conducting tests or field failure analysis will become well aware. The three states of water—gas, liquid, and solid—are of serious concern with regards to their impact as factors on a new product design, whether the design is electrical or mechanical. This chapter will present a number of case studies that demonstrate how these different states can appear and adversely affect the reliability performance of a product design.

By virtue of being located on the planet Earth, very few products will be in an environment that will not be affected by some aspect of water. Thus it behooves the reliability engineer to understand that these factors exist and how they can occur. After achieving this understanding, it becomes necessary to improve the design in various ways to make it impervious to the effects of water.

Typically, any potential problem that is related to water will not usually be addressed during the early design stages—rather it is addressed when a failure occurs during testing or during field service. However, as some of the case studies presented in this chapter will show, a certain amount of precaution or prevention can be incorporated in the design to prevent serious problems.

7.1 HOW THE DIFFERENT STATES OF WATER CAN IMPACT A DESIGN

All three states of water—gas, liquid, and solid—can severely impact a design whether it is operating in an indoor or outdoor environment. Examples of how each state can impact a design will be provided in the case studies later on in this chapter.

Long-term destruction by any of the three states of water is possible, as any of these states can corrode the metallic components of a product. The degrees of corrosion can range from subtle effects, such as a light film that acts like an insulator between two metal surfaces, to as severe as major corrosion causing deterioration of the metal. Another concern is in the area of the via hole where the leads of components are soldered into. This area typically does not have a conformal coat applied, along with the fact that the initial soldering operating may have corrosive elements from the flux. If condensation or ice collects here on the solder joint, there is a possibility of an open circuit in the via hole circuit developing.

In the case of the light film of corrosion, it can act as an insulator between two conducting metal surfaces, and this condition is of major concern with connectors that are used in an outdoor setting or similar-type environment where rain water or humidity is of constant concern. Thus, this becomes an issue for cable TV lines and various radio antenna installations as any break in the conductivity between connections will reduce signal strength. Also, the results of the corrosion process are made worse when there are two dissimilar metals involved.

Also, it is noted that water in its liquid state is capable of conducting electrical current because it contains mineral impurities. Thus when water falls on most types of electrical circuits, particularly analog-based circuits, it can change the way that the unit functions or even cause permanent failure. The case study presented in Section 7.3.1 will discuss in further detail the effects of water on electrical circuits.

When water takes on the form of humidity, it can result in fog conditions, and this can adversely affect different product designs, particularly in the area of display units where glass is used in the design. This can become a major concern to the end-user, whether it is a pilot in a cockpit trying to read a display unit or a consumer trying to read an instrument gauge in the car.

Severe corrosion of exposed surfaces such as the pins of an external connector is also a major concern. Corrosion can be so severe in certain environments that the metal from the pins can break down into metallic salts that can eventually short out adjacent pins (see the photo in Figure 7–1). Mechanical items, metal structures, and connections that are exposed to the rain and humidity also have to be adequately protected by the use of sealant and enclosures. In the case of water in its solid form as ice, it can change the capacitance of components and any structure involved in the area of radio frequency (RF) generation when applied to them. The dielectric constant of water is about 80 at 25°C, and for comparison, plastics are about 2 to 3, glass 5 to 10. Of course, the dielectric of air is very close to 1. This means water can make a very good capacitor. On high-impedance and high-frequency circuits, water across the printed circuit board traces acts like a capacitor and can effectively short-out or load down the circuitry. Thus, certain electrical circuits, particularly analog-based circuits, are extremely susceptible to the effects of condensation and need to be protected adequately by use of conformal coating. Also, while pure water (deionized) is a very poor conductor, any contamination turns water into a relatively good conductor.

Unsealed boxes (anything without a hermetic seal) where condensation may occur caused by temperature cycling need drain holes or weep holes to allow liquid water to escape; otherwise water can build up within the unit. Holes smaller than 3/8-inch will not easily pass liquid water because of its surface tension. Drain holes should be located at the lowest point of the unit when it is mounted in its normal operating position.

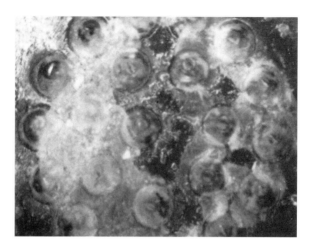

Figure 7-1 Effects of Severe Corrosion
This photo shows the effects of severe corrosion on pins of a connector. This connector was on a unit that underwent humidity testing as part of the qualification program. It can be seen that the effects of corrosion are so severe that the metallic pins have deteriorated into metal salts that are actually shorting out the adjacent pins. Some limited success is realized with metallic-type coatings on pins that are subjected to severe water-type environment, but periodic maintenance still appears to be the most effective solution to fixing this problem.

Another area of concern is the changes between states of water. Much like the area of thermal variations, the changes between water into ice and back to water on a repeated level can induce failures. For example, a pool cover on an outdoor pool may incur damage from the stress of repeated state changes between water and ice.

7.2 REMEDIES FOR EXPOSURE TO WATER

For any new product designs that are used in either an outdoor environment or in an environment where humidity is a concern, a review of the design is necessary for adequate protection against the states of water. Some of these areas of protection include:

- Conformal coating on electronic circuits
- Drain holes or weep holes in the chassis
- Hermetically sealed units (with nitrogen backfill)
- Limited use of dissimilar metals
- Use of paint or sealant to protect exposed surfaces
- Use of grease on exposed threads

A design's susceptibility to water can be discovered during the testing phase of a new design, either during qualification testing or reliability testing. This is the best place to detect such a condition so that potential corrective action can be implemented before full-scale production.

7.3 CASE STUDIES ON THE EFFECTS OF WATER

The following case studies document real-life examples of the effects of water and what was done to address the problem. Some of the examples show how water comes into play when a unit is tested in a thermal chamber and other examples show actual field applications where water is a major issue.

7.3.1 Effects of Water Case Study #1

ITEM: *Air Force transport aircraft electronic control unit undergoing durability testing*

A) Analog circuit board

An electronic control unit that was to be installed on a military aircraft was subjected to both qualification and reliability tests where the affects of humidity was tested. An electrical design engineer assigned to designing this control unit, in which various discrete signals passed through it, had to develop a series of complex voltage divisions for analog components on one circuit board. The engineer needed these voltages to set a reference value for what the proper discrete signal voltage had to be. If a particular discrete value varied more than a few percent of the reference value, the unit's built-in test software would flag a failure.

Although the design engineer was experienced in designing electrical circuits, he apparently had limited awareness of various environmental factors that could affect the way that the analog circuits would work. In particular, he was unaware of the fact that the unit would be subject to levels of humidity and condensation during both testing and field use. Because he used low impedance values for the various analog lines, any water condensation that would form on analog components would change the capacitance of the circuitry to the point that the voltage was changed and hence flagged a failure. He also had hoped that conventional conformal coating on the circuit board should have been sufficient to eliminate the effects of condensation This would not prove to be the case.

The reliability engineer who was initially assigned to this program did not have electrical design experience, and he missed this also. Thus, when the unit entered qualification testing, particularly during thermal cycling where condensation appears during the ramp up from cold to hot and during humidity testing, significant amounts of analog circuit failures occurred.

As a desperate measure, it was decided that the entire eight-inch by ten-inch board had to be protected from water by adding a 1/8-inch layer of silicon! (See Figure 7–2.) Additional testing also revealed that the test point connectors also had to be protected. Needless to say, the addition of this silicon layer added many complications to the production process until a different type of coating from the wax family was used.

Eventually, a thinner-type coating became available and this was effective in reducing the effects of water condensation and humidity. This particular example shows how electrical engineers, while very good at designing circuits, could be caught off-guard by not being aware of the practical aspects of what the design must endure. Thus, practical experience in the environmental area of a design is a very good thing for both electrical and reliability engineers to have!

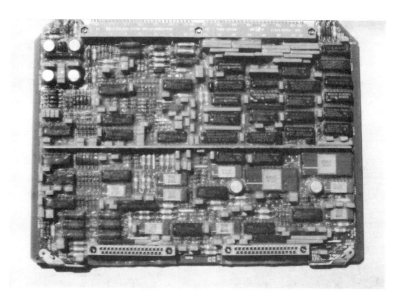

Figure 7-2 Condensation Failures on Analog Board
The photos show an analog circuit board that was made for a control unit used on a military transport aircraft. Both sides of the circuit board are covered with thick protective silicon coating in order to prevent the analog circuits from being affected by water. The process of adding this coating was quite involved, including the need for a complete day for the curing of the coating. A different type of coating was eventually used that was thinner and took less time to cure.

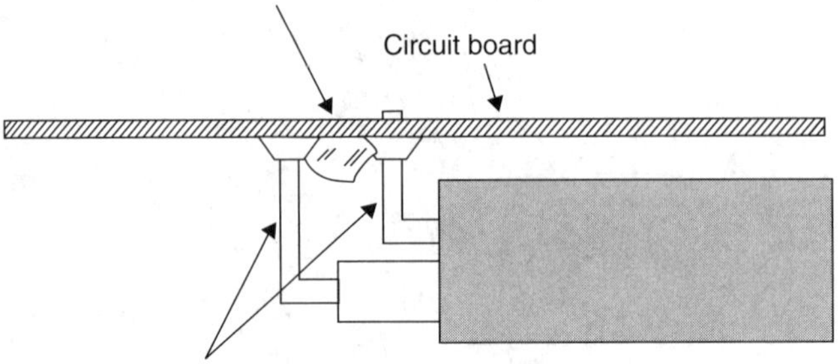

Figure 7-3 Effects of Condensation on Differential Line Pairs

B) Data lines on motherboard

This is a similar case to the analog board example that was described in A. In this case, the board that failed was the motherboard (where all of the other boards plug into) for the same control unit. The motherboard contained a differential pair of signal lines that was used for transmitting and receiving MIL-STD-1553 type data. [The principle of differential signal lines is that the amplitude remains the same proportion on each line under various conditions.] In addition, the lines are spaced very closely to reduce the effects of external noise. However, if one line happens to be affected by the presence of water while the other is not, then the data being processed may be altered.

In this case, the failure was traced to missing conformal coating in a specific area between the two line pairs on the connectors as shown in Figure 7–3. Because the conformal coating was sprayed on, the unusual shape of the connectors precluded ease of spraying coating on the exposed leads in the area of the connectors. Thus, after discovering the failure, the reliability engineer determined that a separate process by hand had to be employed for applying the conformal coating, and this particular failure did not occur anymore in the unit.

7.3.2 Effects of Water Case Study #2

ITEM: Passive electronic components used on cable TV lines

A number of passive electronic components are installed on cable TV lines in the outdoor environment. These include notch filters, external connectors, and splitter devices. All of these devices have potential issues involving water because of how they are exposed to the outdoor environment as well as how they are situated on the house or on the service pole.

Notch filters are often used by cable TV companies in order to cut out certain cable TV channels when the customer is not paying for service to that channel. For cable TV lines that come from aboveground poles, these filters may be installed in the area of the drop to the house from the pole. The filters may be housed in a structure or a box.

This story is based on the experiences of a cable TV subscriber in New York who would experience the following symptoms every year during the winter. Channel 10 was a premium service channel that the customer no longer wanted. The local cable TV company sent a technician to install a notch filter on the pole on the customer's drop line. However, a year later, the customer noticed that Channel 10 was starting to come in and that the sound of Channel 9 was being cut out. The cable TV company was called, and a technician installed a new notch filter to fix the problem. A year later, the same symptoms appeared on the television, and the company sent a technician to replace the filter. Again a year later, the same symptoms appeared, only this time the homeowner was at home to talk to the technician when the repair was performed. Apparently, the filter was located in a box where a tremendous amount of rainwater was accumulating, and this would eventually find its way into the filter over time. During the winter, any water inside the filter would freeze up and cause damage to the filter. The technician employed a simple remedy of relocating the filter so that it would not be in the area where rainwater could accumulate in the box.

The filter is a simple device consisting of a tuned circuit that is packed in insulation-type material. Water and humidity can enter these devices because the mechanical thread connection is not watertight or hermetically sealed. Other devices such as splitters have similar components and are also not in hermetically sealed boxes and can also receive damage from the effects of water and ice. Please refer to Figure 7–4.

Thus, it becomes of paramount importance to properly locate these devices so that there is a minimal amount of intrusion by rainwater. Connectors should be protected as much as possible, away from direct exposure and notch.

A) Notch filter

Water can enter through connector thread and travel internally

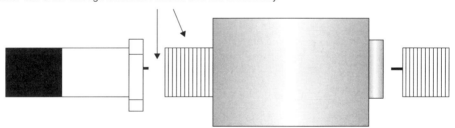

B) External connectors

Film can develop from effects of water on center conductor or on threads of shield

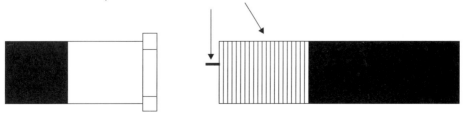

Figure 7-4 Effects of Water on Cable TV Devices in Outdoor Environment

7.3.3 Effects of Water Case Study #3

ITEM: A-10 close air support aircraft airframe components

During the early 1980s, the U.S. Air Force A-10 close air support aircraft was deployed in three operational Air Force bases in the United States, along with four Air National Guard bases. The reliability engineering department for the Fairchild Republic Company, the manufacturer of the aircraft, had to track reliability MTBF values for the aircraft and its various subsystems for each of these bases on a monthly basis. In addition to the monthly tracking of the MTBF value, the data was also examined for failure trends.

The average overall aircraft MTBF value for the A-10 aircraft was in the three- to four-hour range for the operational Air Force bases. Yet after initial deployment at the air national guard bases, the aircraft at one of the air national guard bases in the United States was seeing less than two hours MTBF for the overall aircraft. This was the air national guard based at Martin AFB in Maryland. It did not take a lot of investigation to see why. More than one-third of the type 1 (or inherent) failures that were counted monthly were tracked to corrosion of the airframe components such as the access doors, panels, and other airframe surface components. Indeed, the same specific error code used by the maintenance personnel to indicate corrosion failure was listed against many of the components in the airframe subsystem at this particular location.

Geographically, Martin AFB was the closest to the saltwater of all of the A-10 bases and saltwater film at less than five miles away. It was also common for the aircraft to fly over the water when training. As a result, the corrosion from the saltwater in the air caused numerous maintenance actions where cleaning of the corrosion had to be performed on the skin surfaces of the aircraft. Often, repainting had to be performed after the corrosion on surfaces was cleaned away. Figure 7–5 shows the effects of corrosion on the surfaces, in this case, the tail section of the A-10 aircraft. The photos show that not only are the skin surfaces affected, but also the fasteners and the threads of the rod linkage. For the base at Martin AFB, maintenance had to be performed more often on a regular basis.

Corrosion is an even bigger concern for Navy planes that fly off of the aircraft carriers in the saltwater environment, and many programs address this issue. Not only is saltwater corrosion on skin surfaces an issue, but also its effects on connectors and internal circuitry of electronic equipment used on the aircraft and the carrier.

7.3.4 Effects of Water Case Study #4

ITEM: Chassis material for ground refueling display unit for commercial aircraft

The unit in discussion here is an electronic display unit that was situated in the wing of a commercial aircraft and had an LED display for showing fuel quantity values during ground refueling operation. The vendor that designed this unit opted to use aluminum as the material in the chassis because it was lightweight. However, because a number of plated metal connectors had to be soldered directly to the chassis, and some sort of plating was needed to be applied to the chassis to allow these connectors to be soldered.

The vendor chose nickel plating as the material to be applied to the aluminum chassis and this was painted. The use of nickel plating would eventually prove to be a very poor choice. Why? Nickel and aluminum were on the opposite ends of the chemical scale

A) Airframe surfaces without corrosion

B) Airframe surfaces with corrosion

Figure 7-5   Effect of Corrosion on Aircraft Airframe Surfaces

and that combination of the two materials along with another agent could cause corrosion. This became obvious during field service when the paint was scratched and the metal underneath was exposed. If the unit was then exposed to water, either in the form of deicing fluid that was used on the wings of the aircraft during cold weather or when exposed to salt water, major galvanic action would take place on the exposed areas. The corrosion of the chassis became so bad that switches and glass windows on the unit would fall out.

After ten years of service, the aircraft manufacturer had a new version of the original aircraft in works and at this point, much of the equipment was also slated for a design change. The vendor who made the display unit had the current mechanical engineer conduct a thorough review and trade study on how he could change the chassis material design for the better. After ten years, newer connector designs came along that did not require them to be soldered onto the chassis; rather they used the concept of jam nut locking hardware. This eliminated the need for the nickel plating. Also, in lieu of painting, the mechanical engineer elected to have black anodize applied to the aluminum chassis.

However, the aircraft manufacturer, having been burned previously by numerous replacements and repairs of the previous display unit design, was not convinced initially. The customer asked for the vendor to consider using brass in lieu of aluminum for the chassis material to prevent the occurrence of corrosion problems. The use of brass would have made the design very expensive as well as heavier. Thus in order for the vendor to regain the customer's confidence, a whole series of environmental tests had to be performed on the new aluminum design to prove to the customer that this design would hold up. The tests proved the vendor's concept and the unit has since performed well in the field with minimal problems related to corrosion.

The primary lesson that was learned here is to always be aware of the materials that are used and that they do not have the potential for adverse chemical reaction causing damage. Dissimilar metals that are in contact with each other are always a concern in a design as well as in storage situation. A secondary lesson that was learned was that the customer had suffered through many failures and that in order to win the confidence of the customer back along with not using more expensive metals, several types of tests had to be performed.

7.3.5 Effects of Water Case Study #5

ITEM: LCD for cockpit display unit for Air Force attack aircraft

An aircraft manufacturer was looking for an upgraded cockpit display unit for displaying fuel values that would be in the form of LCD as opposed to the analog tape drive unit that was currently used in an attack aircraft. The vendor that won the contract for building this upgraded display was asked by the manufacturer to perform a fog test for each production unit that was built as part of the acceptance test. Apparently, the manufacturer had some prior experience with other LCD displays used in the aircraft that fogged up during field service and wanted this test to be performed on future LCD designs.

The LCD display had two modes of illumination: back lighting for night use (red lights) and front lighting (white lights) via a glass wedge for dawn/dusk use. The gap between the wedge and the LCD was a few thousandths of an inch. Figure 7–6 shows the setup.

The vendor protested that the fog test was an added expense and that there was no fog problem, but the aircraft manufacturer insisted on imposition of this test. The manufacturer suggested that if there were no problems, each production should be able to pass the test with no problems. Reluctantly, the test was performed by the display vendor. Each production unit was placed in a thermal chamber and cooled for over an hour at temperatures that

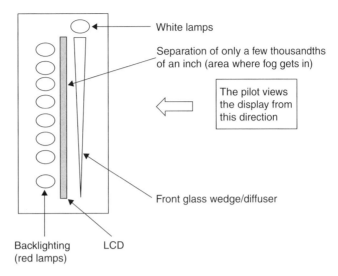

Figure 7-6 Fog Test on LCD Display Unit

were just above freezing. Then the unit was removed from the chamber and an ice cube was placed directly on the LCD itself.

To the surprise of the vendor, fog appeared on the inside of the LCD glass when the first test was performed on the qualification unit. The air between the glass and the LCD had humidity in it, and some sort of bake procedure was needed. The correct way would be to hermetically seal the display itself with the evacuation of all air and back-filling with dry nitrogen. However, this was deemed too expensive at the time (although the vendor would end up paying for several thousands of hours of engineering costs and repairs over the next two years). The vendor proposed a low cost "open" approach that would allow the display to "breathe."

This approach involved assembling the displays in a clean room with controlled humidity. In addition to the assembly process, there would be liberal use of RTV sealing compound over all of the edge joints between the LCD and front glass. The overall problem was never really solved, as the fixes were put in place only to pass the customer-required fogging test. A production unit might require as many as three or four iterations of disassembly and reassembly before the unit could pass the test.

In retrospect, the display module should have been hermetically sealed, baked, and back-filled with dry nitrogen. The customer could have also been persuaded to illuminate the LCD for the dawn/dusk requirement in lieu of using the wedge design. This could have been accomplished by using external lamps or additional rear lamps.

As it turns out, the vendor made over 100 production units, but none were ever installed into the aircraft for field service because of production delays, some because of the fog situation described here. It seems a waste of effort to have made 100 production units that ended up in storage in a warehouse and not installed in an aircraft where the pilot could have benefited with this improved display. But the situation described here points out the delicate situation between the customer and the vendor, particularly if there is a sticking point that exists with regards to a design issue that the customer just won't let go.

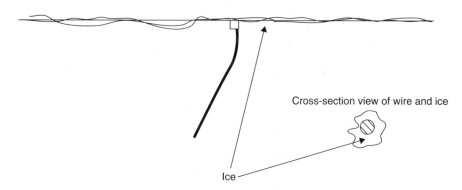

Figure 7-7   Ice on Wire Antenna

7.3.6 Effects of Water Case Study #6

ITEM: Wire antenna

Amateur radio antennas vary from multiple element beam arrays to simple wire antennas. As they are installed outside, there are concerns involving water, whether it is in the form of humidity and rain causing corrosion to feed point connections or in a more severe form, ice.

During a national radio contest that was held in mid-February, the author, a radio amateur in Long Island, New York, was using a simple dipole antenna that was cut for resonance on the 7-MHz frequency (cut for 33 feet in length). With the antenna being resonant, all of the RF power is effectively radiated by the antenna (as measured by a standing wave ratio at 1 to 1).

At the beginning of the contest period, the antenna was resonant, and contacts with other hams were being made. However, overnight a freezing rain occurred and the antenna was suddenly no longer resonant. In the morning, the operator went outside and saw that there was a coating of ice along the entire length of the thirty-three-foot wire antenna as shown in Figure 7–7. The ice had apparently changed the tuning of the antenna. The antenna was cut to a specific length that was based on free-space or air for the electrical wavelength. With ice on the antenna, it is a different dielectric than air, and hence the antenna is not going to tune in the same way as it did before.

Fortunately, that radio operator was able to use a tuner device that was able to compensate for this situation for the time being until it got warm enough for the ice to loosen up off of the wire and be shaken off. The problem becomes more severe in the case of multiple-element array antennas because the bandwidth on this tuning is smaller, and there is little in the way of compensating for ice by using a tuner device.

7.3.7 Effects of Water Case Study #7

ITEM: Hearing-aid design

Hearing-aid design has gone through several iterations in design changes, including size, fit, and in the electronics used. While the trend of hearing-aid design towards smaller size and new electronics have reduced failures, newer failure modes have surfaced because of certain changes.

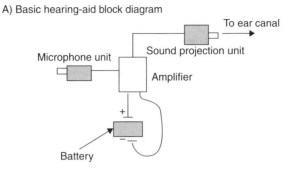
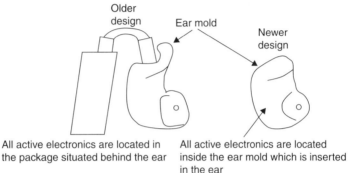

Figure 7-8 Evolution of Hearing-Aid Design and Issues Involving Humidity
In the diagrams, it can be seen that the trend towards smaller hearing aid designs where the hearing aid is located entirely inside the ear leads to additional stresses on the electrical components that were not seen as much in the older design. These stress factors include higher temperature (98.6°F body heat temperature) and increased humidity factors from being located inside the ear canal.

The availability of smaller electrical components has made it possible to make the hearing design increasingly smaller, to the point of being able to fit all of the necessary components into one complete package that can fit into the canal of the ear (see Figure 7–8 for this evolution). When the hearing aid used to be a behind-the-ear design, failure modes that were of concern were stress on the neck of the hearing aid where the sound projection unit was located as well as shock-related failures if the hearing aid should fall off of the ear.

However, as the hearing-aid design has become smaller, thinner low-current thread-style wiring is needed in order to connect the major components. With the introduction of the hearing aid to fit entirely into the ear, there is the new situation of the electrical components now being exposed to humidity factors that are present in the human ear canal. As there is an opening for the sound projection unit to send the amplified sound through, this opening can allow humidity to enter into the internal parts of the hearing aid. Over time, this increased humidity becomes a problem for the low-current wiring. The wiring becomes more brittle and corrosion can set into the areas of the solder connections where the wires are attached.

The exposure of the wires and solder connections to higher humidity cuts into the lifetime use of the hearing-aid setup. The only remedy for this is to optimize the positioning of the wiring as much as possible from the portals of the hearing-aid design.

Thus, this example shows how a generally good trend of reducing the number and the size of components can result in a design change such that a new arrangement is developed and new reliability issues are introduced.

7.4 SUMMARY

This chapter has illustrated to the reader, through the use of case studies, how the different states of water can impact a product and cause failures. Whenever a piece of equipment has to be used in the outdoor environment, any of the different states of water has to be an area of concern. This applies to cable TV, aircraft, automobiles, and other electronic items that may be used this way. There are certain applications where even in the indoor environment, some aspects of water and its states can be an issue that has to be addressed.

Various remedies exist for protecting electrical circuitry such as conformal coating as well as proper location of components that are susceptible to water intrusion. However, conformal coating has a limited lifetime and effectiveness with regards to certain applications. This is where qualification testing of a new product becomes of the utmost importance, particularly the humidity, fog, and rain tests to show any potential susceptibility to water.

7.5 EXERCISES

1. Explain how each of the three states of water (liquid, solid, and gas) can affect electrical equipment that is used in an outdoor environment.
2. Describe the conditions as to how certain products or appliances used indoors are exposed to any of the three states and can be adversely affected by these states.
3. Examine equipment (both electrical and mechanical) that is used in your car, and identify any issues involving water intrusion or exposure that can affect the performance of the equipment.
4. Identify some of the possible fixes that can be employed to correct the issues identified in problem 3.

CHAPTER 8

Failure Analysis and Troubleshooting Methods

INTRODUCTION

It should not be a surprise that some failures will occur on new products and systems entering field service or during the preliminary testing stage. Thus, failure analysis then becomes a major part of new product development where the cause of failure is determined and corrective action is developed to reduce the chances of a particular failure mode from reoccurring. Reliability engineers will often be involved in root cause analysis, as well as other aspects of the failure investigation, whether it is simply in the area of failure reporting to the customer or more detailed failure analysis during reliability test programs. The amount of involvement depends, in large part, on the individual engineer's initiative or the way that the job tasks are assigned in the company.

Failure investigation may often be conducted in a manner much like detective work or forensic engineering in criminal cases. The end result of the failure incident may be a broken component or an internal failure of an inoperative component. After the incident has occurred, the goal then is to find why the component broke or why the internal failure occurred. Sometimes, physical clues may be evident suggesting the way that the failure occurred, such as burn marks from an electrical short on the component. Other times, there may be no obvious clues from the evidence available, and some sort of limited or focused testing is needed to gather more information. All of these investigative methods make up an area commonly known as troubleshooting and will be illustrated throughout this chapter.

The complete reliability engineer should be able to learn some basic failure analysis troubleshooting techniques, particularly if the engineer is involved with reliability tests or with investigating field failures. This chapter covers basic troubleshooting methods that can be used to resolve different types of failures. Methods as to how to isolate these failures are presented in detail to the reader and are the focal point for this chapter. A particular emphasis for this chapter is in the area on how to isolate and troubleshoot intermittent failures.

8.1 BASIC TYPES OF FAILURES

Any one of several types of failures can appear during the course of a product's lifetime, beginning from early testing through field service. As introduced in Chapter 1, the basic reliability life curve or bathtub curve of a component has three basic periods of failure: infant mortality, constant failure rate, and wear-out failures. Infant mortality failures are the types of failures that can be detected during early design stages or during burn-in testing. Wear-out failures or end-of-life failures occur near the end of the product's life cycle, and this may be addressed by preventative maintenance, which will be covered in more detail in Chapter 13.

In the area of the curve where there is a constant failure rate, certain types of failures can occur that require troubleshooting. In fact, a number of failures may actually fall into two categories simultaneously or even may be in a different category than what was originally thought. The different types of failures that will be discussed in this chapter are:

- Hard failures
- Latent failures
- Secondary failures
- Intermittent failures
- Repetitive failures

It is important to be able to categorize the failure, as this helps guide the direction toward isolating the root cause of the failure. For example, if it is known that it is rare for a component to fail in a certain manner, it may be that this failure was actually a secondary failure that was the result of a primary failure of another component.

8.2 BASIC TROUBLESHOOTING METHODS AND TOOLS

When equipment is well into production, full documentation for equipment in the form of schematics and maintenance manuals should be made available. There may also be test equipment that has a diagnostic program that can test individual production units. In addition to these items, the standard tools that a technician may use to fix electrical equipment include multimeters and scopes.

It is important to save any failed parts that are removed during troubleshooting and subsequent repairs. These parts can be dissected for further failure analysis and can be examined under the microscope. In some cases, it may make sense to leave the failed parts intact, bag them, and store them in the event the failure mode appears again in the future. Thus, as more information becomes available through the reappearance of the failure on the same component, a more informed judgment can be made as to what the failure cause is.

8.2.1 Use of the Multimeter for Electrical Troubleshooting

Perhaps the greatest tool toward conducting most types of troubleshooting is the basic multimeter or ohmmeter, particularly a portable pocket-size unit that can be carried to the field or test lab. Even just using the meter in the resistance scale allows the engineer to find the following:

- Opens or high resistance on circuit board traces or wiring
- Electrical shorts to ground
- Blown components (failed open)

The beauty about this type of troubleshooting is that power does not have to be applied to the unit when checking out circuits in the resistance scale.

Often, voltages have to be checked, particularly if the unit has its own built-in power supply. In this case, the unit has to be powered on in order to use the voltage function (either DC or AC, depending on the application). This, too, can be checked by the simple multimeter by setting it in the appropriate voltage scale. Examples of how a multimeter can be used are provided in a number of the case studies throughout this chapter.

8.2.2 Troubleshooting Charts

It is very helpful to create a troubleshooting chart even if one does not exist from the manufacturer of the product or device. It is also useful to have the functional diagram or schematic diagram of an electronic device for troubleshooting purposes. However, lack of this documentation does not necessarily preclude the ability to create at least a simple troubleshooting chart. This is because almost all electronic products will have some of the same basic features. For example, units that plug into house currents will have a transformer and rectifier circuit as part of a power supply. This circuit will take the 110 volts of AC wall current that comes into the power cord and then change the voltage to DC voltage at a reduced level to drive the individual circuits of the product. A similar setup is used for components in aircraft applications where many devices will have internal power supply to step down and convert aircraft voltage from the voltage bus (typically 28 volts of AC).

In addition, most power supplies will have noise filtering of sorts, so large tantalum capacitors will be connected on the voltage input and output lines. One can check the output voltage (if it is not known) of the power supply through the use of a multimeter on these capacitors.

There may be fuses or protection diodes on power supply outputs that go to other circuit boards inside the unit or to an external interface output. Other boards may be in the form of logic circuits, memory ICs, and microprocessors. Typically these circuits will be using 5 volts DC from the power supply, and there may be some analog circuits present that use $+12$ volts and -12 volts. Again, these voltages can be verified through the use of the multimeter. If a product has an LCD display, a unique voltage output will drive the display.

From this examination of any electrical product for which one does not have a schematic diagram or other documentation, a generalize function block diagram can be generated. It is also helpful to generate a FMEA to list potential failure modes and their effects. From the completion of these tasks, it becomes possible to generate a troubleshooting chart or guide. The following case study involves a situation in which a manufacturer made a unit that consisted of a power supply and logic circuit that was used by a bank for their after-hours deposit box, but no documentation was made available by the manufacturer.

8.2.3 Troubleshooting Chart Case Study

Item: Power supply unit used in gas pump transaction unit

A large service station chain had a population of several hundred gas pump transaction units that contained a fuel quantity LCD, small customer CRT monitor, as well as a printer for printing out the receipts when requested by the customer using the keypad. The gas pump transaction unit consisted of a display monitor, a keypad, a receipt printer, and power supply unit, and they were serviced by a field engineering service and repair depot maintained by the gas company. They were generally powered on for 24 hours, 7 days a week.

The repair depot found repairs of certain components performed by the manufacturer of the devices to be very expensive, on the order of several hundred dollars. Therefore, it made sense for the repair depot to have their personnel perform as much of the repairs in the field or in the depot. One of the devices was the power supply, and using the methodology described in the previous section, the reliability engineer for the service station company's technical group was able to determine the output voltages of the unit. From this he was able to construct a block diagram and troubleshooting chart for the device even without schematics being made available by the original manufacturer of the device. Figure 8–1 shows the general functional block diagram and position of some of the key connectors and fuses.

Once the general functional block diagram was completed, the engineer was able to construct a troubleshooting chart on the power supply unit. Certain faults were able to be simulated by removing fuses or disconnecting connectors and cables. A multimeter was used to check the voltage of output lines. As a result, a basic troubleshooting chart was able to be created as shown in Table 8–1. This chart allowed for more repairs to be

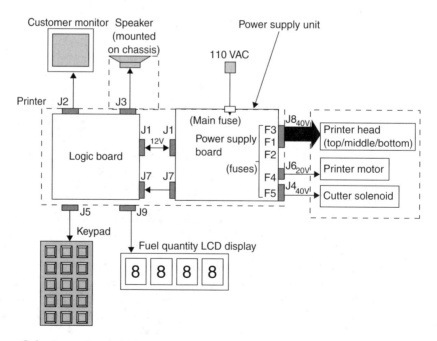

Figure 8-1 Power Supply Unit Functional Block Diagram

Table 8-1 Assembly: Gas Pump Transaction Unit
Subassembly: Power Supply Unit Electrical Failure Troubleshooting Chart

Symptoms	Subassembly	Possible Causes
No power output to any device	Chassis	Main fuse is blown
Distorted print (missing top part of print)	Power supply board	F3 fuse is blown
Distorted print (missing middle of print)	Power supply board	F1 fuse is blown
Distorted print (missing bottom part of print)	Power supply board	F2 fuse is blown
No printer motor response	Power supply board	F4 fuse is blown
No response from the printer cutter solenoid	Power supply board	1) F5 fuse is blown or: 2) J4 connector or ribbon cable assembly from power supply connector to logic board connector is bad
No printing even though the print head moves	Power supply board	J7 connector or ribbon cable assembly from power supply connector to printer unit is bad
Printer head does not move	Power supply board	J6 connector or ribbon cable assembly from power supply connector to printer is bad
Customer monitor blanks out during printing	Power supply board	Bridge rectifiers are bad. 22 volts should be read across and 40 volts should be read on outputs
Printer motor does not advance	Logic board	J6 connector or ribbon cable assembly from logic board to printer unit is bad
Monitor is not powered on	Logic board	1) J6 connector or ribbon cable assembly from logic board to the monitor is bad or: 2) J1 connector or ribbon cable assembly from logic board connector to power supply board connector is bad
Fuel Quantity LCD is not powered on	Logic board	J9 connector or ribbon cable assembly from logic board to LCD is bad
No keypad response	Logic board	J5 connector or cable assembly from logic board connector to keypad is bad
No individual key response	Keypad board	Bad contacts or broken solder joint on individual key assembly
Not able to hold date and time on the monitor	Logic board	Date/Time Chip is bad
Gibberish displayed on the customer monitor as well as gibberish being printed	Logic board	EPROM Chips may be bad
No sound after screen change	Logic board	Broken or loose connector or speaker wires or damaged speaker

performed in the gas company's repair facility as opposed to sending it to the original device manufacturer for repair.

8.3 FAULT ISOLATION TROUBLESHOOTING METHODS FOR HARD FAILURES

Fault isolation troubleshooting techniques can be used on numerous occasions when no formal documentation such as a maintenance manual or electrical schematic is available. This is often the case during the development of a new design, and situations may arise during a qualification or reliability test in which failures need to be analyzed and fixed. Knowledge of the basic theory of operation of the device is generally required. The fault isolation technique may also be employed when the vendor has not provided any schematic or similar documentation, and the customer has to compare a failed piece of equipment with a known good unit of this equipment.

The easiest way to use fault isolation techniques is by swapping components one at a time between a failed unit and a good unit. This is straightforward when the failure is "hard" or still present in the unit. The employment of this method becomes easier in the case of a test program where typically two units are available or in early production when a number of units have been completed.

One would isolate the failure to the circuit board of the failed unit by swapping out parts (circuit boards modules) from a known good unit. A methodology or order is required to keep track of the swapped components. This method is simple to employ, and there may not be a need for a schematic in most cases in order to get down to the failed circuit board level. Once the failed circuit board or module is identified, further failure breakdown requires a schematic and multimeter in order to check out the circuit and isolate the fault to a specific component. Once the failed component is identified, it is replaced with another component, and the failed component is taken off line to be thoroughly diagnosed, usually by microscopic means.

Thus a "drill down" technique is established by fault isolation, going from the circuit board level all the way down to a specific problem inside of a component.

8.3.1 Fault Isolation Troubleshooting of Hard Failures Case Study

Item: Credit card reader

The repair depot operated by a large credit card company had twenty returns of the portable credit card readers used in stores that had failed during field service and were failing test diagnosis. Component failures were suspected in many of the failures. Schematics were available, but because these readers were no longer in production by the vendor that made them, no spare parts were available.

The unit consisted of three main circuit boards, along with a number mechanical components such as a magnetic head, solenoid, and motor assembly. The technician decided to use the swapping part method to try to quickly isolate bad components of each of the failed twenty units. By using a good reader out of stock and taking each known good circuit board from this unit and swapping with each failed unit, the technician was able to isolate the majority of the failed units to a bad circuit board or motor assembly.

Through this process of fault isolation about ten of the readers were restored by using the combination of the good boards. As the unit was out of production, the leftover parts that were tested good were used as spares. This process is not unusual for obsolete or out of production equipment.

8.4 LATENT FAILURE TROUBLESHOOTING METHODS

Latent failures involve components that have failed but are not detected when they occur. Often, they may involve degraded components that are not detected during normal operation. This does not make them any less serious in nature. Often latent failures are only detected when another failure occurs to another component in the device or if an event occurs that exercises the function that has the latent failure.

A very simple example of a latent failure is a newly paved road with poor drainage. When weather conditions are dry, no failure is detected with the road. However, if there is heavy rain, the road is flooded and the failure in the design of the road is discovered. Another example of a latent failure is an explosive charge used in an ejection system. If the charge is no good, there is no detection under normal circumstances, and it would only be discovered during an emergency situation when the ejection seat is used or when the system is tested.

Some latent failures are minor in nature. An example is a ceramic capacitor used for noise filtering. If the capacitor should fail to open, but there are no noise conditions present in the environment, this failure would not be detected.

Latent failures are a particular concern to safety and reliability engineers and are part of the fault tree analysis and a maintainability (FMEA) detectability analyses that are performed during the early stages of product development. A maintainability demonstration will also bring out the conditions that result in latent failures and sometimes will demand some sort of corrective action depending on the failure consequences.

Often latent failures or conditions become noticeable during the occurrence of intermittent failures. This is documented in a case history Section 8.4.1. This case involves a component that was performing at a borderline level that became noticeable when another assembly was changed in the equipment. When there is suspicion of a latent failure, the best way to bring it out is to simulate another failure in the circuit or simulate the conditions that would bring out the failure to the point of detection. Sometimes this requires removal of another component to simulate the failure. This process is part of the troubleshooting methodology that can be used for latent failures.

Loose hardware is another area that can cause latent failures. This is why burn-in of production units is so important as these can be discovered at that point. However, there may also be concerns involving shipping of units where hardware can become loose. This particular area of concern has resulted in the science of packaging engineering, as well as the development of portable accelerometers that have the capability of measuring the highest level of shock and vibration and apply a time stamp against such occurrences.

8.4.1 Latent Failure Case Study

Item: Laser printer failure during field use

With the advent of improved embedded software technology, laser jet printers have been employed in significant numbers in every aspect of modern society. These types of printers will work on the principle of heat transfer of ink particles from toner cartridges installed into the printer. These toner cartridges have some self-contained mechanisms that interface with the printer. Many of these toner cartridges can be reused again by returning empty cartridges to the supplier.

The engineering department of a large company that used many laser printers at many locations was asked to investigate an error code situation that appeared when certain toner cartridges were used to replace empty cartridges in the printer. After the new cartridge was installed, a specific three-digit error code would appear on the LCD display of the printer within one minute of a print attempt. The error code was flagged against the print head of the printer. The problem appeared to be isolated to what brand of toner cartridge was used. One brand toner would work, and another would not.

Closer investigations revealed that the one brand that did not work had a different flag wheel configuration (located on the end of the toner cartridge) than the brand that did work. A flag wheel is a device that is used in many different types of printers and other mechanical devices that use a mechanical means of counting in order to work with the unit's software. The wheel has some slots cut into it that is "read" by a diode-based sensor as shown in Figure 8–2. Each slot represents a count that corresponds with the turn of the axle. This information is fed back to the unit's software so that the correct interface is accomplished and the other devices (such as the print head and the rollers for the paper travel) are properly engaged.

In this particular case, the actual failure was a low-voltage signal output from the print head of the printer. The print head was performing at a borderline level when a change in the brand of toner cartridge caused a change in the voltage settings and this apparently changed the voltage level of the print head. With the change of toner cartridge, the print head failure was no longer in a latent state but in an active fail state.

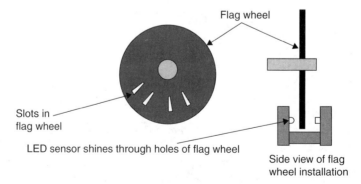

Figure 8-2 Flag Wheel and Sensor Configuration

The diagram shows how flag wheels are used to identify to the product, different types of components (such as toner cartridges) that are installed. In this particular case, a different-type toner cartridge brought out a latent failure condition that was not seen with the original toner cartridge. The latent condition was brought out by the change of settings.

The specific failure of the print head was traced to a failed laser diode in the assembly. The different toner cartridges and different settings caused the failure to appear. Further analysis of this failure, along with three other diodes that failed, revealed that all had suffered electrostatic discharge (ESD) damage, which apparently was introduced during the manufacturing process. It is worth discussing some of the major aspects of ESD at this point.

ESD damage is a major concern for companies that make electronic products. ESD damage is particularly bad on CMOS-type devices and other static sensitive components since degradation to these parts will occur. When handling these components, special grounding straps must be worn by the operator to take any electrostatic charges to ground. Companies typically have regular ESD training meetings to remind their employees about proper handling procedures. Component manufacturers have even more stringent quality controls.

Normally, there are fewer concerns with ESD when these type components are installed in an assembly that may be covered. It is generally advised that external connectors on certain units be capped so that the likelihood of ESD damage being introduced is reduced.

The bad thing regarding the detection of an ESD failure is that they tend to appear sporadically, almost in an intermittent fashion; however, the failure is classified more on the order of a latent failure condition. The failure is basically a degradation-type failure where the part may still be functioning at a certain level but will eventually degrade to the level of nonperformance. Thus, components with ESD can appear as any one of a number of failure types: intermittent, latent, and degraded.

Also when more than one case occurs in which the same component fails in a number of units because of ESD damage, it suggests that there is a manufacturing process error or parts handling problem on a specific log of the failed component. When the reliability engineer sees this happening, questions must be asked of the vendor with regards to how they test components for ESD damage after manufacturing along with their basic manufacturing and parts handling procedure.

The particular failure that was discussed here covered several areas. It is not unusual for the reliability engineer to see this type of situation occur on occasion.

8.5 SECONDARY FAILURE TROUBLESHOOTING METHODS

On occasion, a failed component that is observed in a unit may not have been the original or primary failure. In a case like this, it may be a secondary failure that resulted after another component in the same circuit failed first.

In a certain sense, protective fuses, circuit breakers, and polarity protection diodes that are found to be blown or tripped can be defined as secondary failures. When a fuse is blown or a circuit breaker tripped, there is often an overvoltage condition present in the circuit that was caused by either an internal failure or external failure condition. In the case of the polarity protective diode being blown, it is generally as a result of human error where the polarity was reversed.

There are some basic ways to spot these situations fairly quickly. It is important to know the function of the part that failed and whether it is capable of being the cause of a primary failure. For example, it is very rare to see a resistor or diode to burn up because of an inherent failure of these parts. In most cases, it involves an external overstress situation to these parts, such as high voltage, that caused them to burn up.

8.5.1 Secondary Failure Troubleshooting Case Study

Item: Air force transport aircraft controller unit undergoing durability test

After vibration cycling of a durability test was performed on a test unit, a failure occurred to the secondary (backup) power supply of the unit. Examination of the power supply board revealed that a resistor was physically burnt up. This seemed odd for a resistor to fail as the primary failure, and further examination of the electrical schematic revealed that this resistor was hooked up in parallel with an inductor. Investigation of the inductor revealed a fractured lead on this part, which means that the path across the inductor was broken and that all of the voltage was dropped across the resistor, causing it to burn up from overstressing. A fix for the inductor was subsequently developed in which epoxy would hold it to the circuit board better so that the leads would not break. The point of this example is to show how secondary failures can occur and how they can be spotted.

8.6 INTERMITTENT FAILURE TROUBLESHOOTING METHODS

A large focus of this chapter is dedicated to the area of intermittent failures, which is probably the hardest of all reliability problems for engineers to solve when they occur on equipment undergoing either field service or testing. This is particularly true with electronic components where intermittent failures involve a mechanical aspect of the component, either internally or externally. Intermittent failures may be referred to as "ghost" failures for the obvious reason that they come and go, as well as being hard to be able to reproduce on the bench. It is not always possible to simply swap out circuit boards with a known good unit with another unit that has an intermittent failure.

A typical case of intermittent failures in real life is illustrated by the case in which a car owner brings his or her vehicle in for repair of an intermittent problem, such as a noise, and the mechanic is unable to find the failure, as it is no longer present. An amusing and extreme example of this type of intermittent failure happened some years ago to the owner of a luxury car who would hear a dull thud sound on the right side of the car whenever he made a sharp turn. The owner made repeated trips to the car dealer where he had purchased the car, but the dealer was unable to find any obvious cause for the noise in the suspension or surrounding area. Finally, the dealer made a last attempt to get the problem resolved, and the manufacturer of the car sent two of their top mechanical engineers from the factory to strip the car down to the frame.

When the mechanics removed the rocker panel that was located between the front door and the rear doors on the right side of the car, they saw a small rope tied to the top of the frame. At the other end of the rope was a soda bottle with a note inside, which said, "Ha, you turkeys finally found me!" Needless to say, the owner received a new car, and an intensive internal investigation was conducted by the car manufacturer to locate and dismiss the disgruntled factory worker who did this mean trick.

Unfortunately, in most cases, any engineer who is troubleshooting an intermittent failure will not eventually find the note in a bottle that identifies that the failure has been found. The other point brought out by the car story is that the mechanics from the dealer were trying to troubleshoot the failure by conventional means using standard procedures and repair manuals. They were doomed to failure because they did not once think outside of the box that the failure was something different that would not be covered by the normal repair manual. Only when the car was stripped down to the frame could the root cause of the failure be spotted, but this meant that a lot of man-hours would have to be expended in order to complete this investigation.

The reliability engineer becomes essentially a detective in the investigation of intermittent failures. Indeed, there are elements of forensic engineering when conditions are reproduced or laboratory samples are created from failed parts for microscopic analysis.

An intermittent failure in electrical equipment can be caused by one of three types of factors: vibration, thermal, and humidity.

Intermittent failures of units may appear only during vibration such as a blinking display or lamp failure (as described in Chapter 5) but appear to function when no vibration is applied. This can be caused by:

- cracked solder joints on circuit boards
- cracked component leads but two halves are still touching (Ics, resistor networks)
- lamps with damaged filaments
- damaged wiring (including insulated wiring hiding frayed or broken wiring strands)

One can generally be assured that a failure has occurred and that there is still physical evidence of damage present somewhere in the unit. It may be difficult to find, but one has to start with the information of what function has failed and look initially at the circuit that is involved and trace the path from there, examining all components involved in that particular function.

Thermal-related intermittent failures are typically harder to find than failures that occur during vibration. This is particularly true in the case of many thermal failures that are caused by a temporary set of circumstances related to a combination of thermal expansion of different materials. Thermal expansion may occur internally inside a component, such as an IC, or externally with components soldered onto glass epoxy circuit boards. Condensation and humidity issues are also harder to find as they come and go when moisture is present (particularly during some parts of thermal cycling).

Look for the usual suspects! Start by examining the following components:

- Connectors
- Potentionmeters
- Wiring
- Large components and their mounting to the circuit board

Basic steps to take:

- Try to simulate the exact conditions when the failure occurred. Certain temperature conditions and humidity factors may be involved for the failure to occur.
- Try to isolate the function that is affected. Swap parts to see if the failure goes away.

Because of the general difficulty in finding intermittent failures, a large number of examples from actual case studies will be provided here to give the reader a better understanding of how to troubleshoot this particular type of failure.

8.6.1 Intermittent Failure Case Study #1

Item: Air Force transport aircraft controller unit undergoing durability test

The following case study shows how a simple multimeter using the resistance scale was able to help with troubleshooting, particularly in the test lab environment where limited amounts of test equipment were available. This case study illustrates the methodology to find certain failures, particularly the intermittent failures that were occurring during thermal cycling. The actual failure mechanism will be discussed in-depth in other chapters of this book, however the main intention here is to show the methods used to isolate the failure along with the symptoms that were found to lead to some conclusions.

Intermittent Multi-Layered Circuit Board Failures

During thermal cycling, both of the test units failed consistently during the cold to hot ramp up from −40°C to +55°C. What made the situation worse was the fact that the failures were gone by the time the unit was removed from the thermal chamber and brought to the test bench for troubleshooting. This type of situation required a little bit of a hunch to get started in the right direction. One clue was that a number of the analog functions were affected in one unit and these were primarily concentrated in one circuit board. At some point in the investigation, it was believed that there was a problem involving the circuit board traces.

Point to point on the affected trace measured about three inches in length, yet there would be a resistance reading of 2 to 5 ohms on the multimeter. Typically, one should read 0 or 0.1 ohms on a trace. This was the first clue that there may have been a problem with the integrity of the circuit board itself. As the components were installed and wave-soldered, it was unlikely that the via holes themselves had failed as a consequence of the soldering operation. More likely there was an existing problem with the way that the circuit board was manufactured. Indeed, further investigation revealed a flaw in the way that the circuit board was manufactured with regards to how the via holes were connected to the traces of the circuit board. The actual defect is covered in detail in Chapter 11, along with some mathematical techniques that were used to further isolate the failure. Basically, the process of butting the traces against the via hole was not connection sufficient that could endure thermal cycling and the circuit board manufacturer required a different and more positive connection process.

Figure 8–3 shows the physical setup using the multimeter that was used to find the original failure. A temporary repair was incorporated by installing jumper wires from point to point on the failed traces, but eventually, more and more circuit board traces were failing because of the way that the board was made. The board would be scrapped so that test samples could be made, and this verified the poor connection quality of the trace to via hole connection. This board along with several other multilayer boards of ten layers from both test units had to be scrapped, and replacement circuit boards using a better manufacturing process had to be made and installed into the units. This resulted in additional cost outlay and delays in the completion of the durability test program.

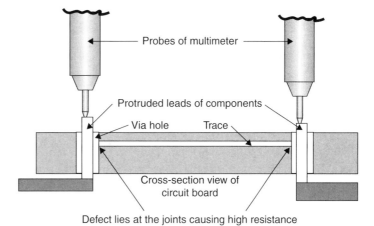

Figure 8-3 Method of Detecting Circuit Board Trace Failure

Power Supply Failure

For finding this particular failure, it was necessary to re-create the actual conditions at the time of the failure and add monitoring to key parts of the circuit. This failure was a particularly difficult one to find as it showed up at the rate of one out of three thermal cycles, always during the temperature ramp-up from cold to hot during the thermal profile. As the failure could not be duplicated on the bench during ambient room temperature, it became necessary to solder a piece of wire to a critical voltage point in the analog circuit on one of the circuit boards. The wire was brought out of the unit as shown in Figure 8–4, and the voltage could be monitored while the unit was being thermal cycled during the ramp-up.

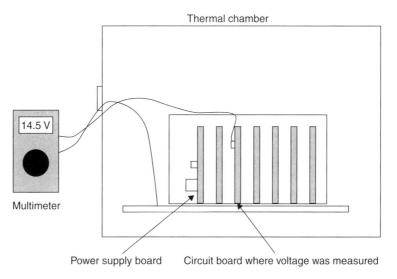

Figure 8-4 Setup that Recreates Failure Conditions

As the failure would only show up during the cold to hot ramp-up, this part of the cycle was repeated a number of times until the failures were reproduced. When the failure did occur, it was noticed on the voltmeter that there was a one-volt drop from the correct value. This was the vital point of information that was needed and not normally supplied by the display of the unit. At this point, it was realized that this condition could only be caused by the power supply. Further analysis isolated the failure to a potentiometer on the power supply board that was opening up during the thermal change. (Please refer to Chapter 5 for a detailed discussion on the failure mode mechanism of potentiometers.)

Condensation Failures

This unit also experienced humidity or condensation-related failures that showed up several minutes after the unit reached hot temperature after ramping-up from cold. The irony was that even though the failure was intermittent, it could be reproduced often with the same timing occurring. At this point, it was believed that water was forming on the circuit boards inside the thermal chamber. As thermal chambers are not perfectly sealed, it is not uncommon to see ice form on the units during cold temperatures, and thus water would result after the unit warmed up. This unit saw two types of failures: one involved analog functions and the other was ice forming on the units during cold temperatures, and thus condensation would result after the unit warmed up. The actual causes of the failures are outlined in Chapter 7. However, it is important to note for the purpose of troubleshooting tips that are covered here is when during the thermal profile the failures occurred.

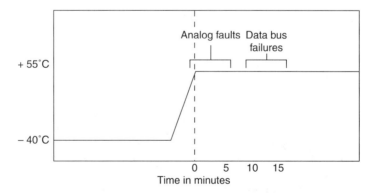

Figure 8-5 Condensation Related Failures During Thermal Profile

The diagram is a thermal profile for a controller unit undergoing reliability testing along with indications of where certain types of faults occurred. The timing of these faults (where they appeared during the thermal cycle) provided to reinforce the notion that specific conditions were present when the failure occurred. In this case, the failures were related to different aspects related to condensation that occurred after thin coating of ice melted inside the unit when the temperature was raised from sub-freezing conditions. This information is critical when trying to determine the root physical cause of the failure.

Because of the timing of when the failures occurred after the +55°C plateau was reached, it had relevance as to what was the actual physical mechanism that was causing the failure. Figure 8–5 shows where these failures occurred, and this timing actually led to the solving of what were the physical causes of the failures. Exposed test connectors on the analog circuit board failed one minute after reaching the 55°C plateau. This was because the ice on the board was just starting to melt and the liquid state of water was causing electrical changes to the analog lines. The data bus failure was caused by ice that had melted completely and shorted out a differential pair at the connector. But this failure took a little longer for the ice to melt into mostly water in order for the short to occur.

8.6.2 Intermittent Failure Case Study #2

Magnetic Stripe Card Reader Failure

The magnetic stripe card reader is one of the more important electronic devices used in our world today as it used for a whole host of products, ranging from badge access entry to automatic teller machines (ATMs). In this particular case study, an engineering group was tasked with finding out the reason why a number of card readers used by their company had a very high misread rate that was much greater than the specified 5 percent value. Those card readers that failed at greater than a 25 percent misread rate were sent back to the engineering lab for analysis. Some obvious problems were found such as dust and foreign object damage, but a consistent number of card readers were failing at a high rate and there were no obvious signs of failure.

The reliability engineer in the group hooked up the failed card readers into test setups in the lab and saw that the readers were reading cards normally, and no problems with the waveform of the card data were apparent. The engineer had some previous experience with certain types of wiring failures in aerospace design, so he decided to investigate the wiring going to the magnetic head device of the card reader. While wiggling the wiring to the head, he placed a multimeter across the two wires that travel into the internal coil of the head and saw that the resistance opened up and went from a nominal value of 35 ohms to several hundred kilo-ohms. This was the basic failure. Because there is an up-and-down action of the magnetic head when reading a card, the wiring is pivoting in one area, just above the head (please see Figure 8–6). About 5 percent of the card readers were suffering from this failure symptom, so this was clearly not a random occurrence, and would require corrective action to fix.

Two areas were identified as contributing to the cause of failure. One area was the fact that the wire traveling into the head was not secured properly, such that it would not pivot. However, the second area was identified as a flaw in the vendor's magnetic head manufacturing process in the potting of the wiring into the head. It was determined that a tinning operation, performed by the vendor, hardened and bonded together the multistrand wire of each lead to a single stress point, in the area of the pivoting action. The vendor discontinued the tinning operation prior to soldering the wires, and this allowed for some added flexibility in the pivot area just above the magnetic head assembly.

Again, the simple multimeter, with the meter set in the resistance scale, was the tool used to spot and verify the failure! Many failures are simple in nature—it's just that it is sometimes hard to find them.

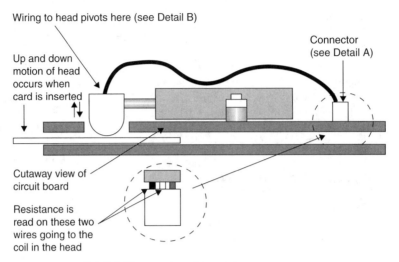

Detail A: Closeup view of connector

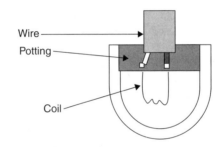

Detail B: Internal view of magnetic head

Figure 8-6 Magnetic Head Failure

8.6.3 Intermittent Failure Case Study #3

Item: Air force transport aircraft cockpit display unit in durability test

Two samples of a cockpit display unit for an air force transport aircraft were undergoing durability testing when after only three thermal cycles; one unit failed a display wrap test that was performed internally by the unit when it was powered on at −40°C. Yet, when the unit was put on the bench, there was no evidence of the failure being present. The failure would show up randomly, perhaps once every thirty or forty thermal cycles. The display wrap test took only 150 microseconds to run and in failing once during an eight-hour thermal cycle, it truly was like trying to find a needle in the haystack.

At first, it was felt that there was a noise transient problem present on the signal lines of which the software was not being tolerant. So the software was changed to increase the built-in test (BIT) tolerance from zero to three iterations before a fault could be flagged. However, the failure reappeared on the same unit at −40°C after twelve more thermal cycles were accrued. A discrete circuit board was replaced, but the unit still failed several cycles later, this time at the high temperature of 71°C.

Suspicion then centered on the display driver chips as these were tested during the display wrap test. The design engineer had a hunch that the driver chips made by the vendor were not up to military quality levels, particularly with regards to the low and high temperature extremes. After the fourth occurrence of this display wrap failure, the display assembly was replaced by another display assembly and a separate test of thermal cycling was performed without failures. It was then decided to replace the chips on the bad display assembly by using another set that was made by another vendor. Testing was allowed to continue until the chips were available. When they were installed, no more failures occurred with the display wrap test for the balance of thermal cycling (approximately 300 more hours of testing).

A number of lessons are brought out by this example. First, you sometimes don't know that you have found the actual failure until you have replaced the suspected parts and then the failures do not occur any more. (In an odd way, this is very similar to certain criminal cases such as arson when the suspected perpetrator is put in jail and the arson fires stop.) The case of the display driver chips described here demonstrates this principle of uncertainty. A hunch was needed in order to move forward.

Second, quality related problems can actually result in intermittent failures, especially when the parts are not durable enough to meet the specified temperature extremes. Inferior material and processes used by the vendor can be a major contributor to quality-type failures.

Finally, with a difficult failure such as the one described here where it only appeared for fractions of a second, one has to take a best guess in trying to isolate the failure to the function that is affected. One must carefully review the functions of all of the components in the circuit involved.

8.6.4 Intermittent Failure Case Study #4

Item: Hospital blood pump unit

A medical equipment manufacturer made a production line of electro-mechanical pump units that were used to pump blood into the patient at a specific pressure. In addition to manufacturing these units, the manufacturer provided the field service support for these units when a service contract was purchased.

The company received a complaint about one unit that was malfunctioning at a hospital in an intermittent fashion and sent a field engineer to investigate. The hospital reported that on occasion, the unit would fail where the CRT monitor screen would blank out shortly after the user pressed the enter button to activate the pump. When this happened, the unit would lose all of its settings, and the user would have to put in all of the basic data again.

The field engineer would test it thoroughly but could not get the unit to fail while in the hospital. The unit checked out, and the field engineer left. But within two weeks, another incident of the failure occurred. Again, a field engineer was dispatched and found nothing wrong with the unit. But because the hospital had lost confidence in this particular unit, the company decided to remove the unit from the hospital and replace it with another unit. The failed unit was returned to the repair depot, and after much investigation, it was realized that the power supply of the computer did not have an engineering change incorporated by the manufacturer of the computer that would have reduced noise on the power lines. It was then reasoned that there was electrical noise present in different areas of the hospital because of other equipment being present along with the presence of air conditioning and heating equipment. Thus, with the power supply of the computer not having the noise protection change incorporated, it was possible for electrical noise to affect the unit's performance, depending on where it was located in the hospital.

It is interesting to see that a unique set of circumstances aided in making this particular intermittent failure difficult to troubleshoot. If the unit was located in a noisy electrical environment when the field engineer originally investigated the failure, the failure might have been able to be reproduced. But the unit was moved around inside of the hospital, thereby changing the environment in which it was situated. Also, the problem came down to the fact that a simple engineering update (noise protection) was not incorporated.

8.6.5 Intermittent Failure Case Study #5

Item: Electric stove

Many electric ovens now have microprocessor controls that monitor many of the oven cooking functions. Items such as digital timing features that control the duration and temperature of the baking and broiling functions are displayed on a LED display panel.

What is also included in this microprocessor control is a built-in test circuit that is able to flag an error code when certain failure conditions appear, particularly those related to the safe operation of the appliance. So imagine when preparing to use the oven in a bake mode one day, and when pressing the bake switch, the characters "F4" show on the LED display. To the ordinary layperson, this does not mean anything (except trouble in the long run). However, any engineer who has worked on electronic equipment with failure-detecting capability would be right in guessing that "F4" is some sort of error code for a failure in the oven.

Unfortunately, the error code is not in the user's manual, probably because the manufacturer does not want various laypeople trying to troubleshoot a stove, particularly with high voltage involved! Calling up the service department brought out the information that the error code is described as "sensor circuit shorted." Examining the sensor wiring and the elements of the stove did not reveal anything except that the fault would go away when the stove was reassembled. The stove would actually work for several months until the dreaded F4 code appeared again—an intermittent failure. Finally a service technician was called in, and as he was well familiar with the stove brand, he knew exactly what the code meant— a failed oven temperature sensor. This would make sense as the stove manufacturer would not want to have the stove operating with this particular sensor because it is safety related. But sensors are used not only in stoves as witnessed by the next case study.

8.6.6 Intermittent Failure Case Study #6

Item: Sport utility vehicle

No chapter on intermittent failures would be complete without the proverbial intermittent automobile problem. Perhaps every person who has owned a car has experienced an intermittent problem when driving the car, but the problem never appears when the mechanic in the shop works on the vehicle.

This story involves an early 1990s sport utility vehicle (SUV). The owner of the car began to experience an annoying failure that appeared at random. The car would start fine for weeks at a time, and then it would not start on numerous occasions. It was particularly troublesome after the car started fine initially during the morning, and then if the engine was turned off for a few minutes, the car would not be able to be restarted.

The owner took the car to the dealer that handled that particular make of the car. Despite the dealer having the maintenance manuals and equipment along with some parts changed, the problem still persisted. Finally, the owner found an independent mechanic who did a little trial and error process on the car. He suspected that one of the sensors was the problem where a bad sensor in the electrical system would cause the computer to prevent the car from starting. One of these sensors was the crankcase sensor. After changing this sensor, the problem never appeared again. Examination of the sensor revealed a simple device that had some internal wiring damage, most likely from age or heat. A reading of this sensor being bad by the car's computer would apparently cause a shut off of the ignition system.

One of the problems with automobiles, particularly in the area of intermittent failures, is that the maintenance manuals for the cars may not cover such cases. Much of the limitations of the manuals are caused by the amount of experience that the company has accumulated with certain components. Newer technology components may have failure modes with no previous history, and thus may not be covered in the manual.

Sensors that feed no-go or go status to a computer of an appliance or car can be a major reliability issue. Sensors are typically simple mechanical devices, sometimes just consisting of thermo-couple–style wiring that is welded together for the purpose of detecting high temperature conditions of a stove or car. When the high temperature is present and the sensor is working properly, the computer receives this information and shuts down the circuitry for the stove or the ignition circuit to the car. If a sensor fails open, it is normally not detected in the circuit unless a second failure in another part of the circuit occurs. However, if a sensor has an intermittent short condition, it can create havoc in the troubleshooting. The rule is to focus on the hardware aspect of the circuit as the software is not capable of failing in an intermittent fashion itself, unless two software processes are involved.

The method of using trial and error where parts are replaced in chasing a failure is one way to address an intermittent failure. However, it would make good sense to do a thorough review of the electrical schematic of the circuit as most sensors are electrical-based devices. The reliability engineer has to be aware that computers require "eyes" or devices that can sense the status of physical conditions and then feed this status back to the computer. But when the sensing device itself fails in the short mode, it seems to cause nuisance-type failures. This is true of BIT circuitry for electronic units used in aircraft.

The failure modes of sensing circuits or devices should be addressed in reliability analysis such as FMEA and FTA. Intermittent failures are a subset of these failure modes and as they are generally physical in nature, can easily cause the random occurrences of failure. Further discussion of software technology that is reliant on sensor circuitry is provided in Chapter 12.

These examples are meant to illustrate how sensors are a critical area in failure investigations as they can lead to single points of failures. It is interesting to see how both software and error code technology has become deeply imbedded in our home appliances. It is startling to those use to seeing the failure detection or BIT technology and associated error codes in aircraft technology, now appear in the electronic appliances used at home! Even in cars, onboard computers have this type of technology that will feed diagnostic information to the service technician at the repair shop via error code reporting.

8.7 REPETITIVE FAILURES

Repetitive failures are defined in either of two ways: 1) a single unit that fails repeatedly in the same manner, and 2) several production units of the same product that fail for the same failure mode. In the first situation, it is the case where the failure mode was not correctly identified during initial troubleshooting and the same failure occurred again later. This can happen frequently during the introduction of a new product and prior to thorough documentation being written and available.

The second case is a serious situation, as it suggests a design problem or a procedural issue in the construction of the product. Sometimes this takes a certain amount of time before the repetitive problem is correctly identified and the appropriate corrective action developed. Along the way, there may be a few red herrings that may divert the effort from finding the actual reason for failure.

A good illustration of the need to correctly identify the failure cause is laid out in the 1958 B movie, *The Giant Behemoth*. The basic story line of this movie involves the efforts of UK scientists investigating the causes for why a large number of dead fish are washing up on the shores of the beaches by the ocean. Analysis of the dead fish revealed an unusually high amount of radiation inside them, so the scientists then focus on finding out where the source of the radiation is coming from. At this point in the movie, a giant dinosaur-like monster makes its appearance and causes major destruction on land. The scientists use their instruments to detect that this monster is emitting very high radiation, and they assume that this monster is the source of the radiation that is causing the death of the fishes. They devise a plan where they use a submarine, armed with a powerful torpedo, to destroy the monster underwater. After the monster is destroyed, and the scientists are returning back to base in their submarine, they are alarmed to hear on the radio that thousands of dead fish still continue to wash up on shore. Apparently, the monster was also a symptom of the failure cause (in this case nuclear radiation) just like the fish and while the monster had to be destroyed, the original problem, nuclear radiation still remained as an issue! Likewise, one has to be careful in the course of a failure investigation and find the right monster (or source of failure) to chase!

The following case study shows an unusual example of repetitive failure that is not unlike the situation encountered by the scientists in the movie, *The Giant Behemoth*.

8.7.1 Repetitive Failure Case Study

Item: Navy fighter cockpit display unit

A new LCD-based display unit that measured engine parameters for a navy fighter was introduced into production and encountered a number of different field failures that appeared after one year of service. A number of failures were repetitive in nature, having appeared in several units. (One of these failure modes was mechanical in nature and discussed in Chapter 5.)

However, one particular failure mode that appeared in two different production units was quite unusual. The failure mode was that the transorb diodes used on the 28-volt input power line for lightning strike protection was discovered to have been blown. The diodes were rated for 35 volts, and it was quite odd that they should have been blown, given that no lightning strikes or voltage surge conditions were observed.

A major meeting was held at the aircraft manufacturer's facility with the vendor and about thirty engineers were in attendance between the two groups to discuss this failure mode and other field failures. When the subject of what could have caused the transorb failure came up, the discussion became very animated. At first, it was thought that there may have been a possible quality problem with the parts manufacturer who made the transorbs, but preliminary analysis by the display vendor had ruled this out. A number of the engineers from the aircraft manufacturer incorrectly assumed that there was a high current condition present in the aircraft that caused the failure. This was not the case since the transorbs were rated for very high current and the problem was apparently voltage-stress related. What could have caused the failure?

The answer came about three months later. A number of production failures were observed during ESS and PRAT testing at the vendor's facility. One of the failures was a blown transorb that apparently had been overstressed. As the units are powered by test equipment with the correct voltage at fixed levels, it was reasoned that the overstress condition could not have occurred from the test equipment. Further investigation traced this overstressing was not due to the test equipment but rather to errors made by the production technician who performed the acceptance test procedure. The technician had apparently used a variable voltage power supply to provide 28 volts to power the unit. Apparently, because of either experimentation or plain incompetence, the technician applied too much voltage on the diode and caused damage to it. The technician had blown out a microprocessor board on another occasion by mixing up 12 volts with the required 5 volts input. Needless to say, he was fired.

It can be seen that this repetitive failure condition could have been classified as a latent failure since the transorb had been overstressed and damaged to the point where it would eventually fail completely during field service. In a way, it was also a secondary failure because it was a consequence of another failure (the application of incorrect voltage). This is an example of the ambiguities that can be encountered in trying to classify a failure. In this case, if one gets past the potential of a quality problem of the component, the only logical conclusion that could be made was that an overstress condition was occurring. It just became a matter of time before the source of this overstress condition could be identified, in this case the technician at the production facility.

8.8 SUMMARY

In Chapter 9, a number of case studies will be provided discussing how the design concept is important in product reliability and how a poor concept can result in certain failure conditions. One of the cases that will be presented discusses a repetitive failure condition involving a LCD product that had air bubbles appearing in the LCD itself after the product had been out in the field for a while. Despite the efforts by the LCD manufacturer to improve the process to reduce the likelihood of air getting into the LCD, the failure mode would appear over and over again. This type of repetitive failure condition is a case where the existing technology is not capable of providing a solution to the problem. Thus, the failure kept on repeating despite much engineering effort to try to solve the problem because of this limitation. In a case like this, it eventually becomes apparent that the product will not be able to meet reliability requirements within the current technology, and what will eventually result is a different type of technology being used to build the product. This becomes another important aspect of reliability engineering where a suitable alternative technology is identified.

Definitions for the different categories of failures have been provided in this chapter to the reader along with some of the ways that can be used to isolate the failure. The case studies provided in this chapter are meant to give the reader a way of thinking in approaching the troubleshooting of failures, particularly in the area of intermittent failures. Special emphasis was placed in the area of intermittent failure because these are the most difficult to find and they appear frequently out of all of the failure category.

Often, when the failure occurs and the cause eventually is found, the engineer responsible for the design may exclaim, "This never happened before." One might be tempted to respond by saying, "That's what the farmer said when his horse died," but the thinking should be to expect anything to happen on a design, particularly the unexpected. The design engineer may often not anticipate certain conditions occurring to the product that can cause a failure.

Another point to emphasize here is that most of the failures shown in the case studies were basically simple in nature. There was generally no complex combination of events that caused the failures—usually it was a physical problem of some sort that was induced while the product was out in the field or during the course of testing. The trick then becomes how to find this simple failure mode in the midst of complex systems consisting of thousands of components.

There may be instances where the failure appears because of a combination of complex processes. This may happen particularly in the area of software issues and examples of these are provided in Chapter 12. In the case of more difficult failures, some sort of brainstorming process may be required to be able to put arms around the problem as well as to start thinking about a potential fix to the problem.

By the same token, the corrective action toward fixing the failure cause should be accomplished in as simple a manner as possible. If too many parts are used or if the fix is complicated, there is a danger of introducing another major failure mode. This is the situation where the cure becomes the disease. Thus, the process of testing or pilot programs becomes a key element in proving out a specific corrective action.

8.9 EXERCISES

1. Describe the methods that you would use to troubleshoot eight electrical units that have returned from the field and all that you have at your disposal is a test station and two good units from stock. There is no schematic or test procedure available.
2. Describe what a latent failure is. Can you identify a latent failure that you may have experienced with a car or an appliance product?
3. Identify an intermittent failure that you may have experienced during your life, either in a car or an appliance product. How was the failure found and what was it?
4. An electronic piece of equipment has an intermittent display that shows up during vibration testing. Describe the methods that you would use to isolate the failure to a circuit board.

CHAPTER 9

The Design Concept's Impact on Reliability Performance

INTRODUCTION

The design engineer has a tremendous amount of power and responsibility in determining the design concept for a new product. The decisions that the engineer makes during the conceptual phase will ultimately have a lot of impact on the field reliability of the product. In formulating the design concept, decisions must be made on mechanical packaging, structural durability and parts selection. Along with this power to make major design decisions, comes a high amount of responsibility and the corresponding pressure to develop something that will work well in the field. It is the job of the reliability engineer assigned to the reliability task to use his or her expertise to point out areas of the design that are of concern and would require the appropriate action prior to the design being finalized.

The following are a few of the major pitfalls in developing the design concept that can result in reliability problems:

"Pitfalls in New Designs"
1. *Unproven Design Concept*—launching a new concept in a product with no prior testing or field experience (i.e., the customer or product user is the guinea pig)
2. *Current Technology Not Able to Meet Design Requirements*—this happens when new technology that comes out is not prove to be reliable yet

3. *Lack of Bulletproofing in Design*—this is the lack of putting in fail-safe protection into the design to prevent catastrophic failures
4. *Failure to Anticipate Differences*—the product may perform differently when used in different environments and in different market areas
5. *Needless Complication of Design*—the design is made more complicated than what design requirements ask for (this is known as the design engineer with the big ego syndrome)
6. *No Margin of Error in Design for Potential Problems*—sometimes referred to as the lack of overdesign to compensate for certain failures

9.1 CASE STUDIES ON DESIGN CONCEPTS

The experienced reliability engineer may be able to spot when a "pitfall" situation develops and can sometimes take certain steps to reduce or stop the damage. Part of this experience is the saving of files and notes from previous design programs. The following are case studies, which vividly demonstrate the importance of the design engineer and the decisions that are made early in the design that could later impact reliability. With each case study presented here, some basic lessons are learned that the reader could use to identify potential design issues with new products.

9.1.1 Unproven Design Concept Case Study

Item: Compact car water pump design

In the early 1980s, a major car company launched a low-cost compact car model as an attempt to be competitive in this particular market. Unfortunately, the company made a number of poor decisions during the design phase in its effort to meet the goal of a low-cost automobile, and this eventually resulted in detrimental reliability performance of this model. The poor reliability performance was inevitable based on the choices that were made.

The problems started with the choice of a small engine size that was a four-cylinder 1.6-liter engine. Small engines by themselves are not a bad thing for a low-weight car, but when numerous items are "hung" off of the engine that make it work much harder, you are dealing with a recipe for a reliability disaster.

Apparently, the engineers felt that a conventional-size water pump was too big for this particular size engine. The water pump for the engine normally would be attached to a bracket on the side of the engine as well as being belt driven by one of the pulleys. For this car, the design engineers chose to make a water pump that would attach directly to the front of the engine block as opposed to a separate unit that hung on a bracket with a belt attached as is common in most car designs. By attaching the water pump directly to the engine, the designers felt it could be made smaller and require less energy to drive it, particularly by a smaller engine size. The water pump had an external gear that was driven by a toothed rubber belt which was also the timing belt connecting the crankshaft and camshaft that drove the overhead valves. This belt and gear arrangement along with the new water pump design would lead to major reliability problems down the road during extensive customer field use.

Because of the unique mounting, the water pump was getting additional stresses from direct engine vibration. The water pump was not designed for this additional stress, and as a consequence, the bearing that the gear and propeller of the water pump was mounted on would eventually go out of round (lose alignment) and eventually seize into the housing of the water pump. When this happened, the water pump would cause the toothed belt to shred and subsequently cause the timing of the engine to freewheel where the camshaft was out of sync with the crankshaft. This would result in every one of the eight valves to collide with the top of the cylinders and bend each of the individual valve shafts. Please refer to Figure 9–1. There was no fail-safe condition or clearance built into the engine design for such a catastrophic failure.

Someone in the design department of the company must have realized that the rubber timing belt was a fragile, weak point and a recommendation that this be changed every 50,000 miles was placed in the service manual. Unfortunately for many owners, the seizure of the water pump preempted this, and the belt would break prior to 50,000 miles being reached on most models of this year. Eventually, the company would issue a service bulletin where the repairs would be covered as part of the warranty. In addition, future engine design would incorporate a margin of safety with better valve clearance above the pistons.

There were additional design problems based on the original design concept chosen. The car did not use any grease fittings for the front axle and wheel assembly. But unfortunately, the front wheels would quickly lose their alignment and the insides of the front tires would wear out quickly. The condition of excessive tow-out was quite obvious where the bottom of the tires were sticking out as much as a two or three inches in reference to the top of the tire. As owners of this car would find out, the only way to fix this condition was not by a front wheel alignment. Instead, the frame of the car had to be straightened out before the front wheel alignment could be performed. So a normal $40 alignment job would cost over $300.

Other poor design decisions made by the company was the unfortunate decision to locate the horn unit on the lower part of the chassis where the horn was subjected to water that came from the road. It would not be uncommon for an owner of the car to replace the horn two or three times during the life of the car because of the poor location used.

The poor design choices in this model could have very well doomed this model for future years. To the company's credit, they did drastically improved the design based on the mistakes of the first model year to the point where the model was eventually named car of the year by a car magazine. The 1.6-liter engine was phased out in favor of a larger-size engine.

In recent years, competitive forces have meant automobile manufacturers, along with the industry itself, have imposed various reliability and durability requirements on vendor-supplied items to cars. The effort to incorporate better reliability into the car design is essentially a requirement driven by the competition to attract consumers.

A consumer has to take on the role of a reliability engineer when shopping for a new car. Is the engine size capable of handling the type of driving by the consumer? Is the car easy to maintain, particularly in the area of regular maintenance items such as oil changes and spark plugs? It is hard to believe that many car manufacturers still have difficulty with water pump designs, both with the size of the unit and the placement with regards to the engine.

The first model year of any item, whether it is a car or amateur radio transceiver, should be scrutinized carefully by the consumer. Automobile manufacturers perform various road tests on new models and from this can sometimes gain the label of high reliability. But this really is not always the true picture—it is only when the model is produced in significant quantity and used by thousands of drivers over an extended period of time.

A) This photo shows the basic water pump unit in question, along with one of eight valves that were bent when the water pump failed in a seized mode. The timing belt that was riding on the pump broke, causing the valves to free wheel.

B) Close-up view of the water pump shows that seizure of the water pump unit was caused by the internal impeller (located at the top in this photo) that had a bearing failure and grabbed into the water pump housing.

Figure 9-1 Water Pump Design Failure Analysis

9.1.2 Current Technology Not Able to Meet Design Requirements Case Study

Item: Aircraft LCD cockpit display

An electronics vendor had been making cockpit displays for a number of military aircraft for several years. Many of these displays consisted of the analog-type drive technology where numbers were printed on a plastic tape that rotated on metal spools. Eventually by the early 1980s, both the U.S. Air Force and Navy were interested in whether the newer liquid crystal display (LCD) technology could be used instead. Trade studies indicated that this was possible and a massive effort to convert the analog tape drive units to LCD-based units began.

The architecture of the design had to be changed somewhat along with new voltage requirements needed to drive the individual segments of the display. There were only a handful of LCD manufacturers at the time with only one located in Canada that demonstrated acceptable product quality. The LCD technology was such that the LCD manufacturer was only to get a 50 percent yield of usable LCD as the process was very complex. This situation would cause delays in getting the parts as well as a number of LCDs not being able to pass the vendor's initial acceptance test.

However, much worse news would occur after six months of field service was accrued on the LCD displays. The appearance of a large bubble would occur on the display, sometimes obscuring the values such that the pilot could not read them. Given that some of these displays were on critical engine instruments, the situation was very serious and needed resolution. How could bubbles or voids appear in hermetically sealed LCDs?

What apparently was happening (and later substantiated by the LCD manufacturer's tests) was that the liquid dye material used in the LCD had dissolved air molecules in it. The high vibrations along with the effects of temperature cycling in the aircraft environment had caused these air molecules to separate from the dye material and collect with other air molecules to form bubbles or voids in the liquid. The manufacturer was then tasked to find ways to improve his processes to try to eliminate the introduction of air into the liquid. This involved some baking procedures for the liquid in order to eliminate air molecules and this added to the cost along with increased amount of lead time for the LCD product.

Even worse news came when the LCD manufacturer announced to the display vendor that its parent company was closing down the facility in Canada as it was not profitable, and they did not want to absorb any more losses. Thus, a last time or lifetime buy was the case in which the vendor had to cough up substantial amounts of money to buy about 100 LCDs for different programs, the projected amount that was needed to carry the programs on for a few years. Since that time, LCD technology has been in a constant evolutionary process going from dichroic LCD to pixel or dot style LCD screens.

A very common situation is that a company can get stuck with a technology that becomes obsolete within a matter of a few years. This is tough for any organization in which substantial investments in personal computers and software packages become obsolete and require major expenditures to replace. There is no marriage that lasts forever in the case of changing technology.

It is not uncommon for companies or organizations to "overreach" in the area of new technology. Recent cases of technical overeaching include the V-22 Osprey, which has the concept of a vertical takeoff on a plane by use of variable thrust engines. Another case is the

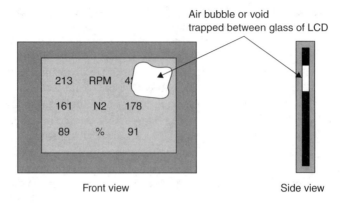

Figure 9-2 LCD Failure
This figure shows how air molecules trapped in the liquid dye for this type of LCD can cause significant problems.

X-33 spacecraft, which NASA ended development in March 2001. Critics of these programs say that the designs are far too ambitious to succeed for the technology that was available. Indeed, it is not just aircraft and spacecraft where the issue of making a reliable product is at the cutting edge of current technology. Automobile designs experience this type of situation often. For example, a recent case involves the area of "smart" air bags where they are deployed to a specific inflation force according to the size of the passenger. In late 2002, three major air-bag suppliers pulled smart-bag technology off of the market for further development, but there were doubts at the time whether a foolproof system could be developed. It was calculated that conventional air bags are 99.99 percent reliable but the smart-bag system was nowhere as reliable. The chief of the National Highway Traffic Safety Administration noted that the use of smart-bag was an extraordinarily expensive rule that is taxing the creativity and ingenuity of auto engineers. Issues that needed to be addressed for this system included misidentification of the size of the passenger due to movement and sitting position along with environmental issues such as water and humidity on the seats.

There can be many issues involved in making new concepts work when they are "pushing the envelope." There is a fine line toward what may be considered as too much for the capabilities of the times and what is truly innovative.

9.1.3 Lack of Bulletproofing in Design Case Study

Item: Magnetic stripe card reader

Credit cards, ATM cards, debit cards, company and student ID cards, transit cards, and similar cards using magnetic stripe technology are widely used in our world. They are read by magnetic stripe card readers that consist of a magnetic head assembly, card slot, and decoding circuitry. The decoding circuit basically consists of logic circuits that can read the high and low signals from the information that are embedded on the magnetic stripe of a customer's card. Many card reader designs allow for two attempts for reading the magnetic stripe on the card, either on the insertion (going into the card reader) or during the removal of the card (going out of the card reader) by the customer (or sometimes both directions where the reader has redundancy).

This study involves a company that made a magnetic stripe card reader that was used for a number of applications such as ATM and transit cards where the reader would be mounted facing up for dipping action. Unfortunately, after introduction into a number of metropolitan areas, there were some incidents of vandalism where coins and other objects were dropped into the reader causing electrical damage whereby units had to be replaced and sent out for repair.

The reliability engineer from one of the companies using this reader investigated the cause of the failure and found that when vandals used metal objects such as coins and paper clips, there was a good chance that these objects could fall onto exposed pins of a connector located inside the reader. These exposed pins on one 20-pin connector had a number of active voltage signals such as $+12$, -12 and $+5$ volts.

It appeared that the manufacturer had some space limitations in trying to fit the circuit boards closely around the card slots so that the design could be used in tight structures such as ATMs and subway turnstiles. In their effort to keep the size of the reader small, the design engineers would make a shortsighted decision of leaving a slot area of leads exposed instead of using standoffs on the circuit boards to the card slots. This was not a good decision, particularly in a metropolitan area where incidents of vandalism may be higher. When metal objects, such as coins or paper clips, made their way to the bottom of the reader where the exposed connector pins were, the object would catch onto the protruded leads, particularly in the area of where the different voltages were coming through.

If the coin happened to rest on one of the active voltage pins, such as the $+5$-volt line, or even the $+12$-volt line, the coin would eventually get so hot that the coin would melt into the plastic material of the card guide and become imbedded in it. (See Figure 9–3.) One could identify the coin easily by the impression that it left behind in the plastic! It is pointed out that the reason why the coin got hot was not that voltage was flowing through the coin, but rather because the coin resting on a pin is not a perfect joint. Because there is a gap of sorts between the coin and the pin, there is a resistance, which will eventually lead to the dissipation of power. Eventually this power dissipation will result in heat dissipation and work its way from the gap into the coin, making it hot and melting the plastic.

What made matters even worse in some instances was that the active pins with the different voltages would be adjacent to each other. So if a coin that traveled down far enough could short both of these two voltages, electrical component damage could result. When the combination of the $+12$-volt and the -12-volt lines were shorted, the bypass capacitor on the -12-volt line would burn completely off the board. This was because the $+12$-volt line had a higher current load than the -12-volt line, and it would overcome the lower current line.

The customer's reliability engineer developed an interim fix of cutting the leads of the pins and then applying electrical tape over the exposed pins. A long-term fix was to use a set of plastic inserts to cover the slot area where the connector pins were exposed.

It can be seen that this failure inconvenienced the users of the card readers, as well as the customer who owned the card reader with loss of service of the device as well as repair service costs. The failure to bulletproof the card reader design by the original design engineer was attributed to a lack of anticipation of potential vandalism problems in a metropolitan market place where volume was very high. However, it is a cardinal sin to have exposed circuitry for any unit that has interface with customers or outside environment. This was an obvious condition that any reliability engineer would have caught at the beginning of the design and

Figure 9-3 Magnetic Stripe Card Reader Failure Analysis
Products that require customer interfaces should be made robust and bulletproof for the worst case incidents of vandalism. The photo shows one side of the card guide of the reader and the results when a penny was forced down the card slot of a magnetic strip card reader with the penny eventually catching on to the exposed leads located at the bottom of the reader. The penny caught on to the +12-volt pin, and this caused the penny to heat up and melt the plastic of the card guides. A cardinal rule is to avoid having active voltage points exposed to potential damage from vandalism, as well as from weather effects.

demanded corrective action prior to implementation into field use. Unfortunately, in this case, there was no such position in the manufacturer's organization at the time that the readers were manufactured. The need to provide bulletproofing in a design is often known as fail-safe designing where a margin of error is put into the design so that certain failures do not cause catastrophic failure conditions of the product. The military is particularly concerned about this situation with regards to aircraft; hence the need for a well-conducted FMEA. This is also an issue in the area of automobile design.

For example, certain car designs have a linkage between the accelerator throttle cable and the transmission in order to synchronize gear shifting with the speed of the car. Yet, there have been designs where a single plastic part that holds the two linkages together can break and then cause catastrophic damage where the transmission burns up as it is running at the wrong speed in comparison to the car. There really is not a cost issue to justify this design since the use of a molded plastic part does not make any difference in price compared to using a more rugged metal part to add more structural reliability.

It should be part of the reliability engineer's job to use common sense and previous experience to look at those areas of the design where a simple component failure can cause major catastrophic damage. The FMEA is one way to accomplish this task if it is done early enough in the program. Also, the reliability should come up with a checklist of important items to look for as part of a "physics of failure" approach to the design concept. This checklist would be based on the experience of the reliability engineer and others, and it would list the pitfalls that can adversely affect a design.

9.1.4 Failure to Anticipate Differences Case Study

Item: HF/VHF amateur radio transceiver

When new amateur radios are introduced into the ham radio market, they face a similar-type situation as when a new automobile model is launched. Just like a car, there is a certain unknown factor out there, particularly if there is market pressure to release the product quickly without the benefit of any type of reliability testing. This was certainly the case in a recent amateur radio transceiver design.

The amateur radio equipment market is a very competitive market with both U.S. and Japanese manufacturers competing for the dollars of hams located around the world. Most of the radios manufactured fall into two categories: HF (high frequency) radios that cover frequencies from 1.8 MHz up to 30 MHz and individual VHF (very high frequency) specialty radios that cover VHF bands such as 50 MHz, 144 MHz, and 440 MHz. Towards the early 1990s, newer transceiver design included radios that combined both the HF bands and the 50-MHz band.

This would lead into the next step when one of the most innovative radios was introduced into the amateur radio market in 1997 that not only included the regular HF coverage, but in addition, the 50-MHz range, the 144-MHz range, and 440-MHz range. The latter two ranges are a very popular frequency range for amateur radio FM repeater work, which is very popular in the United States for operating the radio in an automobile. In addition to the increased frequency coverage, the radio included a large menu of "bells and whistles," such as improved filtering and memory storage.

With the introduction of this particular transceiver, there was obvious pressure on the other major manufacturers to come up with a similar model with even more features. At the 1998 Dayton ham radio convention held in May, one of these companies presented a concept of a similar-type radio design that included the 430-MHz frequency range. All of this would be in a small package where the radio could not only be used as a base station from home but also from a car. Thus, the race was on.

The months dragged on in 1998. First, the radio was to be ready in the fall, then the winter. It became apparent that there were still some issues to be resolved in the design. The company wanted desperately to have this radio available for sale at the 1999 Dayton ham radio convention to be held in May. The radio was sent to one of the staff writers of a ham radio magazine in March 1999. Another radio was purchased by another writer for the magazine, who also did a review that was independent of the first writer's review. Both writers used the radio in various applications as opposed to just conducting bench testing only. This proved very important as a major flaw was discovered during use in the car where the radio would be wired directly to the battery. Both reviewers had experienced the same failure where the transmitter section had failed, shortly after installation and operation of the radio in the cars.

The radios were sent back to the repair depot of the company for failure analysis. It was revealed that one of the voltages inside the radio was dependent on what the external voltage would be. In a base station application, power supplies were regulated at 13.8 volts and no problem would be encountered. However, in many cars made in the United States, the 12-volt battery can be charged even higher than 14 volts, and this higher voltage was responsible for the failure. The designers in Japan did not anticipate this situation in the U.S., as many of the smaller cars in Japan did not typically experience this higher voltage.

In addition to the failures discovered by both reviewers, a number of other hams encountered this same situation where the radio had to be sent back and subsequently modified with a change of about 20 parts. To the company's credit and expense, all radios were fixed under the warranty. This radio did not receive the benefit of enough time of field testing prior to the release in the general market. The amateur radio market continues to be very competitive, where each of the major manufacturers continue to try to outdo the others by introducing more frequency coverage, more features, and smaller radios.

This particular case shows a number of things that can cause problems for a new product. In this case, the highly competitive market of amateur radio equipment added a certain amount of pressure to complete a design in a fast amount of time. This pressure precludes any time for testing, and as a result, the customer discovers design-type failures during certain types of use. Finally, engineers working in companies in other countries may not always be aware of certain conditions in other parts of the world, which in this case was the fact that car battery voltage tends to go over 14 volts for most U.S. manufactured cars.

A more substantial case involving the premise of the failure to anticipate differences in the proposed environment is the Tacoma Narrows Bridge collapse in Washington State that occurred in 1940. The engineers had not anticipated that the high crosswinds in the Puget Sound area would cause vertical oscillations to the point that the bridge reached a point of total collapse only five months after it was opened. The problem might have been solved by using damping mechanisms in the vertical cables, but the engineers failed to do a sufficient survey of the area and then include this knowledge back into the design. This particular problem is detailed through the use of differential equations in the textbook, *Differential Equations and Their Applications* (Braun 1975, Springer-Verlag).

Again, things like a detailed site survey or a detailed market survey that looks at the proposed environment where the product will be introduced is of critical importance and generally requires an engineer to complete.

9.1.5 Needless Complications of Design Case Study

Item: Navy fighter aircraft cockpit display unit backlighting circuit

The sustaining engineering department for a vendor that made cockpit displays was asked to upgrade an existing fighter aircraft LCD-type cockpit display unit from the old dichroic (dye type) to the newer-type LCD technology that did not use liquid dyes. In addition, the aircraft manufacturer asked for fluorescent lighting to be used in the upgraded unit in lieu of the incandescent lamps that were used in the backlighting of the previous design. The aircraft manufacturer asked for a sensing circuit that could be employed that could sense ambient light brightness and increase the fluorescent backlight circuit during bright sunlight conditions.

The vendor's sustaining engineer knew that the customer wanted a simple backlighting sensing design, basically a circuit that ran off of the aircraft bus where all of the cockpit lighting was controlled by a single switch in the cockpit. However, the design engineer who was assigned to the program had other ideas. He had found a logarithm-based circuit design that could allow him to program different levels of backlighting brightness that was detected by an ambient light sensor that was added to the unit. It was more complex than what the customer had asked for, and the sustaining engineer told this to the design engineer but was overruled by both the engineer and the project group leader. What made it more tempting to the design engineer was that over $1 million was allocated for nonrecurring costs. (It turned out that this would not be enough as there were costs overruns.) The customer agreed to the new backlighting design based on the fact that the circuit was supposedly used on other products made by the vendor.

The logarithm-based circuit required further development from another source, the software group, for writing the firmware for the EPROM used in the backlighting circuit. This became a very labor-intensive effort for the software engineers in which each of the different lighting conditions expected in the cockpit had to be programmed into the circuit (it was a nonlinear pattern also).

The end result was that problems with the backlighting were experienced during the test program and during field service. Very harsh lighting spikes were seen by the pilot at odd times during flight. The backlighting never seemed to track well with actual ambient light conditions in the cockpit. It took several iterations over five years by different software engineers to try to smooth the harsh spikes, but the backlighting was never exactly right.

The design engineer moved on to another program, with a slight mark on his reputation. It would take years for the vendor to get the backlighting circuit to work correctly. The original reliability engineer on the program did not have experience in electrical design and did not challenge the design engineer on the excessive complexity that was added. Perhaps another engineer with suitable experience would have asked more questions and thereby, the customer would have gotten what he originally asked for.

It is a violation of the basic reliability law where it is better to have a simple design for better reliability as opposed to making it more complex. The rule KISS (keep it simple stupid) was very appropriate in this example. Finally, while a large amount of money allocated to nonrecurring costs seems enough, it will never be enough money to deal with the extra man-hours to get the design working and for the large amount of troubleshooting afterwards.

9.1.6 Allowing for a Margin of Error in the Design Case Study

Item: The A-10 close air support aircraft program

The Fairchild Republic A-10 Close Air Support aircraft program is an interesting case study in reliability, both in the original design concept and in addressing reliability problems that occurred later on. The following is a narrative account of both the reliability work that went into the design during the concept phase and in a major failure investigation involving the gun gas that became prevalent when the aircraft was deployed in field service.

The design of the A-10 benefited from the experience of the engineers in charge of the various subsystems such as flight controls, avionics, airframe, landing gear, and

Figure 9-4 A-10 Aircraft Design Features
The A-10 Close Air Support aircraft had numerous survivability and durability design features that were key to reliability performance. Some of the ones that can be seen on the outside of the aircraft include spatially separated twin engines that are hung high in the upper part of the fuselage. This feature allows the engines to be protected from runway debris when used in unfinished airfields. In addition, if one engine is not working, the aircraft is capable of being flown with the other engine.

navigation equipment. The overall concept of a simple design with straightforward maintenance would be a very key factor towards the high reliability that this plane would achieve.

The A-10 had a number of major design features that was for improving its survivability in a ground-attack environment. The two turbofan engines were located above the fuselage to protect them from runway debris and during combat from ground fire. (See Figure 9-4.) The pilot and flight controls were protected by a titanium bathtub, which was built into the cockpit structure. There were redundant flight control systems.

The A-10 reached maturity at 70,000 flight hours in 1979. At this point in its service life, the aircraft was achieving a cumulative MTBF value of over 3 hours along with a monthly instantaneous MTBF value of over 3.5 hours. Indeed, when compared with the other aircraft in the U.S. Air Force inventory that used more sophisticated equipment, the A-10 was the most reliable air force aircraft.

However, this is not to say that there was no difficulty encountered for the design. One of the more difficult problems that faced the engineers of the Fairchild Republic A-10 early in the program was the excessive amount of gun gas contamination that was ejected from the 30-mm cannon located in the nose of the aircraft. The cannon could fire at a very high rate, as fast as seven shells per second, and this generated a tremendous amount of gun gas and powder by-products. The high location of the two engines was right in the stream of flow for the gun gas and the engine would occasionally flameout and require restarting in flight. This problem was a unique one as there was no other aircraft made previously that had a high-speed 30-mm cannon and a uniquely shaped plane. Thus, the gun gas problem remained when production started.

Subsequently, the problem became of serious concern when a number of engine flameouts occurred with a few of them causing crashes. In 1981, an ad hoc team was created using engineers from different disciplines, including the reliability department in Fairchild Republic that would investigate the exact causes of the problem along with developing possible solutions. The reliability department was responsible for tabulating all engine flameout events as well as working with engineers to investigate the root causes of the failures. A set of four categories were defined for tabulating events: 1) gun firing flameouts, 2) hardware-induced flameouts, 3) maneuver-induced flameouts, and 4) unresolved events. The additional refinements allowed for a more focused investigation.

It was also found that some engine stalls could be induced by the pilot doing a "throttle chop," (moving the throttle back too fast). In fact, one pilot crashed two planes within a period of two weeks by chopping the throttle quickly, and he was killed in the second crash. A resulting engineering change was introduced by the team to prevent the pilot from moving the throttle too fast. However, more detailed work needed to be done with regards to the main problem of gun gas depositing particles on the engine fan blades and fouling the engine during firing.

One immediate fix was to implement an engine washing procedure after every 100 flying hours. This helped reduce the accumulated buildup of particles. There were several alternatives proposed for a long-term fix for the problem. A number of them involved structural changes such as baffles, extension of the nose section to fully enclose the cannon, a diverter device attached to the cannon, and the introduction of a door in the airframe for removing gas. See the photos of these various fixes in Figure 9–5.

However, the introduction of these structural changes on test aircraft would introduce structural damage such as cracking of adjacent spars in the frame of the nose assembly because there would be more stress added to the airframe during gun firing. The proposed fix of a door in the airframe brought about the concern of a potential door failure in the closed position where inadequate cooling of the gun barrels would lead to a possible explosive situation. The reliability engineering department developed a scaled down FMEA and further proposed change that focused on the relay that controlled the door opening and closing as shown in Table 9–1.

The reliability engineers determined that hazardous conditions could occur in modes B1, B2, and C1, which added up to 59.4 percent of all failure modes. When applying this percentage against the failure rate of 14.31×10^{-6} for the relay, this yielded a λ of 8.5×10^{-6}.

Using this failure rate, the reliability of the door not failing in closed position for a two-hour mission is calculated as:

$$R = e^{-(8.50 \times 10 - 6)(2)} = .999983$$

188 Chapter 9

A) This is a view of the A-10 30-mm cannon with no modifications attached to it. It is a Gatling gun that is capable of firing seven 30-mm shells a second. This high rate of speed causes a lot of gun gas to be generated and cause fouling of the engines.

B) A fix that was tried out was the addition of a gun gas deflection device that was attached to the front of the 30-mm cannon. This device was to break up the gas and deflect it away from the plane.

Figure 9-5 A-10 Gun Gas Flameout Study Program

C) Another change that accompanied the gun gas diverter attachment was the addition of louvers and vents on the front part of the fuselage.

D) Still another design that was considered was the extension of the nose to wrap around the 30-mm cannon along with a diverter door on the top part of the nose assembly (shown in the open mode here) for flushing out the gun gas. Unfortunately, all of the structural changes here could not be used because they induced additional vibration and stresses on the nose assembly structure where cracks developed. Thus, these types of changes and another fix had to be found.

Figure 9-5 A-10 Gun Gas Flameout Study Program *(continued)*

Table 9-1 Reliability Analysis of Ventilation Door Relay

Mission Phase	Description of Failure Mode	% of Failures	Results
A) Operating voltage not applied, control voltage off	1) Relay pulls in	10.8	Door opens
	2) Operating current > 200 ma	1.3	None
B) Operating voltage applied, control voltage applied, relay pulled in	1) Relay does not pull in	24.4	Door remains closed
	2) Relay pulls in, then drops out after time delay	2.7	Door opens briefly, then closes
	3) Operating current > 200 ma	5.1	None
	4) Control current > 20 ma	1.4	None
C) Operating voltage applied, control voltage is applied for 50 ms and goes to zero, while relay is only on for 180 seconds	1) Relay drops out immediately	32.3	Door is open only during gunfire and closes without time delay
	2) Relay drops out, but times out	1.4	Door remains open longer, then drops out
	3) Relay does not drop out as long as operating voltage is applied	19.3	Door remains open for the remainder of flight
	4) Operating current > 200 ma	1.3	None

The expected number of failures of the door remaining closed during gun-firing operations for the A-10 fleet of 660 aircraft, averaging 31 hours per month per aircraft over a 15-year period, is calculated by first figuring out an estimate for flying hours:

$$\text{Flying hours} = 660 \times 15 \text{ years} \times 31 \text{ hours/month} \times 12 \text{ months} = 3{,}682{,}800$$
$$\text{The expected number} = \lambda t = 8.5 \times 10^{-6} (3{,}682{,}800) = 31.3 \text{ failures}$$

It was determined that the amount of expected failures associated with this fix would create too much of a risk as well as introducing potential structural damage to the aircraft because of the radical changes to the forward fuselage structure. Thus, another fix had to be found, and the area that was investigated was activating the ignition plugs whenever the gun was fired so that the engine would be in a start mode.

Some loss of useful life was expected due to increased wear of the ignition plugs, but this was thought to be the least-damaging and most cost-effective solution. Reliability engineering was asked to look at the current failure rate of the igniter plug and estimate how much that it would degrade under two different proposals: a thirty-second duration ignition cycle after the gun was fired and a one-second ignition cycle proposal. At the time, over five years worth of failure rate data was collected on the A-10 and its components. There were two ignition plugs for each of the two engines and there were twenty-two failures during the five years against 227,818 flying hours. Thus the failure rate was:

$$22/(227,8181 \times 4) = 24.1 \times 10^{-6} \text{ failures per flight hour}$$

The normal amount of time for ignition plug activation for the aircraft (during ground start) was an average of 45 seconds during each mission. The 30-second proposal would increase this to 661 seconds per mission while the 1-second proposal would increase this to only 93 seconds per mission. Thus, the degradation factor was applied against the failure rate to get:

1. 30-second proposal: $24 \times 661/45 = 354.2 \times 10^{-6}$ failures per flight hour
2. 1-second proposal: $24 \times 93/45 = 49.9 \times 10^{-6}$ failures per flight hour

Thus, it was reliability engineering's suggestion to go to the 1-second duration design change to deal with the engine flameout problem and ultimately, the igniter fix was the one chosen over the proposals involving structural changes.

The lesson learned here was that even though the gun gas problem was anticipated during the design phase, there was no way that the design engineers could anticipate that it would be even worse than what they thought. Fortunately, there were some options, or a margin of correction in the design that would allow for a fix to be made. This is not always the case for new designs.

Throughout the 1980s, A-10 met many of its performance objectives during simulated wartime exercises and achieving high mission reliability during these exercises. Despite the good performance of the aircraft, it was viewed as a less than glamorous-type aircraft by the Air Force hierarchy, and during the new era of defense cutting in the late 1980s, it was slated to be scrapped from the Air Force inventory. It was hard to keep a specialized aircraft for the close air support role when army helicopters could perform some ground support roles, and the air force was looking at the faster F-16 aircraft to take over the ground attack role. However, hostilities in Kuwait in 1991 would change this situation for the A-10.

During Operation Desert Storm, the A-10 was utilized in a number of unique roles suited to the profile of the aircraft. It was used in various ground support missions such as search and rescue along with it traditional role of attacking ground targets. Even more so, the A-10 demonstrated the value of having redundant systems on a number of occasions. In February 1991, Captain Paul Johnson was able to fly a severely damaged A-10 back to base in Saudi Arabia that had no power in one engine and a large hole in the wing. In the same month, another pilot, Colonel Sawyer, flew an A-10 back to base that had much of the tail destroyed along with one elevator control surface gone. Figure 9–6 shows photos of both of these aircrafts. The redundancy of flight control surfaces and engines proved to be a major plus in wartime reliability. Wartime conditions, where battle damage and rapid turn-around time were daily situations, were major aspects for mission reliability and fully mission capable as opposed to simulated peacetime exercises.

During Operation Desert Storm, the A-10 flew a total of 8,500 sorties, averaging over 2.2 hours per mission. The A-10 sortie total accounted for over 7 percent (19,000 hours) of the total 109,876 sorties flown in total by the allied forces during the 43-day period. Even though each A-10 aircraft flew on average more than one mission a day, a mission capable rate of 95.4 percent was achieved, exceeding the 92 percent average achieved by Air Force

A) The photo below shows an A-10 that was flown back to base during Operation Desert Storm despite a large hole in the wing and the loss of the right engine.

B) The A-10 depicted in the photo below was flown by the wing commander when it was hit in the tail by a SAM missile, yet it was still able to return back to base.

Figure 9-6   A-10 Survival Features Demonstrated During War

planes during the war. The A-10 fleet flew an average of 200 sorties a day, accruing over 400 hours of flying per day. The average total flight time per aircraft during the war was 120 hours for the 43-day conflict. Weapons on the A-10 also achieved high-system reliability with the total reliability of bombs, Maverick missiles, gun and rockets achieving 98.67 percent. The war also showed that the F-16 was not able to replace the ground support roles of the A-10 because it was not designed the same way in terms of flying slow enough and in terms of sufficient protection from ground fire.

After this war, extensive review by both Congress and the Air Force revealed a change in thinking where the aircraft would now be considered an important part of the inventory, particularly in light of reduced defense spending. A common observation was the fact that the plane was overdesigned and it was exceeding its expected service life. Thus, instead of scrapping the plane, the new direction was that the plane should continue to be in service past the year 2020.

Since Desert Storm, the aircraft has seen significant action in the war in Kosovo and in Afghanistan. However, since the aircraft is out of production and spares are limited, the Air Force has to use the aircraft wisely and sparingly for these recent wars since there is no immediate replacement. Thus, in Kosovo, the aircraft was used towards the end of the war as armored targets were flushed out of hiding and in Afghanistan, the aircraft was used in some of the fiercest action in mountain regions where U.S. Army and Marine helicopters sustained heavy damage from ground fire. In the latter situation, it was decided to deploy the A-10 into this theater of action because it had better protection against ground fire when attacking targets.

The experience of the A-10 program showed how a well-thought out design concept provided an excellent performing aircraft for the role that it was planned for. This did not preclude problems from arising, but a good test, analyze, and fix approach was the key in solving production problems such as the gun gas ingestion problem. The point brought out here is that well after the preliminary stages of the program in which reliability does its typical tasks, reliability engineering was involved with FMEA work, predictions and trade studies to help resolve issues such as the flameout problem.

9.2 SUMMARY

Quite often, new product designs are compromised by management decisions and directions, generally as a result of schedule demands. Sometimes a reliability engineer may be relegated to a role of merely a number-cruncher who performs the clerical-type tasks of reliability data items. The reliability engineer who does more than this by putting his or her inputs up front and conducting an aggressive testing program will do more justice to the company and to the customer. For whether certain managers in a company think so or not, a more reliable product is usually the best way to go with less field returns and less angry phone calls from the customer.

The best way to describe the differences in approaches of management is to use the example of Amundsen and Scott in their race to the South Pole that began in 1911. Both men had the same goal of reaching the South Pole, but they had extremely different methods toward achieving this goal. Amundsen had a very aggressive approach and addressed all details, big and small, toward achieving his goal. Some of it was very frank to the level of detail, where he had to plan to kill a number of the sled dogs along the way because of food limitations! Amundsen had to undergo a lot of physical training for the mission also.

Scott, on the other hand, did not do any preliminary work prior to his venture, either in the way of training or in substantially testing out his motorized sleds (which failed shortly after his landing in Antarctica). As stated in the book, *The Last Place on Earth* by Roland Huntford, "That Scott faced temperatures between five and ten degrees lower than Amundsen was his own fault. By taking ponies and consequently delaying his start, he made certain of being on the plateau three weeks after the summer solstice and the turn of the season. Low temperature was a strain he could ill afford. Excepting mittens and boots, Scott had no furs, the absence of which around the face was enough to explain some of the party's persistent frostbite. Poor skiing technique, unintelligent navigation, a badly-laded, ill-maintained and ill-running sledge, inefficient camping routine [such as no floor lining for the tents], the disruptions caused by the last-minute addition of a fifth man; the list of defects was comprehensive. Scott had been so consistently inept as to almost suggest the workings of a death wish."

Amundsen made it to the pole on December 15th, 1911. Scott did make it to the South Pole, but over a month later on January 17, 1912, and now was entering the time of colder weather that begins after the summer solstice. Scott would pay the ultimate price for his mistakes and his fatal delays, when he and the other four men in his team would die on the way back from the pole. Scott had mistakenly believed that the human spirit could override the physical obstacles, and he had grossly underestimated their potential impact, while Amundsen had carefully addressed these obstacles with his careful planning. Scott had made many of the mistakes listed in the beginning of this chapter. These include the use of an unproven design concept (the failure of the motorized sledges), leaving no margin of error in his plans (not planning for delays), and his failure to anticipate differences in the environment (no protection from the elements). Most of all, it is clear that Scott had suffered from the big ego syndrome that was described in Section 9.2. Unfortunately, the big ego syndrome still exists with individuals in many companies and organization and is an issue that has to often be addressed. Sometimes, thoughtful numerical analysis can correct this situation to demonstrate an opposing point of view.

The reliability engineer and the company should both have the same goal similar to reaching the South Pole, that of a reliable product. There are always obstacles to both the reliability engineer and the company towards reaching this goal, but honest assessments at the beginning of the program, along with useful tasks will ensure better results. If a reliability engineer is relegated to only performing paper analysis that has no direct benefit in changing defects in the design, it is meaningless to perform even these paper analysis tasks. But by performing useful tasks, such as physics of failure analysis during the preliminary design stage and later by performing reliability testing, only these tasks will ensure that the goal of high reliability is achieved.

On occasion, the management of the company may prevent the availability of units for reliability tests by allocating all units built to go into production. It is up to the reliability engineer to convince the company of the importance of these up-front tests. In the long run, this is the only way a company can truly stay competitive in today's market by improved productivity and quality. Reliability testing is one process that does this. By removing design flaws before a product goes into full production, the reliability test reduces cost, improves reliability, and improves customer satisfaction.

9.3 EXERCISES

1. What are the most common pitfalls in the early stages of design that were discussed in this chapter?
2. List which one of the six common pitfalls were the basis for the test failures for the case studies presented in Chapter 4, Section 4.6.1.
3. For the two case studies that were presented in Chapter 5, Sections 5.4.1 and 5.4.2, list which one of the six common pitfalls was the basis for the failure for each of the three issues described in each case study. Could any of these failures that occurred have been anticipated early in the design?
4. Take note of the next major product recall that is reported in the news. Which of the common pitfalls described in this chapter fits the reason for the failure that resulted in the recall?
5. List the margin of errors that have been employed in a successful product design that you may have worked on or seen firsthand.

CHAPTER 10

Reliability Graphs and Duane Growth Curves

INTRODUCTION

Reliability graphs are valuable in plotting reliability performance data, as they can provide a visual indication as to the trend that the failure rate or the MTBF is taking. Plotting monthly MTBF data on a graph can be quite useful, and graphs can range from individual equipment up to the level of complex systems such as aircraft. It will be shown how this information is useful during the course of reliability failure investigations.

While linear graphing scales are used for most types of MTBF data tracking, certain types of reliability analyses are plotted on exponential scale paper for clearer visual trends. This chapter will show how the use of exponential scale for these analyses can reveal some important parameters relating to growth trends in reliability performance. The following are the major topics that will be covered in this chapter:

- Reliability MTBF performance graphs and the different types used (instantaneous, cumulative + rolling average)
- Reliability trend analysis and investigation
- Reliability growth curves and exponential plotting

10.1 MTBF GRAPHS

Simple graphs can be used for plotting reliability MTBF performance as determined from failure rate data. These MTBF graphs are useful reliability management tools in that they provide a visual overview of how a program is doing in terms of field reliability performance.

An aircraft manufacturer may track reliability on an overall aircraft level, individual system level, or even as low as the equipment level. The tracking of MTBF is typically plotted on a monthly basis and may be tracked on either an instantaneous level (monthly operating hours divided by total failures for the month) or on a cumulative level (total operating hours divided by total failure accumulated since the beginning of the program.) The type of tracking (instantaneous or cumulative) that is used may often be imposed as part of customer requirements.

For equipment manufacturers, reliability data may be passed down to them from the customers who use their equipment. This data will usually involve monthly operating hours along with the number of monthly field returns so that the MTBF can be calculated. In some cases, as detailed in the next example, certain reliability tests are conducted by tracking field reliability data in a formal manner by the customer. Equipment manufacturers may also track reliability performance of certain components during either long-term reliability testing, by collecting failure information from the results of reliability testing, or cumulative ESS test data.

It is noted that most reliability graphs that track performance are MTBF types rather than failure rate-based. This is not to say that failure rate plotting is not done, however scaling is generally clearer with MTBF graphs.

10.1.1 MTBF Graph Case Study #1

ITEM: Cockpit display unit for air force fighter aircraft

An aircraft manufacturer has imposed a reliability warranty program (RWP) on a display instrument that is installed on the aircraft manufacturer's fighter aircraft for the air force. The aircraft manufacturer and display instrument supplier are to track the failures and flight hours on a monthly and cumulative basis, in order to calculate the cumulative MTBF performance of the display for a one-year period while it is in field service.

The aircraft manufacturer also imposed some additional requirements to the basic MTBF plotting. The aircraft manufacturer recognized that the display unit was on during ground servicing and taxiing, and this was not reflected in the flight hour data. A factor of 1.3 operating hours for every one hour of flight was therefore incorporated into the monthly flight hour data.

The monthly flight hour data and failure data that were collected for this RWP began in October and are as follows:

	Oct	Nov	Dec	Jan	Feb	Mar	Apr	May	June	July	Aug	Sept
Hours	5333	4442	3695	3891	4457	5123	4989	5234	5673	4987	4878	5127
Failures	9	8	13	10	9	12	13	15	12	11	14	17

From this a table of monthly values can be constructed for instantaneous hours, cumulative hours, and instantaneous and cumulative MTBF values. Cumulative hours are the addition of hours from when tracking begins. The following are the basic calculations that are performed for each month:

$$\text{Operating hours} = 1.3 \times \text{Flight hours}$$
$$\text{MTBF (Instantaneous)} = \text{Instantaneous operating hours/ Instantaneous failures}$$
$$\text{MTBF (Cumulative)} = \text{Cumulative operating hours/cumulative failures}$$

Table 10-1 Cockpit Display Unit MTBF Data

Mo.	Flight hours	Operating hours [1.3 × FH]	Cumulative Operating hours	Failures	Cumulative Failures	MTBF (Instantaneous) (Hours)	MTBF (Cumulative) (Hours)
Oct	5,333	6,933	6,933	9	9	770	770
Nov	4,442	5,775	12,708	8	17	722	748
Dec	3,695	4,804	17,511	13	30	370	584
Jan	3,891	5,058	22,569	7	37	723	610
Feb	4,457	5,794	28,363	9	46	644	617
Mar	5,123	6,660	35,023	10	56	666	625
Apr	4,989	6,486	41,509	13	69	499	602
May	5,234	6,804	48,313	8	77	851	627
June	5,673	7,375	55,688	12	89	615	626
July	4,987	6,483	62,171	6	95	1,081	654
Aug	4,878	6,341	68,513	14	109	453	629
Sept	5,127	6,666	75,178	12	121	555	621

From this table of values, graphs can be constructed for the instantaneous and cumulative MTBF as shown in Figure 10–1A and B. The scale used for both graphs is a simple X and Y axis plot. Note that by plotting values on a cumulative basis, the curve is made much smoother. Other methods of smoothing curves include the process of taking a rolling average of three months as shown in Figure 10–1C or as is standard in some industries, a thirteen-month rolling average. This allows some smoothing but also allows recent trends to be spotted. For example, in the area of sunspot data, NOAA, the government agency, tracks the monthly sunspot number through use of a thirteen-month rolling average resulting in the parameter smoothed sunspot number (SSN). This choice is made because of the tremendous spikes and valleys of the raw data that are collected.

From the plotting of MTBF over time, it becomes possible to see if there are any trends in improvement in reliability performance. It is a useful and necessary management tool, on a program level and for reliability warranty programs. In the case illustrated here, a goal of 700 hours was set, and it was hoped that with design improvements implemented during the RWP period, that the goal could eventually be reached. (The goal was not achieved in this case.)

10.1.2 MTBF Graph Case Study #2

ITEM: A-10 close air support aircraft MTBF performance

This case study involves the past experiences of The Fairchild Republic Company, a former military aircraft manufacturer. There was a need to track cumulative MTBF data for a one-year period on the A-10 close air support aircraft built by the company for the U.S. Air

A) Instantaneous MTBF Plot

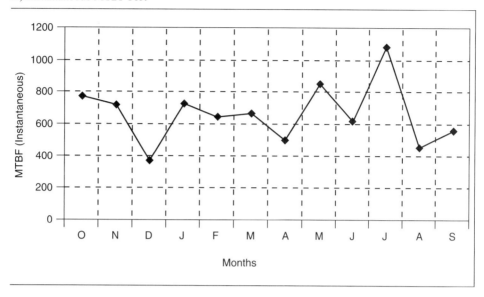

B) Cumulative MTBF Plot

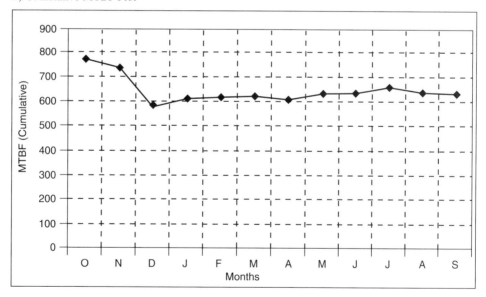

Figure 10-1 Graphs for MTBF Data for Air Force Aircraft Cockpit Display Unit

C) Three-month rolling average MTBF Plot

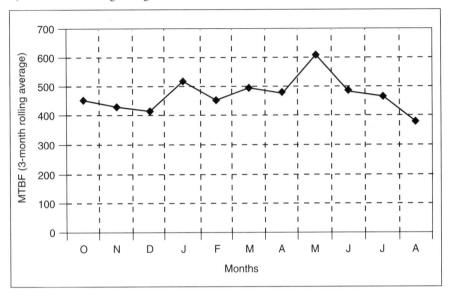

Figure 10-1 *(continued)*

Force. The graph was needed in order to measure the overall aircraft reliability performance, as well as for a number of the aircraft's subsystems. Some of these subsystems include the airframe, the cockpit, landing gear, as well as the flight instruments. The information was needed for the subsystems for a number of reasons that will be discussed later.

First, the company's reliability engineer collected the field data for the A-10 aircraft which was provided by the air force on a monthly basis. The raw data included the monthly aircraft flight hours and the monthly failures reported by subsystem and the overall aircraft level as presented in Table 10–2A. For the purpose of converting this monthly or instantaneous data into cumulative flight hour and failure data, simple summations are performed. If the assumption is made, we assume that the starting point begins in January; the data in the next column (February) is the summation of the instantaneous data of January and February. This summation can be accomplished in a simple spreadsheet program in which the column for the current month's cumulative data is the addition of the current month's instantaneous data added to the previous month's cumulative data. The result of this calculation for this example is shown in Table 10–2B. From this information, cumulative MTBF values can be determined on a monthly basis for each subsystem by dividing the cumulative failures into the cumulative flight hours for each line item. This yields the results as presented in Table 10–2C.

The results from Table 10–2C can be plotted in the form of MTBF charts for each of the subsystems and for the overall aircraft as shown in Figure 10–2. From the graph, it can be seen that the overall aircraft MTBF is relatively flat for the twelve-month period with the MTBF values falling between 3.69 and 3.72. This suggests that the aircraft MTBF per-

Table 10-2A Raw Monthly Failure Data of A-10 Attack Aircraft (Listed by Subsystem)

Months	J	F	M	A	M	J	J	A	S	O	N	D
Flight hours	5,020	4,901	5,221	5,487	5,876	6,103	6,079	5,672	5,548	5,489	5,321	4,894
Airframe	167	156	154	151	163	171	167	156	146	151	143	132
Cockpit	89	91	88	94	101	105	113	104	98	94	92	86
Landing	172	189	192	206	217	238	235	211	212	201	197	199
Flight	154	161	172	178	183	190	181	176	181	174	171	162
Engine	171	183	191	207	216	229	215	201	198	188	192	189
Electrical	47	51	48	53	61	63	57	53	47	47	48	45
Fuel	38	39	38	41	47	46	48	44	37	34	35	34
Instruments	116	111	145	191	178	151	191	199	187	201	178	187
OTHERS	389	354	378	399	403	412	389	376	367	354	347	341
Aircraft (Total)	1,343	1,335	1,406	1,520	1,569	1,605	1,596	1,520	1,473	1,444	1,403	1,375

Table 10-2B Cumulative Monthly Failure Data of A-10 Attack Aircraft (Listed by Subsystem)

Months	J	F	M	A	M	J	J	A	S	O	N	D
Flight hours	5,020	9,921	15,142	20,629	26,505	32,608	38,687	44,359	49,907	55,396	60,717	65,611
Airframe	167	323	477	628	791	962	1,129	1,285	1,431	1,582	1,725	1,857
Cockpit	89	180	268	362	463	568	681	785	883	977	1,069	1,155
Landing	172	361	553	759	976	1,214	1,449	1,660	1,872	2,073	2,270	2,469
Flight	154	315	487	665	848	1,038	1,219	1,395	1,576	1,750	1,921	2,083
Engine	171	354	545	752	968	1,197	1,412	1,613	1,811	1,999	2,191	2,380
Electrical	47	98	146	199	260	323	380	433	480	527	575	620
Fuel	38	77	115	156	203	249	297	341	378	412	447	481
Instruments	123	234	379	570	748	899	1,090	1,289	1,476	1,677	1,855	2,042
OTHERS	394	748	1,126	1,525	1,928	2,340	2,729	3,105	3,472	3,826	4,173	4,514
Aircraft (Total)	1,355	2,690	4,096	5,616	7,185	8,790	10,386	11,906	13,379	14,823	16,226	17,601

formance is pretty much at zero growth or at stabilization. However, when breaking down the data to the subsystem levels, some problem areas can be identified.

Problem areas can be identified when MTBF performance is either negative or erratic, which suggests that an underlying problem may exist with one or more pieces of the equipment. For this example, the instrument's subsystem is an area that requires more investigation. The next step is the drilling down of the failure data on a monthly basis for the specific equipment

Table 10-2C Cumulative Monthly MTBF Values for A-10 Attack Aircraft (Listed by Subsystem)

Months	J	F	M	A	M	J	J	A	S	O	N	D
Airframe	30	31	32	33	34	34	34	35	35	35	35	35
Cockpit	56	55	57	57	57	57	57	57	57	57	57	57
Landing	29	27	27	27	27	27	27	27	27	27	27	27
Flight	33	31	31	31	31	31	32	32	32	32	32	31
Engine	29	28	28	27	27	27	27	28	28	28	28	28
Electrical	107	101	104	104	102	101	102	102	104	105	106	106
Fuel	132	129	132	132	131	131	130	130	132	134	136	136
Instruments	41	42	40	36	35	36	35	34	34	33	33	32
OTHERS	13	13	13	14	14	14	14	14	14	14	15	15
Aircraft (Total)	3.70	3.69	3.70	3.67	3.69	3.71	3.72	3.73	3.73	3.74	3.74	3.73

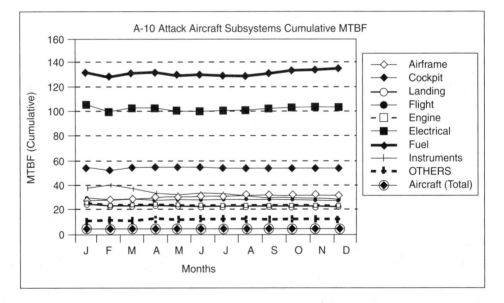

Figure 10-2 A-10 Attack Aircraft Subsystems Cumulative MTBF

of the system; this is done using the method shown in Table 10–2. The major instrument in this subsystem, the heads-up display (HUD), is apparently the culprit towards contributing the negative trend as shown in the series of tables shown in Table 10–3 and in Figure 10–3.

This example was presented in order to illustrate some of the major benefits that come from constructing reliability graphs. Reliability graphs can be used to show trends: positive, negative, or constant. While such trends may not be that easily noticed on the highest level

Table 10-3A Raw Monthly Failure Data of A-10 Instrument Subsystem

Months	J	F	M	A	M	J	J	A	S	O	N	D
Flight hours	5020	4901	5221	5487	5876	6103	6079	5672	5548	5489	5321	4894
HUD	52	49	74	101	96	92	123	132	117	109	104	101
Other Instruments	64	62	71	90	82	59	68	67	70	92	74	86
Instruments (Total)	116	111	145	191	178	151	191	199	187	201	178	187

Table 10-3B Cumulative Monthly Failure Data of A-10 Instrument Subsystem

Months	J	F	M	A	M	J	J	A	S	O	N	D
Flight hours	5020	9921	15142	20629	26505	32608	38687	44359	49907	55396	60717	65611
HUD	52	101	175	276	372	464	587	719	836	945	1049	1150
Other Instruments	64	126	197	287	369	428	496	563	633	725	799	885
Instruments (Total)	116	227	372	563	741	892	1083	1282	1469	1670	1848	2035

Table 10-3C Cumulative Monthly MTBF Values for A-10 Instrument Subsystem

Months	J	F	M	A	M	J	J	A	S	O	N	D
HUD	97	98	87	75	71	70	66	62	60	59	58	57
Other Instruments	78	79	77	72	72	76	78	79	79	76	76	74
Instruments (Total)	43	44	41	37	36	37	36	35	34	33	33	32

(in this case, the aircraft level), they may be spotted on the lower level of data, or the subsystem level (in this case, the instrument's subsystem level). Once a subsystem with a negative or erratic trend is detected, a "drill down" approach can be used to spot the equipment or component that is contributing to this behavior (in this case, the HUD unit).

Another part of this exercise that was illustrated is that performance has to be looked at over a significant period of time to gauge that a trend is occurring. Now, this leads to the concept of reliability growth, in which products improve over time after an initial period of learning or correcting design deficiencies has occurred. The corrective action system is a continuous improvement process that will be reflected in positive reliability growth.

The data that was analyzed here could be used by the aircraft manufacturer in the area of new aircraft proposals, both in the areas of expected reliability MTBF performance for the different subsystems and in the area of manufacturing learning experience where a growth in MTBF values is experienced. This type of analysis can be done on other major systems such as automobiles, railroad equipment, space equipment, and bank equipment.

The concept of reliability growth that was illustrated here leads to the subject of the Duane's growth curve that is covered in the next section.

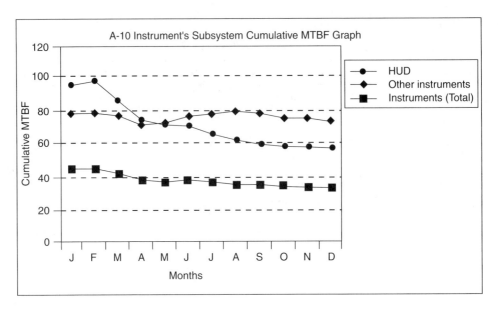

Figure 10-3 A-10 Instrument's Subsystem Cumulative MTBF Graph

10.2 THE DUANE RELIABILITY GROWTH CURVE

Much of the current thinking involving the concept of reliability growth was developed by J. T. Duane, an engineer for General Electric, in his IEEE paper, "Learning Curve Approach to Reliability Monitoring" (1964). The use of reliability growth curves is a useful tool both in tracking reliability during a reliability test and in predicting the long-term field reliability performance for a new product. This chapter will show how the concept developed by Duane has been adopted as a standard tool for reliability programs, initially for military aerospace programs and subsequently for commercial product lines.

The previous sections show the notion of trends in reliability MTBF graphs, and we can now look at more advanced concepts such as reliability growth. According to Duane, the basic concept of reliability growth is based on the observation that several different and complex electromechanical and mechanical systems are shown to have remarkably similar rates of reliability improvement during system development. These similarities can be mapped into a learning curve that can be used to monitor development progress of a program. These curves may be referred to as Duane growth curves or reliability growth curves.

The vast majority of reliability growth approaches require several data observations in order to fit curves that would predict reliability performance for a system or an aircraft consisting of several subsystems. The methodology used by Duane identifies equipment types by general description only (hydromechanical devices, aircraft generators, jet engine), and predicts that the growth curves will plot as a straight line on an exponential scale with cumulative operating time as the X-axis and the cumulative failure rate as the Y-axis. The learning curve or reliability growth curve can then be seen as dependent on two parameters: the slope and the intercept of the straight-line plot.

Duane observed from these various straight-line plots that the cumulative failure rate will vary in a manner directly proportional to some negative power of cumulative operating hours. The basic formula that he developed is expressed in mathematical terms as:

$$\lambda\Sigma = K(\Sigma H)^{-\alpha}$$

where λ_Σ is the cumulative failure rate (cumulative failures divided by cumulative operating hours), K is a constant, and ΣH is the cumulative operating hours. The value, $-\alpha$ represents the negative slope.

Analysis conducted by Duane on the different equipment types investigated showed the negative slope to have a range of 0.4 to 0.5. The intercept parameter depends on equipment complexity and design margins, and thus has a much wider range of variability.

From the basic Duane equation, predicted reliability values at a point in time later in the program in terms of operating hours can be computed. This is done by knowing the slope and cumulative failure rate from an earlier point in the curve and calculated as follows:

$$\frac{\Sigma \lambda_1}{\Sigma \lambda_2} = \frac{H_1^{-\alpha}}{H_2^{-\alpha}}$$

An alternate way of expressing this is in terms of MTBF or:

$$\frac{\theta_1}{\theta_2} = \left[\frac{H_1}{H_2}\right]^{\alpha}$$

The instantaneous MTBF $\theta(t)$, can be determined up to any one point, i, of the curve by the formula below, where

$$\theta_{inst}(t) = \frac{\theta_{cum}(t)}{1 - \alpha}$$

DUANE RELIABILITY GROWTH EXAMPLE

What is the projected slope for a radio installed on an aircraft that has an initial MTBF of 750 hours observed at 1,000 operating hours of testing and has to meet a requirement of 5,000-hour MTBF at 20,000 operating hours of early field service?

Using: $\quad \dfrac{\theta_1}{\theta_2} = \left[\dfrac{H_1}{H_2}\right]^{\alpha}$

We get: $\quad \dfrac{750}{5000} = \left[\dfrac{1000}{20000}\right]^{\alpha}$

$ln .04 = \alpha \ (ln\ .05)$

$\alpha = .63$

Growth curve tracking for values in between the start value and the final value can be accomplished through the following formula:

$$\text{Log } H_m = k (\log H_1 + \log H_2)$$
$$H_m = (H_1 \times H_2)^k$$

Where k represents the percentage or fraction of the in-between point being measured with respect to the final point.

Using the same example shown above, if we wanted to calculate the midpoint or $k = \frac{1}{2}$ and we get:

$$H_m = (H_1 \times H_2)^{1/2} = (1000 \times 20000)^{1/2} = 4472 \text{ cumulative operating hours}$$

This particular formula becomes handy when formal measurement milestones are set up in a program and intermediate growth tracking is required prior to these milestones. The corresponding MTBF value is calculated by plugging back into the main formula to get:

$$\theta_1 / 5000 = (4472 / 20000)^{.63}$$
$$\theta_1 = (.389)(5000) = 1946 \text{ hours}$$

A Duane curve can be constructed from a set of observed data points with a set of X values representing cumulative MTBF values and the corresponding Y values representing cumulative operating hours. From this table, logarithms are taken of each individual X_i and Y_i value and are subsequently plugged into a linear regression or least-squares model. This can be done on all scientific calculators, and from this the slope can be calculated.

Specifically, the data points are set up as:

$$Y_i = \log_{10} \theta_i \text{ and } X_i = \log_{10} H_i$$

where H_i is cumulative operating hours at point i and θ_i is the cumulative MTBF at point i. The basic methodology for linear regressions that is based on the least-squares formula where:

$$S = \sum_{i=1}^{n}(Y_i - a - bX_i)^2$$

Two equations come out:

$$an + b\Sigma X_i = \Sigma Y_i$$
$$a\Sigma X_i + b\Sigma X_i^2 = \Sigma X_i Y_i$$

We can solve for a and b by setting up equations in matrix notation.

$$\begin{bmatrix} n & \Sigma X_i \\ \Sigma X_i & \Sigma X_i^2 \end{bmatrix} \begin{bmatrix} a \\ b \end{bmatrix} = \begin{bmatrix} \Sigma Y_i \\ \Sigma X_i Y_i \end{bmatrix}$$

$$\begin{bmatrix} a \\ b \end{bmatrix} = \frac{1}{n\Sigma X_i^2 - (\Sigma X_i)^2} \begin{bmatrix} \Sigma X_i^2 & -\Sigma X_i \\ \Sigma X_i & n \end{bmatrix} \begin{bmatrix} \Sigma Y_i \\ \Sigma X_i Y_i \end{bmatrix}$$

$$= \frac{1}{n\Sigma X_i^2 - (\Sigma X_i)^2} \begin{bmatrix} (\Sigma X_i^2)(\Sigma Y_i) & - (\Sigma X_i)(\Sigma X_i Y_i) \\ n\Sigma X_i Y_i & - (\Sigma X_i)(\Sigma Y_i) \end{bmatrix}$$

where $Y_i = \log_{10} \theta_i$ and $X_i = \log_{10} H_i$

$$a_{10} = \frac{(\Sigma X_i^2)(\Sigma Y_i) - (\Sigma X_i)(\Sigma X_i Y_i)}{n\Sigma X_i^2 - (\Sigma X_i)^2}$$

$$b_{10} = \frac{n\Sigma X_i Y_i - (\Sigma X_i)(\Sigma Y_i)}{n\Sigma X_i^2 - (\Sigma X_i)^2}$$

The reader may have noted that in the examples above base-ten log scale was used. However, natural logs (ln) can be used also, interchangeably, where base-ten scale log is used. When calculating the slopes of the Duane growth equations, it is not important what base is used. It is straightforward to convert from one logarithmic base to another, as follows:

$$T_i = \ln H_i = kX_i \text{ and}$$
$$S_i = Ln\, q = kY_i, \text{ where } k = \ln 10$$

$$a_e = \frac{(\Sigma T_i^2)(\Sigma S_i) - (\Sigma T_i)(\Sigma T_i S_i)}{n\Sigma T_i^2 - (\Sigma T_i)^2}$$

$$b_e = \frac{n\Sigma T_i S_i - (\Sigma T_i)(\Sigma S_i)}{n\Sigma T_i^2 - (\Sigma T_i)^2}$$

Substituting with T_i and S_i with kX_i and kY_i, respectively, we get:

$$a_e = \frac{k^2(\Sigma X_i^2)(k\Sigma Y_i) - (k\Sigma X_i)(k^2\Sigma X_i Y_i)}{nk^2\Sigma X_i^2 - (k\Sigma X_i)^2} = ka_{10}$$

$$b_e = \frac{nk^2\Sigma X_i Y_i - (\Sigma X_i)(\Sigma Y_i)}{nk^2\Sigma X_i^2 - (k\Sigma X_i)^2} = b_{10}$$

b = slope = growth rate (independent of logarithmic base)
a = equipment constant factor (dependent on logarithmic base)

For the case of a, the slope intercept, it would be different depending on which logarithmic scale is used.

For most new designs, the Duane growth curves follow a similar pattern where there is an initial period of minimal reliability as failure modes are identified and the bugs are being worked out. A sharp period of growth may be achieved as corrective action is implemented until a level of maturity is reached. From that point on, only gradual growth is experienced until wearout or the end of useful life is reached and reliability decreases.

10.2.1 Reliability Growth Curve Case Study #1

ITEM: A-10 close air support aircraft program

TASK: This example involves the experience of Fairchild Republic, a former aircraft manufacturer located on Long Island, where the reliability department had the responsibility of tracking the overall reliability performance of the A-10 attack aircraft that it manufactured for the U.S. Air Force. From this curve, slope values are determined from the empirical data.

The overall reliability for this program is tracked in terms by the cumulative reliability being tracked by using cumulative flight hours and cumulative failures tracked on a monthly basis for the period March 1976 through May 1980. Data was collected by an air force data system, where all inherent failures requiring maintenance action that occurred on the aircraft were collected into the database. The defined point of maturity was 18 calendar months past initial operating capability. This turned out to be at 70,576 cumulative flight hours.

The monthly MTBF data is then plotted on exponential scale graph paper. The results are seen in Figure 10–4 where the initial MTBF values starts well below 2 hours and proceeds up to a value of 3.5 hours MTBF value at the 100,000–cumulative-hour point and over 28,000 failures recorded.

From this plot, the curve can be broken down into two or three main periods of growth. For this particular case, the curve was broken into three portions: the early flight test program portion (I), the early production part of the program (II), and later production (III). There are valid reasons for breaking up the curve and disregarding the first measurement period (I). During the initial phase (I), there was an impact of unscheduled maintenance action coupled with a low number of flying hours, and this produced erratic MTBF values during the initial phase. Also coupled with this was the fact that there was initial maintenance training during which there was incorrect data reported during the initial phase. Thus this period of no growth is excluded from the calculations for the initial slope line.

It is also noted that the curve still has positive growth past the maturity mark, as the effects are realized from a number of corrective engineering changes that are incorporated into production aircraft. At this stage of the program, about 300 aircraft are deployed into the field, accruing an average of 30 to 40 flying hours a month per aircraft. A-10 aircraft production covered a span of ten years with a peak of twelve aircraft a month being rolled into the field. Corrective action implemented into the design early in the program allowed for fewer aircraft requiring retrofit and the prevention of failures in new production aircraft. Reliability growth would continue after the last production aircraft entered field service in 1984, at which point the program had a maximum of over 700 aircraft deployed in field service with over 200,000 flight hours accrued annually. The aircraft fleet has since accumulated well over a million cumulative flight hours since its introduction in field service.

Using the principles established by Duane discussed earlier in this chapter, we now determine the slopes of the linear regression lines that are fitted through the data points for the two regions of interest, Region I and Region II. The slope of the linear regression lines can be determined by either of two ways. A simple method is through the sighting method where a line is drawn by sight through the data points. A more accurate method is obtained by plugging in each set of cumulative flight hours and cumulative failure count recorded at each set of data points in to a calculator that does linear regression analysis. It is noted that it is not unusual to present the overall plot in terms of failure rate; thus, for this particular example, the failure rate slope would be − .08 in lieu of .08 used for the MTBF slope. However, the convention in industry is to track growth curves in terms of MTBF as the parameter.

There is continual growth in the MTBF value, although it is slowing down at the point of maturity. More importantly, from the company's perspective, the slope values collected from this data provided a starting point for their next proposal as seen in the example provided in 10.2.2. There is a major benefit in measuring reliability growth on existing product

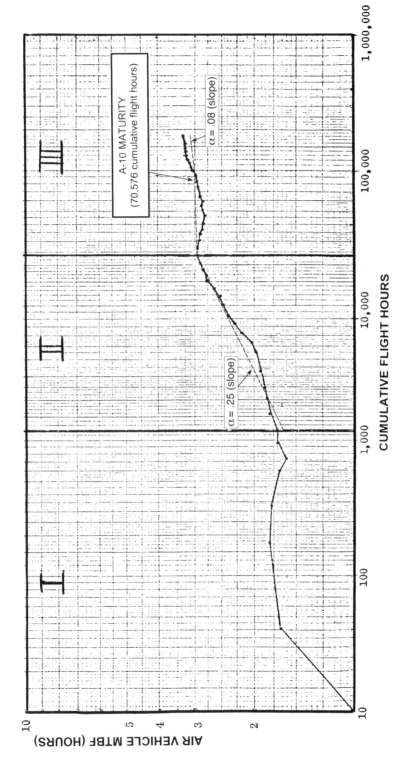

Figure 10-4 Reliability Growth Curve Example Cumulative MTBF Plot of A-10 Aircraft

lines—it provides the basis for projecting future reliability performance on new programs usually while in the proposal stage. It is appropriate for a manufacturer to use the learning curve experience in this way if the processes used are consistent between programs.

10.2.2 Reliability Growth Curve Case Study #2

ITEM: Air force T-46A aircraft program

TASK: A former aircraft manufacturer, the Fairchild Republic company, was asked to develop a Duane growth curve for the overall reliability performance as required for the next generation trainer (NGT) aircraft proposal for the air force (subsequently designated the T-46A). In addition, the engine manufacturer for this aircraft had to construct a similar type curve predicting the reliability growth of the engines in their proposal.

The first part of this exercise involves constructing the predicted growth curve for the overall T-46A aircraft. In order to develop the growth curve for the overall aircraft, Fairchild Republic had to use reasonable assumptions as well as failure rate data of a current aircraft program run by the company, the A-10 program. This was done by examining the growth of the A-10 program that was currently in production at Fairchild Republic during the time of the T-46A proposal. While the A-10 was basically an attack aircraft, it was feasible to use the reliability experience from a number of subsystems on this aircraft for the T-46A trainer proposal. The A-10 had some subsystems that were not being used in the T-46A jet, such as armaments and electronic countermeasures, and the data for these were excluded from consideration for the final curve. This is done by only counting those failures for the subsystems that are under consideration and then computing the cumulative MTBF for each month.

In 1982, at the time of the trainer proposal, the A-10 had reached over 200,000 hours of cumulative flying hours and was considered a mature aircraft. The aircraft was seeing an overall cumulative MTBF of 3.5 hours with the selected subsystems under consideration seeing an overall cumulative MTBF of over 10 hours. The subsystems used for the proposal could be considered to be at the point of maturity as well since many of the design changes and retrofit were completed for earlier failures that had occurred on equipment in each of these subsystems.

Some additional analyses and refinements were performed on the cumulative MTBF curve that was used in the previous example. The resulting curve that was to be used for the proposal was simplified using the concept of two periods of growth for an aircraft program. There would still be a very sharp period of growth in the beginning of the program and a slower rate of growth towards the time of maturity for the aircraft. It was decided by management that any new aircraft program by the company would follow a very similar pattern of reliability growth. Some additional management decisions and refinement of the data was done in order to come up with the following slopes for each portion of the curve that would be used for the proposal:

$$\alpha_1 = .24 \text{ and } \alpha_2 = .12$$

For the T-46A program, the one requirement set in stone by the air force was that at maturity (which was defined as 250,000 cumulative flight hours), the cumulative MTBF

A) T-46A aircraft reliability growth curve

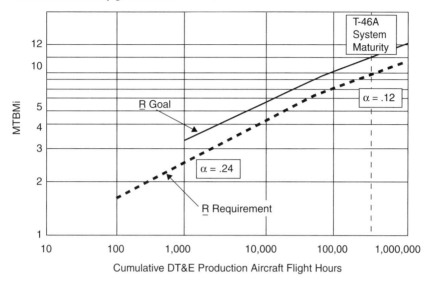

B) T-46A engine reliability growth curve

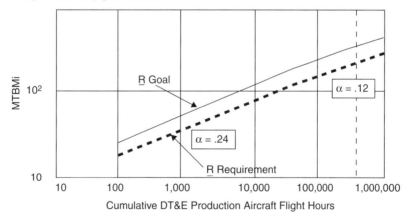

Figure 10-5 Growth Curves for T-46A Trainer Proposal

had to reach 10 hours for the whole aircraft. Now the company analysts had to construct the curve back to the first break point, which was defined as ORA (operational readiness analysis) at 59,000 flight hours. By setting up the Duane formulas, we get:

$$\frac{\theta_1}{10} = \left[\frac{59,000}{250,000}\right]^{.12}$$

$\theta_1 = 10(.841) = 8.41$ hours (MTBF at the ORA break point)

The engine manufacturer of the two T-46A engines was also required to present a Duane reliability growth graph in their proposal. The values used for the slope, $\alpha_1 = .24$ and $\alpha_2 = .12$, were the same values used by Fairchild Republic in their overall reliability growth curve and recommended for the engine manufacturer to use. Again, this was a sharp slope for the first part of the program and a gradual slope for the later part of the program leading up to maturity. The same key break points on the graph at 250,000 flight hours (maturity) and 59,000 flight hours (ORA) were kept with a projected MTBF at maturity to be 250 hours for each engine. The graphs for Duane growth curve on an exponential scale graph for the T-46A aircraft and for the engine are shown in Figure 10–5. Note in this figure, the log-log scale that was used is a 5-cycle by 1.5 cycle. The division lines for the X-axis (time in hours) has been simplified to show just the major division lines for every decibel (power of 10). The examples used in this chapter use different scaling of exponential paper, usually selected to the smallest number of cycles needed to present the graph clearly. For the graph used in Figure 10–5, a Y-scale of less than two cycles was needed, where as five cycles was needed for the X-axis in order to adequately cover flight hours up to the 100,000-hour mark. The selection of the scale is based on the analyst's judgment towards the best presentation of the graph.

The next generation trainer program was awarded to Fairchild Republic in 1984. The reliability performance was tracked during the beginning of the T-46A program in 1986 for the first 500 hours of flight testing until the trainer program was terminated in 1987, well before any major milestones were identified on the growth curve. Since that time, only a few new military aircraft programs have been launched, but they have used the principle of reliability growth as well. For example, for the C-17 military transport aircraft, Douglas Aircraft imposed a slope of .3 as the reliability growth rate for the vendors in the program.

10.2.3 Reliability Growth Curve Case Study #3

ITEM: Navy fighter aircraft engine display unit

TASK: Set up a reliability growth curve for the navy fighter aircraft engine display unit based on results of reliability development test (RDT) program

As part of the requirements for performing the RDT on an engine display unit of a navy fighter aircraft, the vendor had to construct a reliability growth graph on exponential scale graph paper to show that some reliability growth was experienced during the 3,500-hour test. The unit saw a number of failures during the program, many of them occurring during the first 1,000 hours of RDT. For the purpose of tracking the reliability growth, only relevant failures were counted as listed in Table 10–4. Thus, certain test setup failures that were considered as non-relevant failures were not counted for use in this graph. This is one example of data censoring that may be used in the construction of

Table 10-4

Week No.	Cum Hrs Unit A	Cum Hrs Unit B	Cum Hrs TOTAL	Cum Failures	MTBF (Hrs)	Adjusted Failures	Adj MTBF
1	9	23	32	1	32	0	-
2	9	44	53	2	27	0	-
3	9	115	124	3	41	0	-
4	60	157	218	3	72	0	-
5	60	207	267	3	89	0	-
6	177	323	500	3	167	0	-
7	259	406	665	3	222	0	-
8	329	471	800	3	267	0	-
9	442	471	913	3	304	0	-
10	474	588	1,062	4	266	1	1,062
11	474	601	1,075	4	269	1	1,075
12	474	723	1,197	4	299	1	1,197
13	474	824	1,298	4	325	1	1,298
14	474	963	1,437	4	359	1	1,437
15	474	1,070	1,543	4	386	1	1,543
16	474	1,130	1,604	4	401	1	1,604
17	474	1,181	1,655	4	414	1	1,655
18	607	1,181	1,788	4	447	2	894
19	689	1,196	1,885	6	314	2	943
20	822	1,263	2,085	6	348	2	1,043
21	940	1,313	2,253	7	322	2	1,126
22	968	1,313	2,281	8	285	2	1,141
23	968	1,391	2,359	9	262	2	1,179
24	968	1,392	2,360	9	262	2	1,180
25	1,019	1,505	2,524	9	280	2	1,262
26	1,157	1,558	2,715	10	272	2	1,358
27	1,295	1,558	2,853	10	285	2	1,427
28	1,425	1,558	2,983	10	298	2	1,492
29	1,565	1,558	3,153	11	287	2	1,576
30	1,697	1,558	3,285	11	299	2	1,642
31	1,816	1,627	3,443	11	313	2	1,721
32	1,880	1,627	3,507	11	319	2	1,753

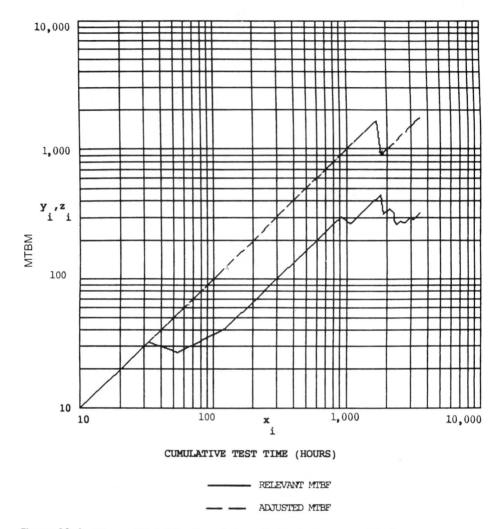

Figure 10-6 Observed Reliability Growth Curve for Engine Instrument Display

a reliability chart. In addition, a second curve was tracked which was adjusted based on what failures had corrective action. Failures that fell into this category were removed from consideration for this curve.

For the RDT program, two sets of growth curve data were being tracked: the raw (unadjusted) data and the adjusted data (where corrective action has been implemented and failure removed). The data from Table 10–4 for the two MTBF curves (adjusted and unadjusted) are presented in Figure 10–6. Clearly, the growth curves for both have very sharp positive slopes and this indicates that the test units are benefiting from the repairs and corrective actions that were made early in the test program. Notice that the scale on this paper is similar to what was used before except that certain division lines are highlighted. It is important for the reliability engineer to be comfortable using different styles of exponential graph paper, as there is no industry standard and the scale and divisions used will often depend on what kind of exponential graph paper is available.

We can calculate some additional things from the observed data. For example, if we wish to calculate the slope for one portion of the curve of the unadjusted data, for example, the period from 32 hours to 1,062 cumulative hours, we use the basic Duane growth formula for MTBF to get the following equation:

$$\left[\frac{32}{1062}\right]^\alpha = \frac{32}{304}$$
$$\alpha \ln(.030) = \ln(.105)$$
$$\alpha = -2.25/(-3.51) = .64$$

It is sometimes useful to calculate the instantaneous MTBF at a data point along the cumulative MTBF curve. For example, if we wish to find the instantaneous MTBF at 913 cumulative test hours, we can use the following formula:

$$\theta_{inst}(t) = \frac{\theta_{cum}(t)}{1 - \alpha}$$

$$\theta_{inst}(t) = \frac{913}{1 - .64} = \frac{913}{.36} = 2536 \text{ hours (Instantaneous MTBF)}$$

This formula helps determine if current MTBF performance is meeting projected MTBF goals.

10.3 GROWTH CURVE THEORY IN OTHER FIELDS

It is interesting to note that the same methodology developed in this chapter is directly applicable to other fields, particularly in the field of biology where there is tracking of growth rates of living organisms. In some cases, some biological systems such as plants have a two-slope growth pattern, an initial growth at a sharp slope and a smaller or gradual slope when the plant reaches maturity.

Figure 10–7 shows an example of a biological growth system. This example involves Codium seaweed, which is a plant that grows rapidly during the summer in the waters of the Atlantic Ocean off of Long Island, New York. It has been a concern for biologists because in some bodies of water such as bays, the seaweed can literary "choke" out other life forms because of its large numbers and rapid growth. Over the years, a number of experiments have been conducted, in which the seaweed's height of a control population on a rack that is set up in certain locations in the ocean around Long Island has been measured on a weekly basis.

One set of data was collected on several plants located in the waters off of Montauk Point at the end of Long Island. Examining the data from one plant with the following X and Y values, where X represents days and Y represents length in centimeters, we can measure the slope or growth rate using the same principles described in the previous section. The points are graphed on exponential scale graph paper as shown in Figure 10–7, and it can be seen that the overall curve is very close to being a straight line. As in the

216 Chapter 10

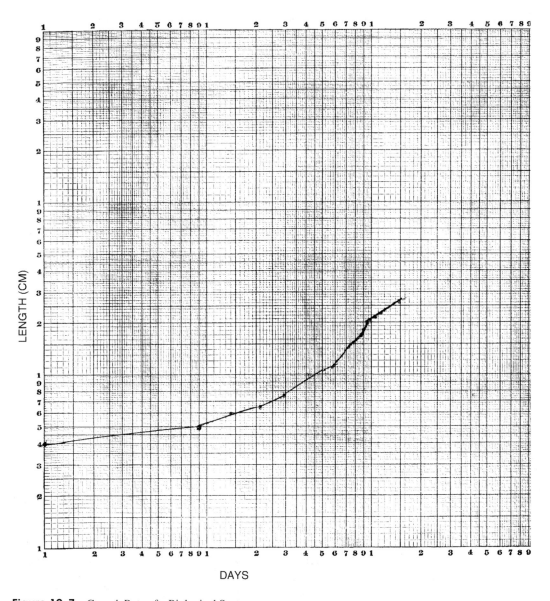

Figure 10-7 Growth Rate of a Biological System

previous examples, an overall slope for the measurement period can be calculated for the curve using the methodology developed by Duane. We use day-1 to day-164 along with the associated length values and insert into the equation:

$$\frac{\theta_1}{\theta_2} = \left[\frac{H_1}{H_2}\right]^\alpha$$

to get:

$$\frac{4}{27} = \left[\frac{1}{164}\right]^\alpha$$

$$ln\ .148 = \alpha\ (ln\ .006)$$

$$\alpha = .37$$

This example of a biological growth system, while not related directly to reliability theory, is presented here to further help the reader understand the basic concept of exponential growth curves. All growth, be it on mechanical devices or living organisms, is for the most part exponentially based, and it can be measured by using this methodology and plotting on exponential scale graph paper. At some point in the life of the mechanical device or living organism, there may be a period of constant growth where the slope of the line is straight.

10.4 SUMMARY

This chapter covered the various types of reliability performance graphs that may be used to track the progress of a program. MTBF tracking and the Duane growth model were presented to the reader as useful reliability management tools. The use of the exponential scale for graphing has allowed us to determine reliability parameters such as the slope of the Duane growth curve.

Why is reliability growth so important, and what is the advantage of performing Duane growth analysis and then plotting it in graphical form? The reader can see that when performing growth rate analysis, it can reveal a lot about how well a company is doing with regards to its design and manufacturing process, as well as surfacing any potential quality issues. For the great majority of the time, a learning curve is part of the process as a company learns to build a product that it has designed. A positive learning curve tracks well with the period of sharp growth seen in the initial portion of the program. As the company gets better and issues are resolved by corrective action, the growth rate changes to a smaller slope.

Negative reliability growth implies intuitively that if a company does not improve, it eventually will reach the point of either walking away from a program or suffering significant financial losses trying to fix its processes. It goes against the law of survival for a program to have sporadic reliability performance throughout the entire program without the customer objecting to poor reliability performance. This is particularly true in the commercial world where customers do not appreciate unreliable products with erratic performance. Thus, in order to survive in a very competitive market, commercial manufacturers have to commit themselves to accepting the principle of continuous improvement that leads to positive reliability growth. In the past, some companies did not want to improve reliability performance because there was no incentive or pressure by the customer to do so, but the laws of competition do eventually come into play where positive growth is required. Consumers who have used poorly performing products have long memories and will leave a product line when an opportunity to go to another product comes along.

At what point should positive growth be seen during a new program, as well as what point does it stop? Positive growth should be seen at least by the midpoint of production. However, there is point in the program when it becomes unrealistic for much growth to occur. This point occurs when the sum effects of component wear-out starts to take over, e.g., the wear-out portion of the reliability bathtub curve that was presented in Chapter 1 in Figure 1–1. There is a real danger that if a company is seeing erratic or negative reliability growth up to the point of cumulative component wear-out beginning, that no real positive growth will be even seen. This situation suggests that there may be many issues with the company's processes, issues with new technology, or even issues with component quality. Negative or erratic reliability growth clearly indicates that there are issues that are needed to be resolved.

For these reasons, then, it makes sense for reliability growth to be tracked as part of the reliability program for new products. Bad trends can push a company into a corrective action cycle to address issues as part of the laws of competition described previously. It is a good practice for companies to evaluate long-term failure trends as part of this growth analysis even if it is not a required contractual item.

10.5 EXERCISES

1. Construct both a instantaneous MTBF graph and cumulative MTBF based on the following monthly failure data for a commercial aircraft:

Month	Operating Hours	Failures
January	30,567	3,501
February	32,678	3,652
March	34,569	3,678
April	34,771	3,701
May	36,823	3,824
June	36,578	3,749
July	34,586	3,455
August	32,101	3,235
September	31,236	3,122
October	30,225	2,998
November	29,998	2,987
December	30,445	3,001

2. What are the three-month rolling average MTBF values and associated graph for the values used in Problem 1?
3. Based on the graph that was constructed in Problem 1, what growth trend can be seen?
4. Construct a growth curve on exponential scale graph paper for an electronic instrument that will be installed in a C-17 transport aircraft using the following values. The predicted mature MTBF is calculated to be 15,000 hours with maturity defined as 100,000

flight hours. The growth curve slope specified by the aircraft manufacture for all equipment is $\alpha = .3$. What is the initial MTFB value at 10 hours? Construct the curve.

5. For the cumulative MTBF graph constructed in problem 1, what is the value of (symbol for alpha)? Calculate by taking the initial MTBF and operating hours values along with the final MTBF and final operating hours value. From this value of (alpha symbol), is the failure rate increasing or decreasing?

6. What would the growth rate (α) be for the following set of values used in Duane's growth curve?
   A. Initial MTBF of 20 hours at 20 cumulative operating hours
   B. Mature MTBF of 1,000 hours at 10,000 cumulative operating hours

CHAPTER 11

Using Hypothesis Testing to Solve Reliability Problems

INTRODUCTION

A number of the principles involving confidence limits that were discussed in Chapter 2 can be developed further into useful statistical tools. These tools can be very useful in investigations regarding field reliability performance of a product and to support a statistical-based conclusion that is based on the data. In this chapter, two useful statistical tools using hypothesis testing for reliability problems are introduced along with the conclusions that can be drawn from the results of the analysis. The two tests that will be demonstrated are the F-test of significance and the chi-square test of independence.

Hypothesis testing involves calculating a test statistic that is based on the reliability data collected from field experience, and then comparing that to an expected value that is a statistic found in a table based on the number of failures (used to calculate the test statistic). Depending on whether the test fails or passes, some statistical conclusion can be drawn involving the confidence of the data.

Both of these tests use the concept of statistical confidence, along with statistical tables for each distribution in a similar fashion to the calculation of confidence limits that were performed in Chapter 2. The chi-square test uses the same chi-square distribution table that was used in Chapter 2, along with the same parameters: degrees of freedoms (based on the number of failures), and probability percentage. Likewise the F-test uses an F-distribution table that follows a similar process for calculating expected values.

The difference here is that instead of performing confidence limits on a single set of data, the statistical tests illustrated in this chapter will generally be applied against two sets of data that are being compared. This is done in order to determine whether there is statistical confidence, or whether the two sets of data are from the same or different populations based on the concept of the expected pattern of behavior.

Some understanding is required when using these concepts, but the application of either test is relatively straightforward by following the steps illustrated in this chapter. A number of real-life examples have been provided in order to help the reader understand. A good understanding of Chapter 2 is required prior to reading this chapter.

11.1 THE CONCEPT OF HYPOTHESIS TESTING

A number of the basic concepts involving hypothesis testing must be discussed first. A population is defined as a whole class of items from which some conclusions are drawn. It is generally not possible, because of size and impracticality, to take measurements from the whole population; therefore a sample is taken from the population. From these sample measurements, general findings are determined in order to draw conclusions about the whole population. In order to get good general findings about a population, a random sample is selected, in which each individual item in the population has an equal chance of being in the population.

Independent random samples can be obtained from the same population if each sample is selected without reference to the way the other sample is made up. Further, two variables are deemed independent when the value fixed for one variance has no effect on the frequency for how the other variable occurs.

Often a sample is drawn from a population for the purpose of testing an initial hypothesis, which we call the null hypothesis about the population. As part of constructing this null hypothesis, a level of significance or a risk of error can be assigned. This level of significance is the percentage probability that even if the null hypothesis does hold true, there is this specified percentage that we reject it. This type of error, where the null hypothesis is rejected, even though it is true, is called Type I error (denoted as α, also known as producer's risk). On the other hand, the failure to reject a false hypothesis is called Type II error (denoted as β, also known as consumer's risk).

The terms confidence limit and degrees of freedom were previously discussed in Chapter 2 with regards to applying them to single MTBF samples. By combining these definitions with the definitions concerning independence and null hypothesis, the stage is set for the concept of statistical testing of two hypotheses—the null hypothesis and an alternative hypothesis. These will typically involve testing two or more samples from a population.

11.2 THE F-TEST OF SIGNIFICANCE

The F-test of significance involves the principle of testing a hypothesis, H_0, also known as the null hypothesis, against an alternative hypothesis H_1 by using an expected value from the F-distribution to set up the comparison. The test will determine whether a test statistic falls in the critical region of the test, i.e., a function that yields the probability of accepting the hypothesis under consideration.

For problems relating to reliability, this may involve whether the failure rate or MTBF experience of two products under consideration are significantly different in their reliability performance such that they could not come from the same population. MTBF values are typically used when comparing these scenarios by taking their ratio and comparing with an expected value that is a statistical value selected from the F-table.

Recall that the MTBF is the ratio of operating hours to the number of failures. The two MTBFs can be compared by using the following representation:

$$MTBF_0 = \theta_0 = T_0/K$$
$$MTBF_1 = \theta_1 = T_1/L$$

The ratio of these two MTBFs can be used (θ_0/θ_1) as a test statistic to either prove the null hypothesis is true or false. Basically if the null hypothesis is false, this ratio will exceed the expected value (a value selected from the F-distribution table) by a greater amount than could reasonably be attributed to chance selection in sampling. Thus, the two hypotheses are used for testing: the null hypothesis, H_0: $\theta_0 = \theta_1$, and an alternate hypothesis: $\theta_0 \neq \theta_1$. The null hypothesis is represented mathematically as:

$$P(\theta_0/\theta_1) > F_{1-\alpha}(2K, 2L)) = \alpha$$

where K is the number of failures used to calculate θ_0 (numerator) and L is the number of failures (denominator) used to calculate θ_1. The portion of the expression, $2K$ and $2L$, represents the degrees of freedom. The expected value that is selected from the F-distribution table is based on the degrees of freedom ($2K$, $2L$) and the confidence value $(1-\alpha)$.

F-table values have been calculated for different confidence levels, but for the problems presented in this chapter we will only use some of the common confidence levels. We will use $\alpha = .05$ (probability value of 95 percent) associated with the F-table values presented in Table 2 of Appendix A and $\alpha = .01$ (probability value of 99 percent) with the associated F-table values presented in Table 3 of Appendix A. The generalized graph for the F-function appears in Figure 11–1.

The process for performing the F-test of significance is as follows:

1. Calculate the actual MTBF values ($MTBF_0 = \theta_0$ and $MTBF_1 = \theta_1$).
2. Calculate the ratio θ_0/θ_1. This is the calculated test statistic value.
3. Use the number of failures that were used to calculate the MTBF values, K and L, respectively, and multiply each by 2 to obtain the degrees of freedom ($2K$, $2L$).
4. Refer to the appropriate F-table for the selected α in Appendix A and look at the top row of the table for $2K$ (numerator value) and then look on the side column for $2L$ (denominator value) to find the expected value.
5. Compare the expected value found in the table with the calculated test statistic value. If the test statistic value is less than the expected value, the null hypothesis holds true. If it is greater than the expected value, the null hypothesis is false.

The degrees of freedom for a reliability problem will generally be the number of failures observed, or a normalized value for failures. Normalization is used in certain cases when the count is too high for the F-table to be used. The following are practical case studies that demonstrate the methodology for applying the F-test of significance for reliability performance studies as well as what conclusions can be drawn from performing the test.

Using Hypothesis Testing to Solve Reliability Problems **223**

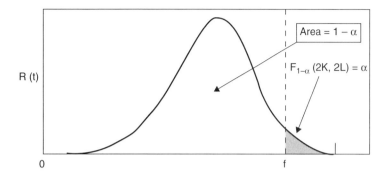

Figure 11-1 General Curve for F-Distribution

11.2.1 F-Test Case Study #1

TASK: Apply the F-test at 95 percent confidence to compare a number of failed ten-layer circuit boards with other boards during reliability testing of a controller unit that is used on an air force military transport aircraft.

During a reliability test of two test units of an electronic controller consisting of eleven circuit boards per units, a number of failures were observed with the more complex types of circuit boards. Because of the complexity of the unit and overall size constraints, all of the circuit boards consisted of several discrete layers containing traces, voltage planes, and ground planes. It appeared from test failures that the more complex circuit boards that consisted of ten layers of internal circuit traces were failing at a higher rate than the circuit boards consisting of eight layers or less. The reliability engineer wanted to add a measure of mathematical relevance that the boards were manufactured using different processes to what appeared to be an obvious situation, so the F-test of significance was applied.

For each of the two units being tested, four boards consisted of ten internal circuit layers, two microprocessor boards, one data communication board, and one analog board. The ten-layer circuit boards were manufactured by one circuit board vendor. The remaining seven boards used in each unit consisted of eight internal layers or less and were made by a different vendor than the manufacturer of the ten-layer boards. The following failure history was accrued during 800 total test hours for the two test units combined:

Type	Quantity	Number of failures
1) Non ten-layer circuit boards	7	1 open circuit failure
2) Ten-layer circuit board	4	11 open circuit failures

Calculating the MTBF for the non-ten-layer circuit boards yields:

$$\text{MTBF}_0 \text{ or } u_0 = \frac{7 \text{ boards} \times 800 \text{ total test hours}}{1 \text{ failure}} = 5600 \text{ hours}$$

The MTBF for the ten-layer circuit boards is:

$$\text{MTBF}_1 \text{ or } u_1 = \frac{4 \text{ boards} \times 800 \text{ total test hours}}{11 \text{ failures}} = 291 \text{ hours}$$

We wish to test the null hypothesis: H_0: $\theta_0 = \theta_1$, (alternate hypothesis: $\theta_0 \neq \theta_1$) that the ten-layer circuit board is from the same population as the other circuit boards (non–ten-layer circuit boards).

Under the null hypothesis, $P(\theta_0/\theta_1 > F_{1-\alpha}(2K, 2L)) = \alpha$, we are using the F-distribution at 95 percent confidence where $\alpha = .05$. In calculating θ_0, we had 1 failure, and therefore we have $2K = 2$ degrees of freedom (two times the number of failures) while for θ_1, we had 11 failures which yields $2L = 22$ degrees of freedom (two times the number of failures). We look up the F-distribution table in Appendix A, with the value 2 that is read from the top line (numerator) and 22 degrees of freedom from the value that is read from the side (denominator). Thus, we get:

$$F_{.95}(2,22) = 3.44 \text{ (critical value from } F\text{-distribution table)}$$

We then calculate the ratio θ_0/θ_1

$$\frac{\theta_0}{\theta_1} = \frac{\text{MTBF}_0}{\text{MTBF}_1} = \frac{5600}{291} = 19.2$$

The value of 19.2 is much greater than 3.44 at the 95 percent confidence level; hence we reject H_0, the null hypothesis. Therefore, we can conclude that the ten-layered circuit boards would not be considered to come from the same population as the other circuit boards based on the reliability performance. This essentially provides some verification that there is a specific problem associated with these circuit boards made by this vendor.

A) Cross Section of Multilayered Circuit Board

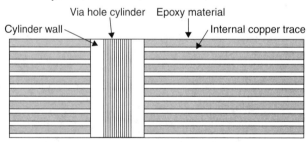

B) Detailed Closeup of Different Types of Internal Circuit Board Connections

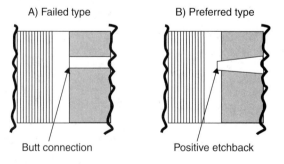

Figure 11-2 Multi-layer Circuit Boards

Subsequent failure investigation revealed that the interior traces of the circuit boards were only butted up against the barrels of the feed-through holes. This is an inferior process in the manufacture of the circuit board used by the vendor. Eventually, the vendor went to an improved manufacturing process known as positive etch-back where the traces were inserted into the circuit board (See Figure 11–2).

It is important to note that the use of statistical analysis does not preclude the need to conduct a thorough failure investigation; rather it is a valuable tool to confirm suspicion of a negative trend in reliability performance as well as reinforce any findings.

11.2.2 F-Test Case Study #2

TASK: Apply the F-test at 99 percent confidence to investigate soldering process on flexible circuit board used in cockpit engine display unit of navy fighter aircraft.

The F-test of significance can be used to identify differences between processes. In addition, it can be used to test more than one scenario or theory while using the same set of data. The following example involves the engine display unit where two units, Unit A and Unit B, underwent reliability testing and apparently two different processes were used in soldering of the motherboard flexible circuit board for which the other circuit boards plug into.

The motherboard flexible circuit board of one unit had all holes soldered by hand, both the holes inserted with component leads and the various feed-through holes that did not have component leads installed (these holes are part of the internal network of the circuit board layout). The flexible circuit board of the other unit under test only had solder applied to those holes with component leads installed and no solder applied to the feed-through holes. The following failure data was observed during the reliability test:

	Unit A	Unit B
Number of feed-through holes	150	150
Number of failed feed-through holes	0	8
Number or regular holes	720	720
Number of failed regular holes	0	0
Test hours accrued	1,626	1,880

TEST 1:

Comparison of feed-through holes with the regular holes (component leads inserted) on Unit B.

The null hypothesis that we set for this example is that there is no difference between the feed-through hole configuration and the regular hole configuration in Unit B. We then apply the null hypothesis to see if the feed-through holes and the regular holes of Unit B are from the same population. For the purpose of this exercise, one failure is assumed for the regular holes in order to be able to compute a nominal MTBF value for the purpose of comparing the two groups. Assigning one failure is in a sense, setting a 60 percent confidence limit value and should not affect the results. Eight failures will be used for the MTBF calculation of the feed-through holes.

$$\text{Regular holes MTBF}_0 = \frac{720 \times 1626 \text{ hours}}{1} = 1{,}170{,}720 \text{ hours}$$

$$\text{Feed-through holes MTBF}_1 = \frac{150 \times 1626 \text{ hours}}{8} = 30{,}488 \text{ hours}$$

Using the same methodology that was used in the previous example, the null hypothesis is then determined for the F-test critical value using the following parameters:

$\alpha = .01$, $K = 2$, $L = 16$, $F_{.99}(2,16) = 6.23$ (critical value from F-distribution table)

The test statistic θ_0/θ_1 is set up as:

$$\frac{\text{MTBF}_0}{\text{MTBF}_1} = \frac{1{,}170{,}720}{30{,}487} = 38.4 \text{ which is much greater than } 6.23$$

Because the test statistic θ_0/θ_1 is greater than the expected critical value that was found in the F-table, we reject H_0, the null hypothesis (the alternate hypothesis is accepted). This suggests that the feed-through holes are not of the same population as the regular holes, leading to the suspicion that the manufacturing process of the two types of holes is different.

By the way, if we had chosen to add the experience of Unit A to this query, it would have "softened" the difference between the two MTBF but the test for the null hypothesis would still have failed. When performing calculation for these types of problems, it is best not to dilute the data with extra information that may have a bearing on the test. This is a form of censoring data so that the results are clear for the test being performed. The censoring of data is based on the judgment of the analyst.

TEST 2:

Comparison of the feed-through holes of Unit A with Unit B.

The null hypothesis that we set for this example is that there is no difference between the feed-through hole configuration of Unit A and the feed-through hole configuration of Unit B. Again, one failure is assumed for the feed-through holes of Unit A in order to do the F-test of significance at 1 percent.

$$\text{Unit A feed-through holes MTBF}_0 = \frac{150 \times 1880 \text{ hours}}{1} = 282{,}000 \text{ hours}$$

$$\text{Unit B's feed-through holes MTBF}_1 = \frac{150 \times 1626 \text{ hours}}{8} = 30{,}488 \text{ hours}$$

As we did in Test 1, the null hypothesis is applied with the following values determined:

$\alpha = .01$, $2k = 2$, $2L = 16$, $F_{.99}(2,16) = 6.23$ (critical value from F-distribution table)

$$\frac{\text{MTBF}_0}{\text{MTBF}_1} = \frac{282{,}000}{30{,}488} = 9.25 \text{ which is greater than } 6.23$$

The null hypothesis of this query is rejected with the results showing that the feed-through holes of Unit B are not from the same population of Unit A. Armed with this in-

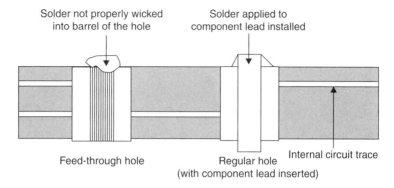

Figure 11-3 Cross-Section View of Two Different Types of Hole Solder Process

formation, it is necessary to check what was done during the solder operation of the motherboard flex of each unit. Through further investigation, it was found that the operator who applied the solder to the feed-through holes did not wick the solder to fill the hole as the operator who soldered the flex on Unit B. See Figure 11–3 for this picture. Further corrective action was developed where it was decided to eliminate the soldering process for the feed-through holes altogether—it was determined that the solder that was added made no difference in making the hole any stronger.

In the application of the 99 percent value, note that the value for comparison, 6.23, is greater than if the 95 percent value was used for the same set of parameters, or the value of 3.63. Using the 99 percent value made the results in Test 2 a little bit closer, but the same conclusion regarding the null hypothesis was achieved. The graphical representation comparing the two confidence levels (95 and 99 percent) on the MTBF ratio results of Test 2 is shown in Figure 11–4.

11.2.3 F-Test Case Study #3

TASK: Apply F-test at 95 percent confidence level in comparing field performance of two types of card readers in an ATM design that is deployed in two market areas.

A number of more advanced concepts will be illustrated in this particular example. First, it will be demonstrated that the F-test of significance can be used multiple times on a large set of data to draw individual conclusions for each query. Second, a method of data reduction will be demonstrated in this example to be able to use the F-table for valid comparisons. Finally, it will be demonstrated that an approximation of values can be estimated when specific table values are not available.

The application of the F-test was applied to a particular study of two different types of card readers used in the drive-up ATMs of two different areas of the United States with an ATM network. The test was used to test different sets of hypotheses of independence. The parameter used as the criteria for the study was the number of failed card dips where the card reader could not successfully read the magnetic stripe data off of the customer's ATM card. Card reader data used for this study was taken from East Coast areas and Midwest areas for both type readers used in the drive-up ATM for the time pe-

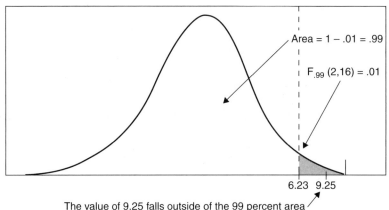

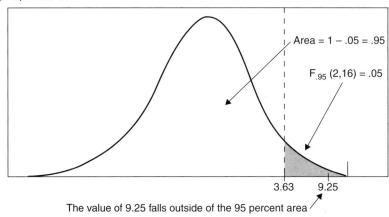

Figure 11-4 Comparison of F-Test Values for Test 2 in Example 11.2.2

riod January through April. This time period covers the colder months of the year which represents worst case conditions and the data over four months represents a significant amount of time for the data base.

The overall results are shown in the Table 11–1. An F-test of significance was applied to the MTBF value (MTBF here represents mean transactions or card dips between failures) of the two types of readers to verify that their performance was statistically different. However, the performance of Type 2 readers from the East Coast area and from the Midwest area was virtually identical such that they could be considered to be statistically from the same population. The same results hold true for Type 1 readers from both areas where the performance is identical and thus considered to be from the same population.

The F-test of significance was be applied to the various MTBF values calculated for Type 1 and Type 2 readers for both East Coast and Midwest. The null hypothesis is a test

Table 11-1 ATM Card Reader Performance Data

Location	Type	Quantity	Total Dips	Total Bad Dips	MTBF	AVG Daily Bad Dips (each reader)	AVG Weekly Bad Dips (each reader)
East Coast	Type 1	24	268774	52191	5.15	18	126
East Coast	Type 2	5	54052	2824	19.14	5	35
Midwest	Type 1	82	591788	109710	5.39	11	77
Midwest	Type 2	16	63812	3357	19.01	2	14

where the ratio of the various MTBF values of the card readers will be compared to the value listed in the F-table for the following probability equation:

$$P(\text{MTBF}_0/\text{MTBF}_1 > F_{1-\alpha}(2K, 2L)) = \alpha$$

A 95 percent probability factor will be used (or $\alpha = .05$). Since the degrees of freedom (number of failures) are quite large, the values for K and L will be a value close to infinity and the expected value that is taken from the F-table to be compared with the MTBF ratios is almost 1. This test would generally be valid except for certain cases where the comparison would be too close to call. It would make more sense to use another method.

Thus, in this example, we will use a process known as normalizing data. The K and L values are much too large to properly use the F-table, but we can reduce the data and maintain the basic ratio of failure count between each group being measured. Inherently, it can be seen that from Table 11-1, the MTBF values are pretty consistent between the two groups (for example, the Type 2 readers for both areas have similar MTBF values). It would not be reaching too far to keep these MTBF values while reducing the failure count data using a different scale. The data was collected over a four-month period (120 days), so it is possible that we could obtain a daily and a weekly failure count for each group in an effort to reduce the failure count, yet still maintain the relationship between each group.

Thus the column "weekly failures per reader" is calculated by taking the total number of failed dips for the 120-daily period and dividing by the number of readers in the population and by 120 days. This yields an average daily failure count per reader. This number is then multiplied by seven to yield an average weekly count. It is noted that we are dealing with averages, but this will not adversely affect the results.

TEST 1:

Perform a statistical comparison of the two types of readers in the East Coast area using the F-test.

$$\text{MTBF}_0 = 54{,}052/2824 = 19.14 \text{ (MTBF of Type 2 Readers)}$$
$$\text{MTBF}_1 = 268{,}774/52{,}191 = 5.15 \text{ (MTBF of Type 1 Readers)}$$
$$\frac{\text{MTBF}_0}{\text{MTBF}_1} = \frac{19.14}{5.15} = 3.72$$

Using the weekly failure values that are listed in Table 11-1 for our degrees of freedom, where $K = 35$ and $L = 126$, we look up the value for $F_{.95}$ $(2K,2L)$ and we get $F_{.95}(70,252)$ $\cong 1.3$. We are able to get the value of 1.3 by doing a rough interpolation estimate between the table values presented (the range is from 1.43 and 1.23). It is acceptable to use an approximated value since we know the upper and lower limits from the table. The ratio of $MTBF_0/MTBF_1$ of 3.72 is higher than 1.3, hence the null hypothesis is rejected with the conclusions that the two MTBFs are not from the same population.

TEST 2:

Perform a comparison of the two type of readers (Type 1 and Type 2) in the Midwest area using the F-test.

Using the same methodology that was used in Test 1, we look up Table 11-1 to get the following values:

$$F_{.95}(28,147) = 1.5 \text{ (critical value that is interpolated from the F-table)}$$

$$MTBF_0 = \frac{63,812}{3,357} = 19.01 \text{ (MTBF of Type 2 Readers)}$$

$$MTBF_1 = \frac{591,788}{109,710} = 5.39 \text{ (MTBF of Type1 Readers)}$$

$$\frac{MTBF_0}{MTBF_1} = \frac{19.01}{5.39} = 3.52, \text{ which is greater than 1.5 (the critical value)}$$

Hence, the null hypothesis is rejected, and the two MTBFs are not from the same population.

TEST 3:

Perform a Type 2 readers comparison between two markets.

From Table 11-1, note that the MTBF of the East Coast–area Type 2 reader compares favorably with the Midwest-area Type 2 reader. By applying the F-test, we obtain the following:

$$F_{.95}(70,28) = 1.75 \text{ (interpolated from the F-table)}$$

$$\frac{MTBF_0}{MTBF_1} = \frac{19.14}{19.01} = 1.01, \text{ which is less than 1.7 (the critical value)}$$

Hence, the null hypothesis is accepted as the two MTBFs are considered to be from the same population as the Type 2 readers compare similarly in performance.

TEST 4:

Perform Type 1 readers comparison between two markets.

Likewise, from Table 11-1 it is noted that the MTBF of the East Coast–area Type 1 reader compares favorably with the Midwest-area Type 1 reader. The F-test yields:

$$F_{.95}(154,252) = 1.2 \text{ (interpolated from the F-table)}$$

$$\frac{MTBF_0}{MTBF_1} = \frac{5.39}{5.15} = 1.05, \text{ which is less than 1.2}$$

While the null hypothesis is accepted for Test #4 with the two MTBFs probably being from the same population, the margin of error is very close and this fact should be noted when presenting this conclusion. The closeness of this test was driven by the very high number of degrees of freedom.

In this example, the F-test was applied four times to the same set of data to test out four different queries. This analysis shows that in this example, the remarkable fact that performance of readers of the same type was identical, even when installed in different areas. Due to the significant amount of data that was used in this exercise, there is a large confidence in the results and the conclusions that can be drawn from them. Other concepts of data reduction and estimating values between presented F-table values were also illustrated.

The conclusions obtained from our testing suggest a number of factors at play. First, they show that the different environmental factors of each area (East Coast and Midwest) did not play a significant difference in the performance of the two card readers as similar MTBF values were obtained for each type reader as used in the two areas and substantiated by hypothesis testing. (Environmental factor for this study showed that the temperatures were generally colder in the Midwest compared to the East Coast during the months January through April.)

But application of the F-test did suggest that certain differences in the design would affect performance in the field. It certainly appeared that the Type 2 reader was performing better on the field than the Type 1. Some design differences were noted between the Type 1 design and the Type 2 design, with the Type 2 being the more complex-style card reader in the area of electrical circuit design, magnetic head design and logic design. Initially, it was thought that the extra features of the Type 2 reader aided in more consistent reading of cards.

However, deeper investigation revealed that the software for the two types used different criteria for counting a bad card dip. The Type 1 counted all misreads, where as the Type 2 did not count the first two bad attempts of a customer's card and counted the third bad attempt because the reader performed an additional action to try to read the card. (It held the card and moved the magnetic head). Thus, the application of the F-test proved valid as it pointed out that there was something significantly different involved. In this case, it was the software and eventually, the software was changed in the Type 2 to count all bad read attempts.

When comparing data that uses time-based parameters such as MTBF, it is best to use the F-test of significance to compare the data. However, there is an alternative equivalent test that can be used in comparing different scenarios, often using the same set of data. This test is known as the chi-square test of independence where expected values are used. This technique has an advantage in that it is capable of extending to situations involving more than two categories for each basis of classification. The results that are computed from this test will be just as valid as the F-test and in many cases both tests can be applied.

11.3 THE CHI-SQUARE (χ^2) TEST OF INDEPENDENCE

The chi-square (χ^2) test of independence is another test that can be used to compare two populations and conduct hypothesis testing. The basic concept of this test involves the goodness of fit of the statistic to the population that is being measured.

As with the F-test that was demonstrated in the previous section of this chapter, this test involves testing a null hypothesis by using a critical probability value for testing this null hypothesis. As with the F-test, the chi-square distribution value is determined by the number of degrees of freedom (based on failure count) for a selected α value.

The chi-square, or (χ^2), distribution describes the expected pattern of behavior for the sum of a specified number of squares, n, of independent normally distributed variables, each of which has been selected from a population with the characteristics of a mean equal to zero and standard deviation equal to one. The distribution is tested by comparing observed frequency (fo) of occurrences with the expected frequency (fe) of occurrences for each of the individual observations. The value $(fo - fe)^2/fe$ is calculated for each set of fo values. This term represents the square of a normally distributed variable with a mean equal to zero and a standard. The degrees of freedom for the chi-square distribution is equal to $n - 1$.

CHI-SQUARE TEST EXAMPLE

For example, supposed we wish to perform a null hypothesis test on a set of values to determine if they meet the criteria for the chi-square distribution. Let the following values be the observed number of failures (fo) that are accrued by a fixed population of units in the field during a six-month period of measurement. We wish to determine this value at a confidence level of 95 percent.

Month	Observed Number of Failures (fo)
January	150
February	187
March	171
April	201
May	168
June	173
Totals	1050

The mean or average is equal to 1050/6 = 175 (this is also known as the expected value). This value is the expected frequency or fo.

Now a calculation table can be set up to come up with the values for $(fo - fe)^2/fe$ for each of the six observations.

Month	fo	fe	fo − fe	$(fo - fe)^2$	$(fo - fe)^2/fe$
January	150	175	−25	625	3.57
February	187	175	12	144	0.82
March	171	175	−4	16	0.09
April	201	175	26	676	3.86
May	168	175	−7	49	0.28
June	173	175	−2	4	0.02
Totals	1050				8.64

The chi-square value is at 95 percent confidence or $\alpha = 1 - .95 = .05$ for $n - 1$ degrees of freedom or 5 degrees of freedom for this example. Looking up the tables yields the result:

$$\chi^2_{0.05}(5) = 11.07$$

Thus the calculated value of 8.64 is less than the chi-square value at 95 percent confidence of 11.07 and the null test is passed, meaning that the observed values are within in the expected range of the expected value.

One aspect of using this test is that rather than using MTBF values as we did in the previous example using the F-test, we can use the number of failures observed for a population. This may come in handy for cases when we do not know the number of operating hours accrued by the units in the population.

Some additional considerations to remember are that a minimum of 50 observations is required if the theoretical chi-square distribution is to provide a reasonable approximation of the expected sampling distribution. Also, in general, the expected frequency for each category should be at least 5. On occasion it may be necessary to combine categories in order to achieve this objective. When the expected frequency for a category is less than 5 the chi-square approximation may be poor; however in some cases, this approximation may be sufficient to prove a case as is shown in the first case study.

CHI-SQUARE CONTINGENCY MATRIX

The beauty of this distribution is that the chi-square test of independence can be set up and can be applied to two populations, each with a two or more sets of observed occurrences, and this can be done by setting up a matrix in a row by column setup. This is also known as contingency testing. This process will be demonstrated using a two-by-two contingency matrix setup by the following steps:

1. Create a two row by two column table matrix (known as a r by c contingency table), where the scenarios to be tested are listed:

	Condition A	Condition B	Row Totals
Scenario 1	X_{11}	X_{12}	$\Sigma(X_{11} + X_{12})$
Scenario 2	X_{21}	X_{22}	$\Sigma(X_{21} + X_{22})$
Column totals	$\Sigma(X_{11} + X_{21})$	$\Sigma(X_{12} + X_{22})$	$\Sigma(X_{11} + X_{12} + X_{21} + X_{22})$

X_{11} represents the observed results for items meeting scenario 1 and condition A, X_{12} represents the observed results for items meeting both scenario 1 and condition B and so on.

2. Calculate the expected values for each cell of the matrix using the general formula:

$$(fe)_{ij} = \frac{\Sigma(\text{row i}) \times \Sigma(\text{column j})}{\text{Total number of observations}}$$

In nonmathematical terms, this can be expressed as the product of the row total and the column total and divided by the total number of observations.

For the two-by-two matrix setup, we then calculate the expected frequency values for the four cells in the matrix:

$$(fe)_{11} = \frac{\Sigma(\text{row 1}) \times \Sigma(\text{column 1})}{\text{Total number of observations}} = \frac{\Sigma(X_{11} + X_{12}) \times \Sigma(X_{11} + X_{21})}{\Sigma(X_{11} + X_{12} + X_{21} + X_{22})}$$

$$(fe)_{12} = \frac{\Sigma(\text{row 1}) \times \Sigma(\text{column 2})}{\text{Total number of observations}} = \frac{\Sigma(X_{11} + X_{12}) \times \Sigma(X_{12} + X_{22})}{\Sigma(X_{11} + X_{12} + X_{21} + X_{22})}$$

$$(fe)_{21} = \frac{\Sigma(\text{row 2}) \times \Sigma(\text{column 1})}{\text{Total number of observations}} = \frac{\Sigma(X_{21} + X_{22}) \times \Sigma(X_{11} + X_{21})}{\Sigma(X_{11} + X_{12} + X_{21} + X_{22})}$$

$$(fe)_{22} = \frac{\Sigma(\text{row 2}) \times \Sigma(\text{column 2})}{\text{Total number of observations}} = \frac{\Sigma(X_{21} + X_{22}) \times \Sigma(X_{12} + X_{22})}{\Sigma(X_{11} + X_{12} + X_{21} + X_{22})}$$

3. Calculate the chi-square statistic by setting up the following table of values:

Cell	(fo)	(fe)	(fo − fe)	$(fo - fe)^2$	$(fo - fe)^2/fe$
1,1	X_{11}	$(fe)_{11}$	$X_{11} - (fe)_{11}$	$(X_{11} - (fe)_{11})^2$	$(X_{11} - (fe)_{11})^2/fe$
1,2	X_{12}	$(fe)_{12}$	$X_{12} - (fe)_{12}$	$(X_{12} - (fe)_{12})^2$	$(X_{12} - (fe)_{12})^2/fe$
2,1	X_{21}	$(fe)_{21}$	$X_{21} - (fe)_{21}$	$(X_{21} - (fe)_{21})^2$	$(X_{21} - (fe)_{21})^2/fe$
2,2	X_{22}	$(fe)_{22}$	$X_{22} - (fe)_{22}$	$(X_{22} - (fe)_{22})^2$	$(X_{22} - (fe)_{22})^2/fe$

The last column of the above table [$(fo - fe)^2/fe$] is then summed as the value to be used when comparing with the chi-square value.

4. Establish the decision rule by determining the number of degrees of freedom. For a r by c contingency table, this is calculated by the product, $(r - 1) \times (c - 1)$, which for the two-by-two version shown above, yields one degree of freedom. The number of degrees of freedom, (1), is used when looking up the chi-square table, usually using a probability of 90 percent or $(1 - \alpha) = .10$. For one degree of freedom, this is:

$$\chi^2_{0.10}(1) = 2.71$$

This can be represented pictorially as shown in Figure 11–5.

The chi-square value from the table is then compared to the calculated chi-square statistic (from Step 4). If the calculated value is less than the table value (in this case, 2.71), then the hypothesis of independence is accepted. If it is greater, then the hypothesis of independence is rejected and there is a relationship between the two hypotheses being tested.

Basically, the value 2.71 is pretty much all one has to remember for many application of this test, since most examples involves the use of two by two matrices and a probability of 90 percent. If there is a different matrix size or critical value, then the value has to be looked up in the table. A major advantage of setting up problems using the chi-square method is that one only has to remember this criteria value 2.71 is used for comparing data, as opposed to having a F-distribution table on hand for doing the F-test.

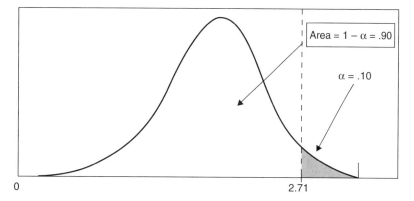

Figure 11-5 General Curve for Chi-Square Distribution

11.3.1 Chi-Square Test Case Study #1

Task: Study of Flexible Circuit Board Situation Used in Case Study 11.2.2

We can apply the chi-square test to the first part of the situation presented in Case study 11.2.2 The premise that is being tested is whether the eight failures observed on the motherboard flexible circuit board of Unit A is a chance occurrence and that it is not related to a specific issue. As per the step described in the previous section, a table is constructed by listing the different entries:

Number of failed feed-through holes observed on Unit A's motherboard flex = 0
Number of feed-through holes that did not fail on Unit A's motherboard flex = 150
Number of failed feed-through holes on Unit B's motherboard flex = 8
Number of feed-through holes that did not fail on Unit B's motherboard flex = 142

In this case, test hours accumulated for each unit are close enough for a valid statistical comparison to be made: 1,626 hours for Unit A and 1,880 hours for Unit B.

	Failed Holes	Holes without Failures	Total
UNIT A	0	150	150
UNIT B	8	142	150
TOTAL	8	292	300

Using the table above, the expected values are calculated:

$$(fe)_{11} = \frac{150 \times 8}{300} = 4 \qquad (fe)_{12} = \frac{150 \times 292}{300} = 146$$

$$(fe)_{21} = \frac{150 \times 8}{300} = 4 \qquad (fe)_{22} = \frac{292 \times 150}{300} = 146$$

to yield:

Cell	(fo)	(fe)	(fo − fe)	$(fo - fe)^2$	$(fo - fe)^2/fe$
1,1	0	4	−4	16	4.0
1,2	150	146	−4	16	0.11
2,1	8	4	4	16	4.0
2,2	142	146	−4	16	0.11

The sum of the four values of the last column is: Σ = 8.22.

This value will be compared with the chi-square value that we computed by establishing the decision rule by determining the degrees of freedom for the two-by-two matrix above to be (2 − 1) × (2 − 1) = 1

As we determined in the previous section, the 90 percent confidence limit for this two-by-two contingency table is = $\chi^2_{.10}(1) = 2.71$ and this is less than 8.22 which was computed in the table above. The graph would be the same as presented in Figure 11-5. The hypothesis of independence is rejected as the discrepancy between the expected frequencies and the observed frequencies is too great to be reasonably attributed to chance selection in sampling. Hence, there is a specific issue concerning the number of failures that have been occurring in the flex of Unit A.

It is noted that even though that two of the fe values were only 4, this did not affect the overall results despite the caution earlier about using small fe values. This example worked because here was a significant difference between the fe values of these rows with the fe values from the other rows. Even if we were to arbitrarily substitute the number 5 for 4 for $(fe)_{11}$ and $(fe)_{21}$, the test would still fail. This is why it may pay to perform different tests to verify the conclusions reached so that some level of confidence is achieved. In this case, the conclusion reached by performing this test is essentially the same conclusion reached by performing the F-test of significance in Case study 11.2.2 and we can feel reasonably confident with the results.

11.3.2 Chi-Square Test Case Study #2

TASK: Perform a comparison between two types of receipt printers used in the ATMs of a major bank by using field performance in chi-square test

A major bank is studying the field reliability performance of two different receipt printers used in their ATM design: the old simple-style printer and the new smart-style printer with self-correcting software over a one month period. The bank owns 1,200 ATMs, with the breakdown of 1,020 simple printers and 180 smart printers in one marketplace. Initial observations show that one of the major error codes related to dirty sensors affects the smart printer more than the simple printers such that multiple failures (more than three occurrences per month per unit) are observed.

Number of Simple Printers with multiple failures due to specific error code = 9
Number of Simple Printers that did not fail due to specific error code = 1111
Number of Smart Printers with multiple failures due to specific error code = 104
Number of Smart Printers that did not fail due to specific error code = 76

	Failed Units	Units with No Failures	Total
SIMPLE PRINTER	9	1111	1020
SMART PRINTER	104	76	180
TOTAL	113	1187	1200

Using the table above, the expected values are calculated:

$$(fe)_{11} = \frac{1020 \times 113}{1200} = 96 \quad (fe)_{12} = \frac{1020 \times 1187}{1200} = 1009$$

$$(fe)_{21} = \frac{180 \times 113}{1200} = 17 \quad (fe)_{22} = \frac{180 \times 1187}{1200} = 178$$

yielding the following table:

Cell	(fo)	(fe)	$(fo - fe)$	$(fo - fe)^2$	$(fo - fe)^2/fe$
1,1	9	96	−85	7,225	75.2
1,2	1,111	1,009	−102	10,404	10.3
2,1	104	17	87	7,569	445.2
2,2	76	178	−102	10,404	58.4

The sum of the four values of the last column is: $\Sigma = 589.1$.

This value will be compared with the chi-square value which we compute by establishing the decision rule by determining the degrees of freedom for the two-by-two matrix above to be $(2 - 1) \times (2 - 1) = 1$. The 90 percent confidence limit = $\chi^2_{.10}(1) = 2.71$, which is less than 589.1 which was computed. Thus the hypothesis of independence is rejected. The discrepancy between the expected frequencies and the observed frequencies are too great to be reasonably attributed to chance selection in sampling. Hence, there is a specific issue concerning the specific error code on the smart printers that makes it less reliable than the older simple printers.

Investigation revealed that although the newer smart printer had more recovery features, it had a design where the sensors were located in the lower half of the printer and this was more susceptible to dust accumulation. This shows that newer technology may appear to be more reliable as per paper analysis, but the basic design feature of the sensor locations doomed it from being more reliable than the older printers. As a result of the finding, a monthly maintenance program was introduced to clean the sensors of the smart program as well as consolidating all of the smart printers into specified sites where this cleaning could be monitored as well as providing the necessary training to the branch people.

11.3.3 Chi-Square Test Case Study #3

TASK: Verify a number of premises involving a manufacturer's yield of capacitors

This example is set up to show that there are applications where a matrix bigger than two-by-two can be constructed in order to conduct various chi-square tests.

A manufacturer of capacitors has three separate manufacturing runs of tantalum type capacitors over the course of the year. A number of capacitors had failed during the burn-in screening process at the manufacturer's facility with two failure modes occurring, short or open. The following chart is the results of each run:

Production run #1	10,000 good	10 failed short	9 failed open
Production run #2	25,000 good	20 failed short	25 failed open
Production run #3	40,000 good	15 failed short	17 failed open

A) Conduct the chi-square test of independence to determine if there is independence between each production lot with regards to good and bad components. For this example, the two failure modes will be combined for each lot into one number for failures to get the following three-row-by-two-column table:

	Failed Capacitors	Good Capacitors	Total
RUN 1	19	10,000	10,019
RUN 2	45	25,000	25,045
RUN 3	32	40,000	40,032
TOTAL	106	75,000	75,106

As in the previous example, expected frequency equations are set up:

$$(fe)_{11} = \frac{106 \times 10019}{75106} = 14 \quad (fe)_{12} = \frac{75000 \times 10019}{75106} = 10005$$

$$(fe)_{21} = \frac{106 \times 25045}{75106} = 35 \quad (fe)_{22} = \frac{75000 \times 25045}{75106} = 25010$$

$$(fe)_{31} = \frac{106 \times 40032}{75106} = 56 \quad (fe)_{32} = \frac{75000 \times 40032}{75106} = 39976$$

yielding the following table:

Cell	(fo)	(fe)	$(fo - fe)$	$(fo - fe)^2$	$(fo - fe)^2/fe$
1,1	19	14	5	25	1.79
1,2	10,019	10,005	14	196	.02
2,1	45	35	−10	100	2.86
2,2	25,000	25,010	−10	100	.004
3,1	32	56	−24	576	10.29
3,2	40,032	39,976	56	3136	.08

The sum of the four values of the last column is: $\Sigma = 15.04$.

For a three-by-two contingency table, the calculation $(3 - 1) \times (2 - 1)$, yields two degrees of freedom. The number of degrees of freedom is used when looking up the chi-square table, usually using a probability of 90 percent or $(1 - \alpha) = .10$. For two degrees of freedom, the value from the table is:

$$\chi^2_{0.10}(2) = 4.61$$

This is less than the 15.04 that was calculated and thus the hypothesis of independence is rejected and the results observed are in the area of what is to be expected between each run.

B) Conduct a chi-square test of independence to determine if there is independence between each production lot with regards to the failure mode of open and short for the capacitors. For this example, the two failure modes will be compared against each production run to get the following three-row by two-column table:

	Failed Short	Failed Open	Total
RUN 1	20	9	29
RUN 2	20	25	45
RUN 3	15	17	32
TOTAL	55	51	106

Expected frequency equations are set up as follows:

$$(fe)_{11} = \frac{55 \times 29}{106} = 15 \quad (fe)_{12} = \frac{51 \times 29}{106} = 14$$

$$(fe)_{21} = \frac{55 \times 45}{106} = 23 \quad (fe)_{22} = \frac{51 \times 45}{106} = 22$$

$$(fe)_{31} = \frac{55 \times 32}{106} = 17 \quad (fe)_{32} = \frac{51 \times 32}{106} = 15$$

Cell	(fo)	(fe)	$(fo - fe)$	$(fo - fe)^2$	$(fo - fe)^2/fe$
1,1	20	15	5	25	1.67
1,2	9	14	-5	25	1.79
2,1	20	23	-3	9	0.39
2,2	25	22	3	9	0.41
3,1	15	17	-2	4	0.24
3,2	17	15	2	4	0.27

The sum of the four values of the last column is: $\Sigma = 4.77$.

For a three-by-two contingency table, the calculation is $(3 - 1) \times (2 - 1)$, yields two degrees of freedom. Two degrees of freedom is used when looking up the chi-square table at a probability of 90 percent or $(1 - \alpha) = .10$. For two degrees of freedom, the value is:

$$\chi^2_{0.10}(2) = 4.61$$

It can be seen that this value is greater than the 4.77 value that we calculated which means that the test for independence has failed. However, since this was very close, it suggests that further analysis my need to be performed.

C) Compare the failure results for production run 1 and run 2 to further investigate possible differences in the data results. For this example, we can revert back to the two-by-two matrix setup.

	Failed Short	Failed Open	Total
RUN 1	20	9	29
RUN 2	20	25	45
TOTAL	40	34	74

Expected frequency equations are set up:

$$(fe)_{11} = \frac{40 \times 29}{74} = 16 \qquad (fe)_{12} = \frac{34 \times 29}{74} = 13$$

$$(fe)_{21} = \frac{40 \times 45}{74} = 24 \qquad (fe)_{22} = \frac{34 \times 45}{74} = 21$$

yielding the following table:

Cell	(fo)	(fe)	(fo − fe)	(fo − fe)²	(fo − fe)²/fe
1,1	20	16	4	16	1.00
1,2	9	13	−4	16	1.23
2,1	20	24	−4	16	0.67
2,2	25	21	4	16	0.76

The sum of the four values of the last column is: $\Sigma = 3.66$.

For a two-by-two contingency table, we get one degree of freedom. The number of degrees of freedom is used when looking up the chi-square table, usually using a probability of 90 percent or $(1 - \alpha = .10$. The 90 percent confidence limit = $\chi^2_{.10}(1) = 2.71$, which is less than the value of 3.66 that was computed. Thus the test for independence has failed, and the discrepancy between the expected frequencies and the observed frequencies is too great to be reasonably attributed to chance selection in sampling.

This particular case study shows a matrix that has some scenarios that are larger than two-by-two, is meant to show the reader that any set of combinations and matrix table size can be set up for applying the chi-square test for analyzing the data. One has to be very careful, particularly when the margin of difference is close. When this happens, it is a wise idea to investigate further and conduct further statistical tests.

11.4 SUMMARY

This chapter shows the importance of conducting statistical-based hypothesis testing to solve reliability problems. As illustrated in the various case studies in this chapter, it can be seen that some problems can be solved by either method (F-test or chi-square) with similar results. The type of test that is used will generally depend on what parameters are available. The F-test is perhaps easier to set up, but it will generally require an F-distribution table handy to select the value for comparison. On the other hand, the chi-square test requires a little more effort to set up. However, since most setups involve a two-by-two matrix, the same value of comparison that is taken from the chi-square table for one degree of freedom of 2.71 (at 90% confidence), is used for most of the examples. Thus a table is not needed if one can remember this specific value to use in the comparison.

Other statistical tools exist that can be applied in similar fashion as the two described here, however, these two are fairly straightforward to apply with the same general results achieved if either method is used. It is important to realize that the tools are meant to support findings in failure investigation, not necessarily discover the cause of a failure. Sometimes, the cases are so obvious prior to applying a statistical test that the mathematical exercise becomes trivial, as in Case study 11.3.2. However, it is still important to perform these statistical tests because it is more valid to say that observations or conclusions are supported by statistical analysis as opposed to the subjective approach of just saying that it looks a certain way (and varying depending on the observer's perspective). It is an important aspect of reliability engineering to be able to support observations with sufficient statistical analysis.

It is also recommended that when providing this type of analysis as part of a report or presentation, it is best to present it either as an appendix or as a separate supporting document. The analysis can be quite impressive, but manager-types in general may not be able to connect with the mathematical concepts and may even be intimidated by them. It is important not to lose the audience's attention by throwing too much mathematical analysis at them. Thus, the main part of the report can refer to the fact that a statistical test has been performed to back up conclusions stated in the report and refer the reader to the detailed statistical analysis provided in the appendix. This is the way that things can work when reliability engineers interface with nontechnical and nonmathematical people in the real world.

This is the last chapter in this book that features the significant use of numerical reliability analysis. The secret to using these math tools is to not be intimidated by them, but rather treat them as common everyday analysis. It is critical to understand that these tools can be used to draw limited conclusions within a defined set of parameters. It would be wrong to draw out anything more than what the calculations mean, as they should be used to answer a specific question. These are not "magic numbers" that solve all reliability-type problems, and management has to be made aware of the specific scope involved with this type of analysis and the associated limitations.

11.5 EXERCISES

1. The field performances of two different populations of units are being compared. One population of 20 units does not have a corrective action fix and has accrued 30 failures. The other group of 30 units have had corrective action implemented and has accrued 5 failures. All units entered service at the same time and ran for 60 days with power on

continuously. Perform an F-test of significance at 95 percent in order to compare the field performance of these two populations by performing the following steps:
   A. What are the number of degrees of freedom for each population group?
   B. Calculate the MTBF (in hours) for each population.
   C. Calculate the ratio of the larger MTBF value to the smaller MTBF value.
   D. Using the number of degrees of freedom (calculated in A), select the 95 percent confidence level value from the F-table in Appendix A.
   E. Is the null hypothesis accepted or rejected as to whether the two groups of units would be considered to have come from the same population?
   F. Can one conclude that the corrective action was effective?
2. Two units are undergoing reliability testing. The process for making the circuit board has changed. Perform a F-test of significance on the following set of data at the 95 percent confidence level to determine if the units were made in similar fashion:

   Unit A 1,450 hours, 21 failures
   Unit B 1,650, 9 failures

3. Perform a chi-square test of independence at the 90 percent level on two vendors who make the same component for a customer. Use the test to determine whether the vendors are using similar processes and quality. The data collected is:

   Vendor A   1,500 components built, 50 failures recorded when used in customer's unit
   Vendor B   900 components built, 15 failures recorded when used in customer's unit

4. Using the same numbers as used in the case study in 11.2.3, Test 2, construct a contingency table and perform a chi-square test of independence at the 90 percent level. Perform the following steps:
   A. What is the size of the contingency table?
   B. Set up the contingency table and perform role and column calculations.
   C. Set up a table of expected frequency values for each cell of the table.
   D. Compute the chi-square statistic by constructing a table of values.
   E. Determine the number of degrees of freedom [Hint: use the formula $(r - 1) \times (c - 1)$].
   F. Look up the chi-square value for 90 percent from the chi-square table in Appendix A.
   G. Is the chi-square table value larger than the calculated chi-square statistic?
   H. Are the results obtained by this method similar to the results obtained by the F-test method?

5. An LCD manufacturer, using a certain process, made a production run of 500 LCDs with 230 LCDs failing during the production run. After making some design changes in the construction of the LCD, the manufacturer made a second production run of 500 LCDs with only 115 LCDs failing. Is this change in the yield significantly different? Solve this problem by first applying the F-test of significance and observe the results. Then set up the chi-square test of independence and observe these results. Assume that each LCD is tested for 10 hours during burn-in testing.

CHAPTER 12

Software Reliability Concepts

INTRODUCTION

It would be somewhat accurate to say that the software field has changed many times over since it started becoming a major part of product design during the 1960s. The use of software in products has allowed products to perform more functions in a reliable manner as less parts are needed in the design. It would be even truer to state that many software reliability concepts and models developed over the years have been made obsolete by the rapid changes and the significantly expanded capabilities of software. Thus, the field of software reliability has to be a dynamic field that is capable of adapting to these rapid changes in industry.

The reduction of the size of software-based equipment, from mainframe systems to smaller PCs, and the reduction in the size of integrated circuits (chip) packages have brought about a revolutionary process. Perhaps the biggest change in the use of software is the integration of software packages into smaller-size product designs. Software has made it possible for many types of products and subsystems to be made smaller and more manageable. Through the use of software in these products, there is a significant reduction in the number of parts as certain components that contain software programs can send commands as opposed to developing a set of discrete circuits for similar functionality.

Some basic concepts involving software have remained remarkably consistent throughout the many changes and updates. While we will discuss the effects of software in this chapter, we will limit our discussion on the actual coding of the software or any discussion on different software languages and packages. This is because, in the area of reliability, typical software faults are more often not the result of the actual language or coding that is used, but rather of programming logical errors, sometimes very subtle, that are used to instruct the device to operate. Basic coding errors would usually be detected during early software development, prior to implementation into hardware.

Even though different computer languages may be used over the years, basic concepts involving software reliability do not change. These basic concepts have reliability concerns, and we will discuss these concepts in detail in this chapter.

12.1 BASIC RELIABILITY SOFTWARE CONCEPTS

As we all know, software is present in practically every phase of our lives, so much so that much of it is transparent to us most of the time. Perhaps it is fair to state that the more transparent the software is in the products that we use daily, the greater the probability that the product has high field reliability that is made possible by the way the software is written.

It is important to state up front that the actual software itself does not have "reliability" or a "failure rate" in a practical sense. Unlike hardware reliability, software reliability does not follow a bathtub curve, nor does it wear out. But one should not be misguided into thinking that there are no reliability issues involving software or that the software is always written correctly and working properly.

The general definition of software reliability is that the software program functions without an external failure of a specific time period in the system that is used, under actual working conditions. The trick is to verify that the software was properly written to address all conditions that will be seen during field implementation.

12.1.1 The Three Different Types of Software

Any engineer who deals with reliability software issues on individual self-contained equipment, or a complete system in the form of a network, may encounter software issues in one or more of the following areas:

- Embedded software: This is the software program that is imbedded in a programmable memory device (EPROM, EAPROM, E^2PROM, Flash) inside a unit. Embedded software is sometimes known as "firmware" because it is rarely changed once the unit is delivered, if at all.
- Application software: This is software that is used on a system level in which different devices or equipment communicate with each other. This may use a higher level of computer language, and it will be installed on the hard drive of the computer in most cases.
- Support software: Typically this is software that is used in test equipment or numerical control–type machines. Like application software, this would be written using a high-level computer language.

12.1.2 Reliability Issues with Software

Many issues concerning reliability should be reviewed and tested by the engineering department during the software development of the product. The issues involving software faults that are commonly called "bugs" include:

- Errors in specific instruction coding
- Incomplete instructions
- Ambiguity in instructions

- Collision of two sets of processes that cause "lockup" with no way out
- Dead-end paths or "AWOL" conditions in modules
- Infinite loops (instructions in a module that have no exit, and therefore it repeats)
- Reset errors
- Initialization errors
- Built-in test or error code annunciation do not perform correctly
- All situations that will be seen during field service are not properly addressed
- Inadequate documentation of the software package
- Obsolescence and sunset process of software program packages
- Insufficient support of software releases throughout the life of the product
- Hardware failures of the EPROM or the hard drive that houses the software
- Lack of sufficient data storage capacity or data transmission traffic capacity

Some of the subtle aspects of failures in software packages are that most of the programs have several subroutines and modules that are inside the main program. Of concern is if a module of software instructions is entered into, is there an appropriate line of code for an eventual exit? Otherwise, an infinite loop situation will exist, and an exit cannot be made and the software will appear to "hang up." In a situation like this, a conditional statement such as "IF N = [some value]," or a GO TO statement, has to be inside this loop in order to allow for an eventual exit from the module and back into the main program.

Another trick that software engineers use is to simulate timing by having a series of simple counting instructions performed inside a module (such as $N = N + 1$) and when a certain count is reached, an exit is made from the module. Timing issues surface with software applications, and they can be addressed often by putting in simple delays such as a simple counting sequence in a module located inside the software package.

There is a tremendous amount of market pressure placed on the vendors who develop software application packages to continuously improve their product. Because of the near-exponential growth of software capabilities in the world, it becomes necessary to keep track of software releases with numerical revision numbers, e.g., release 4.1 replaces previous software release 4.0. Over time, older releases may eventually not be supported by the vendor.

In recent decades, software reliability has become more of a reliability concern with many products, particularly with those electronic products that make intensive use of components with embedded programming capabilities. The time to watch for the possibility of incorrectly written software instructions is at the beginning stages of a new program. The reliability engineer or design engineer assigned to verify the reliability performance of the program will need to keep a watchful eye on the way the software behaves in a unit, particularly during qualification and reliability testing. It is important that the software is kept as concise as possible. Extra complexity by means of additional commands in how the software performs is not helpful, particularly when troubleshooting needs to be performed in the event of hardware failures.

The obvious errors are usually spotted right away during early development of a new product. The errors that are less obvious are those that might be embedded in a subroutine or module that are not called upon that often, such that they would be easily spotted. These errors tend to slip through until they are spotted during field service. Other software bugs may result from less subtle conditions—timing errors or collisions of two or more processes.

There is a hardware presence to software in the form of the physical devices into which the software is installed. For embedded software, this is the form of an EPROM (erasable programmable read-only memory) or EEPROM (electrically erasable programmable read-only memory). The EPROM contains a window for which the program for running the device can be installed or erased. A metallic-based sticker is often used to cover the window after the programming is completed to protect the contents from being erased accidentally. There is a limit to how many times an EPROM can be erased and reprogrammed, and this number of reprogramming cycles depends on the cost. Low-cost devices may only be reliably reprogrammed a few times, whereas higher-cost devices may be reprogrammed on the order of hundreds of times. Excessive erasures and reprogramming beyond the capability of the device may result in the failure of the EPROM.

Another issue is the time the program in the device will stay as originally programmed. Most programmable devices have a defined program retention time, usually specified in the data sheet. Some low cost devices may only maintain the program in them for five years; better devices will last for fifty years or more. The lifetime will be affected by the applied voltage, temperature, and other factors. With software, the changing of only a single bit in a program as large as 64-K bytes can yield catastrophic results.

A well-designed system will, on power-up, calculate a procedure known as the checksum and check to see that it matches the original checksum that was programmed into the device initially. The checksum is the hexadecimal addition of all of the bits for a program, and it should always match between EPROMs if the same program and loading process are used. Since a single incorrect bit resulting in a different checksum will be detected, this test will verify that the programmable device has not had a failure. Normal system response with a checksum fault should be to send an error flag to the output and then stop. However, adding checksums takes time, effort, and code space. During development, self-tests like checksums are often disabled. With all that goes on during the hectic last-minute changes readying a product for production, enabling the self-tests may be forgotten, or purposely omitted to meet a delivery date. Sometimes the code grows too big for the memory available, so the first thing to go is some self-test code. The reliability engineer can work to insure that these self-test features are included and are tested for proper operation. As the EPROM is a hardware device, it can fail just like any other component with regards to hardware-type failure modes.

Another feature often used in software applications is the watchdog timer. Watchdog timers can be hardware-based, software-based, or both. The basic idea is that a timer runs for a prescribed time, and if the time ever runs out a system reset is forced. The hardware approach uses a re-triggerable one-shot or the equivalent timer logic that must be reset by software. Special integrated circuit chips (IC's) are available to perform this function independently of the microprocessor and its software. Under normal operation the software constantly re-triggers the watchdog timer, allowing the software to continue to run. The watchdog timeout duration is set relative to the speed needed by the system. For most electronic devices, the trigger rate will be in the order of 50 mS to a few seconds. For high-reliability systems where failure may result in loss of life, there should be at least two independent watchdog systems; so if an unanticipated event disables one watchdog, another is available to reset the system. The reset system should default to a safe condition.

Applications programs are installed into a hardware device, typically known as a disk or hard drive. The hard drive is electronically altered to retain the bits for a software program. The application software makes use of error reporting and system messages.

When investigating software failures, issues tend to be with the coding or with timing. In more advanced systems, generally some sort of event recording or log of important messages is usually in a software file created by the application. If the reliability engineer is fortunate to have access to this type of log, it can be very useful in the area of troubleshooting software-related failures. Some of the case studies presented in this chapter provide a detailed look into products that use software and some of the means for troubleshooting failures.

12.2 SOFTWARE IN GENERIC CUSTOMER TERMINAL DESIGN

Customer terminals are used in great numbers and can take on many forms: automatic teller machines (ATM), post office stamp dispensers, gas pump terminals, ticket dispensers, etc. A study of the generic customer terminal design can reveal how the different types of software applications are used and work together. The infrastructure of the customer terminal software requires that the engineer have a good working knowledge of how the different modules work together when reliability issues occur.

The major hardware devices that comprise most customer terminal designs are the card reader (or key pad), the computer, deposit unit (for taking cash in a stamp dispenser or for receiving envelopes in an ATM), cash/stamp/ticket dispenser, and receipt printer. Each device has a control board or control circuitry that has embedded software in the EPROM that will provide the functionality of the device as well as allowing it to communicate with the computer.

Application software is typically loaded into the hard drive of the customer terminal computer. This application software is basically the operating system, and it allows the computer to communicate and control each device. In addition, the software allows the computer to flag errors that occur in any of the customer terminal devices. These errors may be capable of being transmitted either on the customer screen or a maintenance screen (the latter being available only to the maintenance personnel when they arrive to troubleshoot the failure on the customer terminal). The level of capability for the ease of troubleshooting is determined by how much programming was performed in the application software.

Support software is usually in the form of a network that communicates directly with the customer terminal computer. One application of this will allow a customer terminal such as an ATM to access account information and the amounts available. In this case, from the customer's card number that was read, the software determines from internal tables if there is a surcharge to be applied (for a user who is not a customer of the financial institution that owns the ATM).

Any errors that are flagged by the computer of the customer terminal are recorded by the internal software and may be transmitted over the network via the support software that is loaded into the monitoring system computers. Monitoring may be performed by a help desk or support group on a daily basis much in the same way alarm systems are monitored. Figure 12–1 shows the basic software design of the typical customer terminal and the different software items involved.

Because of the complexity of the software, it is not surprising when reliability issues arise between the different modules of software that are used in customer terminals. Certain failure conditions can occur when different software processes collide (as discussed in the case study in 12.2.1), and these can cause major failure conditions. Another concern is

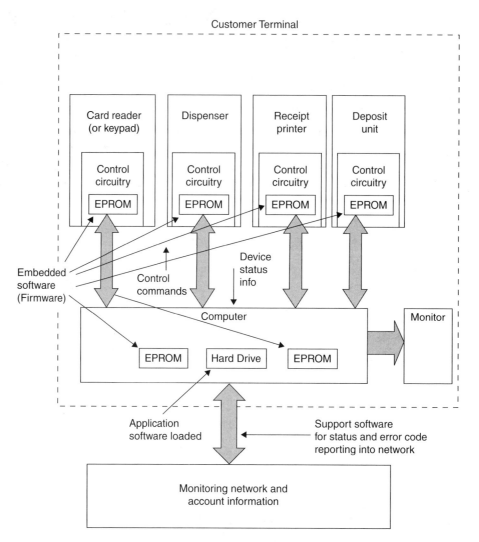

Figure 12-1 Generic Customer Terminal Software Design

that correct failure code information is being passed along in the monitoring network. Also, software can be used to help correct certain failure conditions.

Other issues involve the hardware interface between the devices and the software world. For example, the mechanical devices, such as the cash dispensers, use a flag wheel that either has cutouts or marks that allow a mechanical count to be made on the movement of gears or rollers of the dispenser. A sensor detects changes in the marks or cutouts and feeds this information back into the microprocessor that controls the overall unit software. Quite often, perceived software errors are really situated in the area of flag wheel/sensor interface between the mechanical hardware world and the software world. It becomes a challenge for the engineer to isolate and then correctly identify the true source of the error.

Indeed, when new changes are made in the software that are based on a particular configuration of the hardware interface, there is a potential for unanticipated failures to come into play. There are also concerns whenever there are changes in the electronic circuitry of a product that uses software, particularly if there are value or timing changes. Again, this is an area engineers have to be involved with whenever hardware change takes place in the hardware design. Thus, a watchful eye has to be maintained whenever software revisions are incorporated into a product.

Another reliability issue of concern in customer terminal software is in the area of data handling capacity by the monitoring network. The monitoring or reporting software has to be capable of handling significant amounts of data that are transmitted from the customer terminal as transactions are performed. Concerns regarding capacity can arise when additional customer terminals are added to the network, so it is generally a wise idea to perform feasibility testing prior to any expansion of the network. It is necessary to verify that the additional customer terminal data will not overload the network, particularly if there are time periods with high customer terminal usage and demand (such as a holiday weekend). Failure to perform this feasibility study and testing prior to implementation may result in the unpleasant surprise of a network crashing with individual customer terminals not being available to the customer.

The areas discussed here usually fall into the responsibility of the software quality engineer with the reliability engineer playing an important part by conducting the FMEA analysis along with tests being performed. The effort does not stop there as the reliability engineer will become involved in field failure investigations and trade studies involving software reliability issues.

12.2.1 Generic Customer Terminal Software Case Study #1

ITEM: "Lockup" condition or "Frozen" state

As the customer terminal design becomes more and more software intensive, the timing of the processes becomes extremely important. There is a concern when individual software processes overlap, thus making collisions of these processes possible. When collisions of processes occur, a "lockup" condition could occur on occasion, and the customer terminal would become "frozen" or stuck in one display screen. This is similar to when a personal computer gets "stuck" or nonresponsive. When the customer terminal falls into this condition, it will not communicate to the network, and the maintenance personnel are unable to go into diagnostics on the maintenance panel. In addition, there is usually no way for the network monitoring to flag the fact that the customer terminal is in this state and not available to customers.

After reviewing the event logs of an affected machine in our case study, it could be seen that no activity was occurring during the time that the customer terminal was in this frozen state. Essentially, this condition was undetectable to the monitoring network.

The only way to exit from this particular failure condition was to turn the entire customer terminal off and restart the computer. A major investigation was conducted into the problem, and some special logging software tools were used where it showed at what point the failure occurred. The analysis showed that two processes (or "threads," as such processes are often called) were colliding in the customer terminal where there was an overlap of timing sequences, and once stuck, there was no path for the software to follow to exit out of a locked-up situation. For example, if the computer was running a special software application, and a customer was inserting a card at the same time, this could cause a "lockup" situation within a specific timing window of opportunity.

The problem proved to be too complex to solve by any simple change in the timing of the processes. It was observed that adding some enhanced logging features altered the timing such that the event did not occur as frequently. Eventually, the manufacturer's software department of this terminal developed a simple test in the application software to see if this condition occurred, and if the test failed, the software would trigger the computer of the customer terminal to reboot.

This case study demonstrates how two different processes in the application software can actually collide and cause a software-type failure. When a lockup condition occurs, software—specifically application-type software—should be suspected first as the culprit.

12.2.2 Generic Customer Terminal Software Case Study #2

Item: Using software to correct certain field failure conditions

With the advent of more control features of physical devices by application software (sometimes known as physical device handlers), there are more capabilities for using this application software to fix certain field failure conditions of the product in which the software is housed.

Many devices typically have an internal initialization procedure when first turned on. As a customer terminal is a system that has a collection of individual devices, the application software can turn on these devices whenever the software is rebooted or restarted. Some individual devices have features that can clear certain failures. For example, the receipt printers of generic customer terminal designs have their own imbedded software that controls the printing of receipts and the transporting of these receipts to the exit area where customers can take them. A common problem regarding printing records is that they can get stuck in the transport area and create a jamming condition. Some of the printers used in generic customer terminal designs have correcting features for these jams.

In this particular case study, the help desk that monitored a network of customer terminals noticed that if they observed a jammed condition on the printer, they could clear the problem over the network by rebooting the customer terminal. It seemed that when they reinitialized, any printer that reflected a jam condition previously would now be cleared. Thus, a manual practice evolved in which the help desk would manually send a remote reboot command over the network to the address of the affected customer terminal, and this action would be used to clear jammed printers and other error conditions with devices in the customer terminal. However, the problem was clearly a work-around situation, and it resulted in a lot of confusion with regards as to how effective it was; it also affected dispatching of field engineers and branch personnel to fix the problem.

An engineer investigated this problem and simulated a jam condition on a printer of a lab unit that would be picked up by the software. It took five minutes to reboot the customer terminal, as the software was reloaded and each device of the customer terminal was initialized. Upon initialization of the printer, a self-test procedure started by the firmware of the printer was performed. One of the tests that was performed in this procedure was an ejection test, in which the rollers in the exit area of the printer were cycled several turns along with the ejection mechanism being exercised. By observing this initialization process for the customer terminal printer during the reboot process, the reliability engineer could see that the self-test was able to clear most paper receipt jams in the exit area. This self-correction happened to be coincidental to the original purpose of the printer self-test. However, this self-correction process was noticed by many of the nontechnical people of the bank, including the help desk, and they would reboot the customer terminal from a remote location outside the branch in order to try to clear paper receipt jams. They did this even though they never knew the precise mechanical reasons as to why the jams were cleared.

The process of performing this reboot action from a remote location was labor-intensive, and in the cases in which it was not successful, it would cause extra delays in dispatching either field engineers or branch personal to manually clear the jam condition of the printer.

The reliability engineer recognized that it would make sense if the software developers of the customer's terminal software could incorporate the printer initialization self-test process whenever a jam condition was detected. By making the software capable of performing an automatic recovery process of sorts, it would reduce the need for most manual intervention. The only time that the help desk would be involved would be in case the automatic recovery process did not clear the fault after one try. The reliability engineer wrote up an extensive trade study and reliability analysis that by incorporating this recovery process, all sorts of failures, downtime, and repair time would be reduced. This trade study had to go through a thorough review by management, who, although not convinced totally of the projected savings, still liked the idea of labor-time reduction.

As the software for the customer terminal is a complicated package, it took a few months for the software developers to incorporate this change. After the change was installed on all of the customer terminals in the network, a reduction of the printer jam error condition was realized to the tune of over 80 percent reduction in failures. In addition, there was a savings of several hundred hours of downtime on the printer along with reductions in the man hours being expended by both the field and the help desk to correct the failure condition. By having the software perform an automatic recovery, all of these savings were realized.

The lesson learned here is that some of the embedded software inside devices may often have initialization procedures that have the ability to fix certain failure conditions. An alert reliability engineer can spot these opportunities to use self-correcting features when the device initializes. The engineer can then make a strong case for implementation of these features through the use of reliability-oriented trade studies that show improvements in the way of failure reduction and the reduction of servicing time.

12.3 SOFTWARE IN AIRCRAFT APPLICATIONS

With the advent of EPROM and embedded software technology, it is standard practice for both military and commercial aircraft equipment to make significant use of embedded software in many of the aircraft subsystems. Several subsystems such as engine monitoring and fuel-quantity displays make use of embedded software. This software is used for both controlling vital functions of the equipment, as well as for the reporting of errors into the aircraft monitoring system (typically the data bus). In addition, the reporting of errors by software can be used during the repair process to adequately isolate and troubleshoot the failure. The reporting of errors at a local level is done by the use of built-in test (BIT) software that requires additional components situated at various points in the design to aid in this feature.

The reporting of individual internal errors is available on both the local level and often on the system level. Both military and commercial aircraft make use of a data bus where information is collected, such as the up or down status of each device that feeds into the system, along with any relevant error codes. Depending on how the software is written and the severity of the error code, some of these failures may be reflected to the flight crew; otherwise a less-serious failure may be read from the data bus to the ground maintenance crew when the aircraft is checked after flight.

The introduction of software into aircraft equipment has created a major field of software engineering as well as a whole host of reliability issues. Reliability issues in this area center around the effectiveness of the BIT features, as well as any errors that are inadvertently brought into the software during the coding process. Sometimes errors of interpretation are introduced based on an erroneous assumption by the software engineer about the customer or design engineer requirements (an example is provided in section 12.3.1).

There is an occasional tendency for design engineers to try to increase software functionality as a "magic bullet" to solve all major design issues. This can unfortunately complicate simple situations and create more problems that may not be able to be solved. An example of this was provided in Chapter 9 involving the adjustable backlighting of a cockpit display in an air force fighter aircraft. Again, the basic goal of reliability of keeping it simple still applies when using software as an integral part of the design.

12.3.1 Aircraft Software Case Study #1

Item: Navy fighter aircraft engine cockpit display unit

A pair of test units of LCD-type engine display units underwent reliability development testing (RDT). During thermal cycling, the same failure symptoms would consistently appear after cold soak for one hour at $-54°C$, at the point that the temperature was ramped up to $27°C$. The failure would manifest itself when the BIT message on the LCD would be illuminated for one minute and then stop illuminating and then repeat this process one or two more times. As the failure appeared in both units, it was apparent that it was not a random part failure but rather a problem with an aspect of the design.

Failure investigation revealed a problem with the way that the unit's software was written. The failure condition was traced to the on-and-off cycling of the two thermostats used on a circuit card for warming the LCD display during cold temperature. The BIT circuitry that monitored the heater circuitry consisted of three distinct circuits: two individual thermostat heater BIT circuits and the heater watchdog circuits that monitored the status of the two-thermostat BIT circuits. A latent software error in the unit's original software program monitored all three circuits and would indicate a fault condition if any one of the three circuits indicated an "on" condition for more than three minutes. However, as the effects of temperature stabilization at 27°C from the ramp-up from cold was still occurring, one thermostat was cycling on and off while the other thermostat remained closed. It was later determined that this was not a true failure condition and that the software had to be revised to monitor the heater watchdog circuit only for any switched-on state remaining for more than two minutes, and then flag a BIT error. The other circuits would be used only for intermediate maintenance.

The error was the result of an erroneous assumption that was made by the software engineer when the programming was being written as to how the BIT logic was supposed to work. This particular error would have only been discovered during thermal testing (when the thermostat circuits were activated) and the reliability test that was conducted in this example was successful in unearthing this failure mode, resulting in a corrective action change of the software.

12.3.2 Aircraft Software Case Study #2

Item: Business jet fuel quantity system signal conditioner

As discussed in the beginning of this section, one of the subsystems of an aircraft that makes extensive use of software is the fuel-quantity gauging system. The fuel quantity gauging system of an aircraft requires these main items: the fuel probes, a display unit for the cockpit and for ground refueling along with a signal conditioner (or computer unit). The signal conditioner is used to translate the analog voltages received from the fuel probes into data that can be read on a display. Aircraft fuel-quantity systems may use either the capacitance method or ultrasonic method to measure fuel in a tank. When a fuel system is developed initially, the design engineers must determine a table of precise voltage values that correspond with the associated fuel value in the tank. This is based on analytical data and early testing in the lab. The voltage values are then processed inside the signal conditioner unit by the EPROM/microprocessor combination into useable data values that are sent out on the data bus and can be read by the cockpit display or refueling display unit. Please refer to the functional block diagram in Figure 12–2.

A manufacturer of fuel-quantity gauging systems followed the same process by developing a table with a set of values for a fuel-quantity system on a business jet. The process started with testing in the software lab during the prototype stage, and then into the customer's test aircraft when all sorts of problems were discovered. None of the fuel values read correctly, and the business jet manufacturer was very angry and demanded that an engineer from the vendor be assigned on site full-time for two weeks until the problem was fixed.

The problem was caused by a number of factors that were unforeseen by the vendor, most noticeably some miscalculations concerning the size of the tanks and some other field environmental factors not previously considered. It took a number of man-hours to complete, but a revised set of values for the software table had to be created for the signal conditioner's software to process correctly.

This example shows the need to test software in actual real-world situations after it has been developed in the lab environment and before full-scale production begins. When a customer offers this opportunity prior to actual flight testing, it is in the company's best interest to install a unit in an aircraft or system to test out the software.

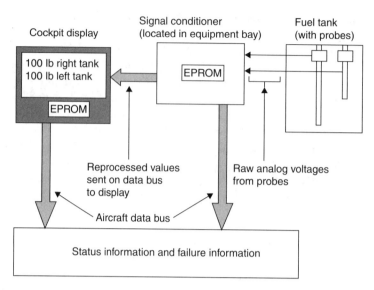

Figure 12-2 Aircraft Fuel Quantity Gauging System Functional Block Diagram

12.4 MEDICAL EQUIPMENT AND SOFTWARE

A walk into any hospital will reveal a world of electronic equipment that is driven by software used to monitor and control patients' critical life functions. Most of this equipment has CRT, LCD, and LED displays to interface with technicians, nurses, or doctors. As in the customer terminal example described earlier, these units contain embedded software, application software, and support software.

Medical equipment reliability is a major field that continues to grow. Along with this growth, there is a heavy dependence on the equipment design for software. Those engineers charged with the responsibility of ensuring high reliability in the equipment design must have knowledge of potential software issues that can arise. In addition, within the software that runs the unit, many items have built-in test software that can flag failures inside the unit.

Hospital and medical organizations have a significant responsibility to select reliable equipment that has some self-test capability. The implementation of straightforward self-test or built-in test circuitry does not necessarily raise the cost of the unit that much. Self-test circuitry will have sensors and transistor switching as part of the design and this does not require expensive components. There may be a dedicated microprocessor and EPROM combination with program code for self-test and failure indications. Insurance companies may insist that the medical organization use equipment with self-test features; certainly doctors should insist on using equipment with the best quality and features. This is not an area for companies to take the cheap route—medical equipment should have these features for both patient safety and user safety. Patients are most vulnerable when they are in the hospital and hooked up to electronic equipment that is regulating and monitoring the patients' major life functions. It is desirable to have virtually 100 percent reliability for medical equipment.

12.4.1 Medical Equipment Software Case Study #1

Item: Eye pressure unit for cataract operation

The following is a very sad story that illustrates why it is so important to have some type of self-test or BIT feature in medical equipment that can spot catastrophic failure situations.

A company that manufactured both aircraft equipment and a line of medical equipment made a unit that was used to maintain the pressure in the eye during cataract operations. This equipment design incorporated a series of pumps, pressure lines, and electrical relays that turned the pumps on and off. For such a delicate medical procedure, it seemed strange that the manufacturer never considered the use of BIT circuitry in this design for added safety, but the manufacturer did not do so because of concerns about overall costs.

In addition to the streamlined design, the company did not perform rigid quality checks on production units (i.e., a formal fault-tree analysis or FMEA on the initial design). Thus, there was an opportunity for a major failure situation to occur and no way to check for it. As Murphy's law would have it, this happened on the production line when a worker in the manufacturer's plant installed a relay incorrectly in the unit. The relay, even though installed wrong, still appeared to be working.

This cataract pressure unit went to a clinic and was used for a cataract operation. When the doctor used the unit to maintain the eye pressure, the unit worked in reverse; instead of maintaining pressure, it actually sucked fluid out of the eye to the horror of the doctor. Needless to say, the patient lost sight in the eye, and a major lawsuit was instituted against the equipment manufacturer.

One could have made a case for some sort of self-test to be implemented in the design. The technology was there to do so, but the management elected to go low cost all the way when developing the unit. It seems sad that for only a few dollars of added circuitry and software, this tragedy could have been avoided, saving millions of dollars in legal costs.

Another lesson learned from this example is the fact that many commercial companies, such as those in the medical field, do not know of the existence and benefits of having some aspects of reliability engineering available. Items such as the FMEA, while not required by the customer, should be performed as it adds a level of confidence in the design. Also, BIT is another feature that should be considered as mandatory in this particular field!

12.4.2 Medical Equipment Software Case Study #2

Item: X-ray machine exposure duration

Modern x-ray machines offer automatic exposure duration. The duration is calculated based on sensors below the x-ray film to determine correct exposure. The sensor data is typically amplified and converted to digital via an analog-to-digital converter and then read by a microprocessor. The microprocessor runs an algorithm combining operator setting for high voltage, x-ray current, and/or the sensor data, to calculate the duration of the exposure. The x-ray technician initiates the exposure, and the microprocessor decides when to terminate the exposure.

With all the sensors, software, and hardware in the x-ray machine, many things can and do go wrong. If, because of a fault, the x-ray does not terminate, the patient could be exposed to a very high (or even lethal) dose. This has happened when companies do not have an extra level of redundancy for this catastrophic failure mode. To insure patient safety, at least two independent ways to limit exposure duration should be incorporated into the design. Typically this can be accomplished if a software watchdog timer monitors the control loop, and a maximum exposure hardware timer trips if the software watchdog should ever fail. In this example of an x-ray machine, the reset will be in the mode of "x-ray off" in the event of a failure.

A cost is associated with adding this change into the design, primarily in the engineering involved to incorporate the hardware and software aspects. But considering that the payoff is in the area of patient safety and less risk of litigation, it would appear that in both examples regarding medical equipment presented here, it makes commons sense to incorporate these features!

12.5 AUTOMOBILES AND SOFTWARE

All modern automobiles have a software module that provides vital engine controls. In addition, there are software modules for the instrument panel and various annunciator lights located on these panels. Strict environmental laws have required that environmental sensors be used to detect the quality of the emissions coming from the automobile. In some cases, the computer in the car can also control some aspects of the emissions. All of this is done by software and microprocessor technology.

As in the other products discussed previously in this book, the automobile requires inputs from sensor devices to provide status into the microprocessor. Depending on the values that are provided, the software guides the system into certain decision factors. Like other products discussed previously, a data bus inside the automobile causes certain faults to be reflected to the driver in the form of a warning light. Specific error codes for these warning lights may be read by the mechanic when the computer is interrogated.

12.5.1 Automobile Software Case Study

Early automobile computer designs had growing pains with regards to field performance. One case was a compact car that had some unusual failure symptoms when car reached about fifty thousand miles.

The car would be able to start without difficulty, but after fifteen seconds, the engine would stop working. Initially, it would appear the problem was related to a fuel flow problem, such as a malfunctioning fuel pump or clogged fuel filter. But by pulling off one of the spark plug wires and observing the spark, it could actually be seen that the spark itself ceased at the fifteen-second mark. Even a failed distributor system could probably not cause this type of failure at exactly the same time after the engine was started.

Suspicion started focusing on the computer module that controlled the ignition to the spark plugs. This was probably the only item in the system that had this capability to cut off the ignition in this particular fashion (after 15 seconds). The module was replaced, and the engine worked normally after this. Examination of the failed module did not reveal any external electrical damage, although it was noted that the interior of the module was packed with a gelatin; this led the car manufacturer to be concerned with environmental factors (water and engine vibration) affecting this module. The module itself had a microprocessor and EPROM.

Since the introduction of small computer modules, larger on-board computers located in automobiles are handling more and more functions. These computers have a hardware presence and, unlike home computers, have some additional concerns regarding the environment in which they are operating, where wider temperature ranges and higher vibration levels could cause stress. Thus, packaging to reduce these effects on the on-board car computer is a major concern in the design.

12.6 AMATEUR RADIO TRANSCEIVER AND SOFTWARE

Another area where software has made a significant impact is in the area of amateur radio design. Prior to the development of embedded software technology being used in radio designs in the early 1980s, radios were moderately large with weights on many units averaging about thirty pounds or more. Software has not only simplified amateur radio designs in terms of the number of parts that are used but has also increased the functionality of radio designs through expanded function menus. Current designs have many features that could never have been implemented easily in the past without tremendous amounts of electrical circuitry.

12.6.1 Amateur Radio Software Case Study

An important factor in amateur radio usage is the correct matching of the radio to the antenna and the feed line. In the past, a high standing wave ratio (SWR) of over 3 to 1 indicates a mismatch between the antenna and the feed line and this could cause serious component damage in the final stages of the radio, as the power is not being efficiently transferred to the antenna via the feed line. A poor SWR reading can be caused by an error made by the radio operator in terms of the antenna design or some failure condition of the antenna (broken wire, corrosion in the connector of the feed line). When a high SWR reading occurs, the RF power is being reflected back into the radio, and this causes high stress on the components. If the high SWR is ignored and transmission is continued, components will eventually fail and repairs can be quite extensive.

With more software control features, new radios now have the feature of software-controlled shutdown in the event of a high SWR being sensed by the radio. In addition, an icon shows up on the LCD screen indicating "High SWR," while at the same time that no power is allowed to be transmitted from the radio.

One of the drawbacks to increased software capabilities is that it is currently impossible for most radio amateurs to troubleshoot their own equipment if the problem lies within the software module. This was a hobby that grew out of radio amateurs being able to build and repair their own equipment. However, the increased desire for more features and functions has overridden the previous goal of being able to repair equipment easily with the implementation of proprietary software modules into radio designs.

The advent of technology has brought software capabilities into the area of handheld amateur radio walkie-talkie units. But in addition to the amateur radio field, there is tremendous usage in the consumer market, as shown in the next section.

12.7 HANDHELD ELECTRONIC UNITS

Many portable handheld electronic units have LCD displays, and they use embedded software. In addition to radio walkie-talkies, there are calculators, pocket organizers, cellphones, global positioning system (GPS) handheld units, and the like. All of these units have embedded software that is burned into the EPROM of the device. Some of these devices may actually have an interface port for accepting external software downloads. For example, one can download a file of names and addresses from a PC into an organizer. Another case is where the current phone listing for a cellphone network can be downloaded through the external interface port.

12.7.1 GPS Handheld Unit Case Study

A handheld GPS unit has an antenna that receives signals from a network of twenty-four satellites orbiting about eleven thousand miles above the earth. When the GPS unit is able to get a lock on a minimum of three satellites, it is able to determine where one is on the earth in terms of longitude and latitude. Each satellite is assigned a number by the GPS and is typically displayed on a miniature map on the LCD display.

A reliability problem concerning some GPS units is in the area of the battery power. When a GPS unit is not used for a while and the battery runs down to no voltage, the unit does not automatically work when a new set of batteries are installed. What happens when this failure condition occurs is that the GPS has to relearn its settings, and the unit starts from the beginning by initializing from satellite number one. What is particularly bad about this condition is that it takes up to five minutes to search for each satellite in incremental order, and a whole set of batteries may be used up to complete this initialization process. Another problem is that there is no caution note in the owner's manual about this condition. The condition is avoidable by periodically turning on the GPS to renew its last setting.

Some types of handheld appliances may have a backup battery to save certain settings, but many do not. A car for example does not have a backup battery in the radio, so when the twelve-volt battery in the car is removed and replaced, the clock functions in the radio have to be reset manually by the owner. Initialization failures related to loss of battery voltage would be best addressed by caution notes on the unit and in the owner's manual.

12.8 SUMMARY

We have discussed how software has expanded the functionality as well as the reliability of many products. Reliability issues can arise when new software is introduced, but a well-prepared engineer can adequately address them. There are some major milestones for verifying that the software works correctly, including the initial design stage and during field service.

What determines whether a piece of equipment has self-test or BIT software capability to check internal failure modes? Often, this is a customer requirement, particularly in the area of military equipment. In the commercial field, self-test is a major selling feature. For example, in the area of printers hooked up to computers, error messages are discerned by internal software and presented to the user via the LCD display of high-end models and LED indicators for the less-expensive models.

A number of engineering organizations have invested significant efforts into developing software reliability models, but in terms of practical analysis, the topics covered here should provide a good set of tools for a reliability engineer to use in any new product line that has some sort of software capabilities.

The reliability engineer can play a very important role with regards to software functionality when a new product is being developed. The functionality can be verified with a thorough FMEA being performed on all potential failure modes. In addition, the engineer can conduct reliability testing where failure modes are introduced into the product and software reaction is observed.

Witness the case of the Marine Corps V-22 Osprey tilt rotor aircraft that crashed in December of 2000. The initial failure that caused the crash was a rupture in the hydraulic line caused by the effects of chaffing. However, a weakness in the software made it harder for the crew to control the aircraft after the rupture occurred. Officials investigating the crash recommended that a complete review of the software be performed, along with possible redesign of the hydraulic system. It would appear that a more in-depth FMEA would have to be developed where failure modes are actually simulated and subsequent software actions are observed, as opposed to just paper analysis being performed.

This particular case illustrates the difficulty that software programmers have in anticipating product failures when they are actually deployed into the field environment. It is difficult for the programmer to test certain scenarios using just a small software lab environment. The software engineer may actually be in a vacuum or isolated environment with respect to the end effects of a particular failure mode. Witness the frozen screen problem that affected the generic customer terminal in the case study presented in 12.2.1.

Thus a reliability engineer becomes a critical part of the team by the timely development of the FMEA document as well as a checklist of critical items that should be tested in either a test lab or in a field test. Thus, it is not important that the reliability engineer actually knows the individual coding of a software program for a product but rather the different ways to thoroughly test the software as part of the exercise to properly complete the FMEA or similar-type reliability checklist analysis. This would involve working with the software engineer, as well as understanding the functional description of how the software is designed to work in the unit.

12.9 EXERCISES

1. What are the three types of software that can be present in a piece of electronic equipment? Which one of these types has a hardware presence?
2. What is the purpose of performing a checksum of the software used in a product design during the power-up cycle of a product?

3. Describe how a lockup or frozen-state condition can occur with software. What type of product designs might experience this condition?
4. Describe the types of devices that control how software is able to interface with the hardware in order to coordinate commands.
5. Identify electrical appliances and electronic units that you use on a regular basis that use embedded software. What functions are most likely controlled by internal software?
6. Select a major electrical appliance in your house that you can open easily. Can you locate the EPROM that contains embedded software on one of the circuit boards? Are there any ports on the circuit board or in the rear chassis of the unit that is used for testing purposes during troubleshooting?
7. Can you identify a product that should have a watchdog timer to prevent unsafe operation? How is the fail-safe mode of the watchdog timer set up?

CHAPTER 13

Maintainability Concepts

INTRODUCTION

It is very important that the reliability engineer be aware of some basic maintainability concepts since reliability and maintainability overlap in many areas. As reliability values are used to calculate many maintainability values, it becomes important for the reliability engineer to realize the importance of MTBF values that are calculated. Many of the calculated MTBF values are used in the maintainability field in the various maintainability calculations that are used to determine availability and spares.

A number of other concepts will be discussed in this chapter, such as the idea of preventative maintenance, along with a few real-world situations that illustrates its importance. Also, this chapter will illustrate examples regarding how a new product can accommodate a design that is easy to maintain.

13.1 MAINTAINABILITY TERMS

There are many acronyms associated with maintainability, more so than what is used in reliability. The major ones will be covered in this chapter along with the subtle differences between terms.

The following terms are the general definitions for maintainability terms referenced in MIL-STD-778, "Maintainability Terms and Definitions" and are used as a guide.

13.1.1 Mean Time to Repair (MTTR)

The mean time to repair (MTTR) is the statistical mean or average of the distribution of time to repair. The MTTR value is calculated by taking the cumulative totals of active repair times during a specific period of time and dividing it by the total number of malfunctions performed during this period of time. Mathematically, MTTR is expressed as:

$$\text{MTTR} = \sum_{i=1}^{n} Ti / n$$

where n = the number of malfunctions that are repaired and each Ti value represents the time to repair the ith malfunction.

MTTR EXAMPLE

During a one-month time period, a repair shop serviced five electronic units that were removed from aircraft belonging to a major airline. The times to do the total repairs for the five maintenance actions were 2 hours, 2.5 hours, 1.8 hours, 2.1 hours, and 2.2 hours. What is the MTTR?

The solution to this question becomes a simple average as follows:

$$\text{MTTR} = (2 + 2.5 + 1.8 + 2.1 + 2.2) / 5 = 10.6/5 = 2.1 \text{ hours}$$

Typically, each session of repair time can be further broken down into the following tasks:

1. Preparation
2. Localization
3. Isolation
4. Disassembly
5. Interchange
6. Reassembly
7. Alignment
8. Verification or final check-out

Using the above list, the time for disassembly is usually equal to time for re-assembly because the same components are involved. Some of the individual times for performing an individual maintenance task can be derived from using standards such as MIL-HDBK-472, a standard used for many years, in which times are presented for routine tasks such as removing a screw or bolt. Some organization may elect to do an industrial engineering-based time study of various tasks to come up with MTTR values for specific items and programs.

13.1.2 Mean Downtime (MDT)

The term downtime represents the time that equipment is not functioning satisfactorily. The mean downtime (MDT) is the average downtime during a specific time interval when repairs are being accomplished. This includes supply and administrative downtime during the same time interval as well as response time (for field service). This is calculated as simply:

$$\text{MDT} = \frac{\text{Total downtime}}{\text{number of actions}}$$

13.1.3 Mean Time Between Maintenance (MTBM)

The mean time between maintenance (MTBM) represents the average time between maintenance plus the ready time during the specific time interval. When preventative maintenance downtime is equal to zero, MTBM becomes MTBF. Another term that is close to this definition is mean time between unscheduled removals (MTBUR). Some organizations may use the term mean time between corrective maintenance (MTBMc).

13.1.4 Inherent Availability

The inherent availability is the probability that a system or piece of equipment when used under the specified conditions operates satisfactorily at any given time. This is done without consideration for scheduled or preventative maintenance or that there is an ideal support environment where tools, parts and manuals are available. The inherent availability of an item is based on both the MTBF and MTTR values that we have defined previously and is expressed as:

$$A_i = \frac{\text{MTBF}}{\text{MTBF} + \text{MTTR}}$$

INHERENT AVAILABILITY EXAMPLE

What is the inherent availability of an aircraft display unit that achieves an inherent MTBF of fifty hours and has an inherent MTTR of five hours?
This is calculated as:

$$A_i = 50/(50 + 5) = 90.9 \text{ percent availability}$$

13.1.5 Operational Availability

The operational availability is the probability that a system or piece of equipment when used in an actual supply environment shall operate satisfactorily at any given time. The operational availability of an item can be expressed in a couple of different ways.

A. One way to calculate operational availability is based on both the MTBM and MDT values that we have defined previously and are expressed as:

$$A_o = \frac{\text{MTBM}}{\text{MTBM} + \text{MDT}}$$

B. Another way to calculate operational availability is based on number of down hours and total operating hours for a specific time period. It can be expressed as:

$$A_o = \frac{\text{Up time}}{\text{Total Operating Hours}}$$

or as:

$$A_o = 1 - \frac{\text{Downtime}}{\text{Total Operating Hours}}$$

OPERATIONAL AVAILABILITY EXAMPLE #1

What is the operational availability of an ATM product line that achieves an inherent MTBM of eighty hours and has an operational MDT of three hours?

This is simply:

$$A_i = 80/(80 + 3) = .964 \text{ or } 96.4 \text{ percent availability}$$

OPERATIONAL AVAILABILITY EXAMPLE #2

What is the operational availability of a group of five ATMs at a branch location for a one-week period where there were a total number of twenty down hours for the location?

This is calculated as:

$$1 - [20/ (5 \text{ ATMs} \times 24 \text{ hours} \times 7 \text{ days})] = 1 - (20/840) = .976$$

or 97.6 percent availability

13.1.6 Mission Capable Rate

This is a maintenance measurement that is expressed in percentage and it indicates the flight or mission status of an aircraft or other systems. Four primary terms are associated with this measurement:

- Mission capable (MC)
- Fully mission capable (FMC)
- Partially mission capable (PMC)
- Not mission capable (NMC)

For both military and commercial aircraft, there is a minimum equipment list (MEL) of what equipment is needed for the aircraft to be able to fly a certain mission. This equipment will be checked out by maintenance crew and the pilot prior to takeoff. Any equipment that is on the MEL and is not functioning for the mission that is to be flown will cause the cancellation of the flight and result in the aircraft being in NMC status. Otherwise the aircraft is in fully mission capable (FMC) status. Equipment that is not functioning but not on the MEL will allow the aircraft to be flown, but the aircraft is in PMC status and this is calculated into the MC rate. The equipment will be assigned as an action item for repair when the aircraft is back on the ground. All essential equipment that is functioning will keep the aircraft in PMC status. Any action that prevents leading to the ground abort of an aircraft will prevent the aircraft from flying and relegate it to NMC status. Other failures that need to be corrected but otherwise will not prevent the mission from occurring would be factored into the PMC rate.

The concept of mission capable or a minimum equipment list needed for a mission is not necessarily confined to military or commercial aircraft. These concepts could easily be translated to other areas as well. For example, a bank may define the ATM status as mission capable or "up" status if it is able to read customers' cards and give out cash. The receipt printer that gives a printed receipt on the transaction may not necessarily be considered by the bank or operator of the ATM to be mission essential for customer success. A car is another case. Obviously a working engine or transmission is needed for a car to travel where as a burned-out bulb in the tail light may be tolerated for a single trip until it can be fixed.

MISSION CAPABLE RATE EXAMPLE

During Operation Desert Storm in 1991, the A-10 fleet of 153 aircraft flew a total of 8524 sorties (missions) and had the following events affecting mission status:

- 9,912 planned missions
- 8,524 missions that were MC
- 8,454 missions that were FMC
- 1,387 missions that were NMC
- 69 missions that were PMC

From this information the following mission capable rates can be calculated:

- MC rate = 8,524/9,912 = 86.0%
- FMC rate = 8,454/9,912 = 85.3% FMC
- NMC rate = 1,387/9,912 = 14.0% NMC
- PMC rate = 69/9,912 = 0.7% PMC

13.2 MAINTENANCE CONCEPT

The maintenance concept defines the criteria covering the repair philosophy of the equipment and the definition of the maintenance support levels. Typically, the maintenance concept for a new product must be done during the early stages of the program. One of the most important tasks is to define the levels of maintenance. There are three basic levels: organizational level, intermediate level, and depot level.

13.2.1 Organizational Level

Organizational level tasks are those simple repair and maintenance tasks that can be done on the end item. At this level of maintenance, the organization-level personnel are generally concerned with the operation, preventative maintenance and use of the equipment, along with some minimal amount of checkout. Personnel at this level usually do not repair removed components, as they will be forwarded to the intermediate level.

Often, the end-users may perform some of these tasks as in the example of an office person who maintains a copy machine in the office. However, a field engineer will, typically perform those types of repairs where components or assemblies are removed from the copy machine. This is a similar situation with organizational tasks that are performed on an aircraft by ground personnel.

13.2.2 Intermediate Level

Intermediate level is where higher levels of repairs are performed. It can either be performed in a specialized shop located at the customer's facility or by mobile-type setups. The latter can be seen in examples such as a cable TV repair truck, an oil burner service truck, or a field engineer's vehicle. Many parts are carried in the mobile unit in order to perform common repairs and maintenance actions.

On the aircraft level, the airline or military organization may have a repair shop facility that is a sub-stock for common equipment that is removed and replaced. Repairs at this level may be accomplished by the replacement of major modules or assemblies.

13.2.3 Depot Level

The depot level represents the highest level of maintenance. It may be located at the facility of the vendor who makes a particular component or assembly. In the military, special shops exist for depot-level repair for most electronic equipment. In some cases, the vendor's facility is designated as an authorized repair deposit for the military.

At the repair depot, individual components are available from stock to complete most repairs on the circuit board levels. At this level of repair, certain failed components may be classified as throwaway components as they cannot be repaired or the costs are too high to repair them.

13.3 MAINTAINABILITY DOCUMENTATION

Documents used for performing maintenance tasks on a product may require a number of items to ensure that maintenance tasks are performed and recorded correctly. The following are a sampling of some of this documentation and how they are used.

13.3.1 Discrepancy Reports and Failure Tags

One might wonder how reliability engineers are able to calculate field MTBF values. This is made possible by the inputs provided by maintenance personnel and field engineers, written on special forms or tags while working on both military and commercial equipment.

The process starts with inputs from the customer. For a military aircraft, the customer would be the pilot who will generate a list of discrepancies discovered during flight on a "crab" sheet and give this to the maintenance crew chief. In the commercial world, a field engineer may be dispatched to a location to work on a piece of equipment based on a user's complaint. This complaint may have been initiated via a call into a help desk setup or via an automatic trouble ticket system that is based on a failure reporting network (using the equipment's monitoring software).

After the item is repaired on site or repaired and removed, the discrepancy report is completed. A failure tag of some sort is attached to the item itself when it is sent down to the next level (either intermediate or depot level) for further action. In the military world, this data is entered into a computerized system that is used to track failures. The air force uses a 66-1 system and the navy uses a 3-M system. The commercial world uses proprietary software for compiling this data on a monthly basis, as reliability engineers at either the customer or subcontractor level can compute MTBF values, as well as spotting significant failure mode trends and occurrences.

Likewise, any parts that are used in the repair are noted on the reports. This is important for stocking information, as well as flagging any failure analysis that needs to be performed.

13.3.2 Component Maintenance Manuals and Parts Lists

A component maintenance manual or service manual is a book that is originally written by the vendor who makes the product, providing details on how to do most repairs. Typically the manual will contain an illustrated parts breakdown (IPB) that shows an exploded view of the parts along with assigned part numbers. The IPB along with reliability failure rate data may be used to help generate the spare parts list for a product. The writing of the component manual and the development of a spare parts list are maintainability engineering tasks that may often be performed by the reliability engineer.

COMPONENT SPARES EXAMPLE

Item: Printer used in office building

A certain brand of printer was used in several offices building locations for an insurance company that had a repair facility for its equipment. A total population of 200 units was being maintained. A spare parts list was needed to be developed for keeping sufficient quantities of parts that would be needed for repairs of failed units.

The engineer at the repair facility would make a number of judgements based on available failure rate data and a number of documents such as the IPD. From this, a generalized list of high-failure-rate items would be created as follows:

Item	*Rationale as High-Failure-Rate Item*
Switches/Keypad	Wear-out from high usage
Covers	Abuse from users
Levers for opening panels	Abuse from users
Print head	Wear-out from high usage
Rollers	Wear-out from high usage
Springs	Wear (loss of tension)
Power Supply	Heat from high-duty cycle

The engineer would conduct further analysis with failure rate data and calculate the expected number of failures for a given time period to determine the number of spares needed for the list.

For example, if it is known that the switch used for the "copy" button has an average of 50,000 cycles between failures, the engineer could calculate an average number of cycles per switch. From this, the number of spares needed for a one-year period could be determined. If it is determined on a daily basis that the "copy" button is cycled 250 times a day, this would translate to 1,250 cycles a week for a five day work week or 65,000 cycles a year.

The switch failure rate is = 1/50,000, and applied against 65,000 cycles yields 1.3 as the expected number of failed switches a year. This applied against a population of 200 printers would yield a sparing requirement of 260 switches to be put into spares stock on a yearly basis.

13.3.3 Repair History Sticker

Located on a product or in a convenient place in a system (like on the inside door of a printer) is a sticker that shows a quick repair history. It will list the date and initials of the field service engineer and sometimes what repair action was taken (depending on customer requirements).

13.4 DISPATCH RELIABILITY

Dispatch reliability is an important area for the use of reliability values. Dispatch reliability is used heavily in the commercial airline business and often in similar fields such as truck and bus routes. This chapter will present some of the basic concepts and calculations used in this particular aspect of reliability.

The dispatch reliability for a system (or vehicle) is defined in mathematical terms as:

$$R_D = 1 - D,$$

where D represents the mechanical delay rate. The mechanical delay rate is based on the number of times that mechanical issues exceeding a specified value and is expressed as:

$$\text{Mech Delay rate} = \frac{(\text{\# of mech. delays} > \text{specified value}) + (\text{\# of mech. cancellations})}{(\text{aircraft actual departures}) + (\text{number of mech. cancellations})}$$

We bring in the definition of unreliability from Chapter 1:

$$Q(t) = 1 - e^{-\lambda t}$$

If we substitute for the above in the original equation for dispatch reliability, we get:

$$Rd = 1 - Q(t) = 1 - (1 - e^{-\lambda t}) = e^{-\lambda t}$$

DISPATCH RELIABILITY EXAMPLE

Item: Commercial airline flight dispatch

An airline has the requirement of canceling a flight if any mechanical delays should exceed 20 minutes. For a given month, the airline had 320 flight scheduled that saw 21 cancellations along with 17 mechanical delays. What is the dispatch reliability?

First we see that there were $320 - 21 = 299$ actual departures. Then we calculate the mechanical delay rate and by using the formula on the previous page, we get:

$$D = \frac{(21 + 17)}{(21 + 299)} = .119$$

$$R_D = 1 - D \text{ or } 1 - .119 = .881$$

13.5 OTHER MAINTENANCE ENGINEERING TASKS

A number of maintenance engineering tasks should be performed at the beginning of a new program in order to verify the fault isolation and troubleshooting methods that are used to correct failures. Depending on the size of the organization in the company, these tasks may be typically performed by the reliability engineer. If the organization is small such that there is no reliability engineer, then the design engineer must take charge of these tasks.

13.5.1 Maintainability Demonstration

This can either be a qualitative or quantitative exercise depending on customer requirements. For a quantitative-based maintainability demonstration, a list of potential faults are generated between the customer and vendor (based on maintainability data item submittals). From the list, faults are inserted into a test unit one at a time with the technician not present. This may involve the lifting of a component lead or shorting out a pair of leads to simulate the fault. After the fault is inserted, the technician is brought in to try to be able to isolate the failure using test points and schematics. Typically for the purpose of the demonstration, the time to isolate would be recorded by the customer to make sure any time requirements to isolate to the fault are not exceeded.

For a qualitative maintainability demonstration, the customer's maintainability engineer will examine the product for a number of maintainability design features. This includes the following shopping list:

- Ease of maintenance
- Commonality of hardware
- Modular assembly
- Murphy-proofing

The last item is defined as the process in eliminating the possibility of the incorrect hardware being inserted in the wrong place or a part being installed incorrectly because of ambiguities. A cover that must go on a certain way may have a unique hole pattern such that the cover can only go on one way.

A simple example of "Murphy"-proofing that the reader may be aware of concerns the oil filter for car engines. The thread size on the mounting stud is different for different-size engines, such that it is generally not possible to put a small filter on a larger engine of an upgraded model made by the same car manufacturer. This is a simple example, as there are more complex situations that can arise when many parts are present in products installed on aircraft such as in the following case study.

13.5.2 Maintainability Demonstration Case Study

Item: Emergency oxygen system for navy fighter aircraft

The maintainability review was held at the vendor's facility with the customer's maintainability engineer (the ejection seat manufacturer) present. A shop technician took apart and reassembled the emergency oxygen system. He was not timed as the customer was more interested in the ease of maintenance for the overall design as opposed to the times of completing individual maintenance tasks.

After this the maintainability engineer looked at the design for two major aspects: Murphy-proofing and commonality of hardware. Murphy-proofing means that any hardware item cannot be inserted in the wrong hole or that the item can be inadvertently assembled wrong (refer to the medical equipment failure discussed in Chapter 12, where a relay was installed incorrectly resulting in major failure consequences). Commonality of hardware involves using as few different sizes as possible with regards to screws and Allen-head screws. This is done for reasons of making it easier to stock fewer parts, as well as needing fewer tools to do maintenance.

The review on the survival seat showed the need for improvement in both the area of Murphy-proofing and in the area of common hardware. The customer's maintainability engineer pointed to three areas where one size fastener could be used. It is important to simplify maintenance tasks with using the least number of different sizes possible for the hardware. By doing so, the incidence of human error and potential failure of the product during field service is minimized.

13.5.3 Testability or Test Point Analysis

Another maintainability task that may be shown during a maintainability demonstration is the test point analysis. The purpose of this demonstration is to verify the effectiveness of the existing test points in the area of troubleshooting and fault isolation. System schematics are used with this exercise and the customer or military organization wants to verify the ease of maintenance of the product for future maintenance that will be performed by their field service or maintenance personnel. Figure 13–1 shows a portion of an electrical circuit that has a test point that is used during troubleshooting and repair.

13.5.4 BIT Demonstration

Typically, a minimum percentage (usually 80 percent or greater) is set by the customer for the percentage of faults that are successfully detected by the unit's built-in test (BIT) software or via a visual indication (LED warning light or message on LCD display).

Maintainability Concepts 271

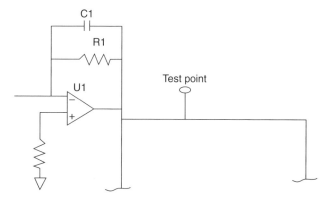

Figure 13-1 Test Point Example

The figure is a schematic of a portion of a peak detection circuit typically used in an electronic avionic unit that has specific voltage values that determine whether the data is being processed correctly. The test point that is located after the linear IC U1 is used during troubleshooting to see that the correct design voltage is being generated by the combination of the IC and the resistor R1 and capacitor C1.

13.5.5 Bit Demonstration Case Study

Item: Controller unit for air force transport aircraft

This test was performed on a unit that had completed reliability testing and was still in working order after the test. Faults from a previously assembled list (determined by the vendor and the customer) were inserted into the unit. After the fault is inserted (usually the removal of a component in the area being verified), it is observed to see whether the unit's BIT software is able to detect the failure and then flag it to a display or via external means.

13.6 PREVENTATIVE MAINTENANCE CONCEPTS

Preventative maintenance is simply the scheduling of regular maintenance actions to prevent the occurrences of failures to a product. This regular maintenance may be in its simplest form, such as minor cleaning, or in a more involved procedure, such as replacement of worn parts. An example that many of us experience is the normal oil change and replacement of the engine oil filter on our automobiles. Often on new product designs, there may not be an immediate recognition by the design engineers that some sort of periodic preventative maintenance has to be performed. Another case was covered in Chapter 9 regarding the washing of the fan blades on the engines of the A-10 military aircraft after a specific amount of flying time had been accrued and after being used during gunfire exercises. See Figure 13–2. This is a form of preventative maintenance that is used to prevent the occurrence of engine flameout.

Figure 13-2 A-10 Aircraft Engine Maintenance

As discussed in Chapter 9, the A-10 aircraft had a failure condition where the fan blades of the engine were fouled by the gas residue from the firing of the cannon. The fouling of the blades could cause engine overtemperature or engine flameout. A preventative maintenance action was incorporated in which the blades would be cleaned by an engine-washing procedure for every 100 hours of flight. This case illustrates the fact that preventative maintenance actions may not always be anticipated prior to field service.

13.6.1 Preventative Maintenance Case Study #1

Item: Receipt printer for gas pump

The modern gas pump is a combination of electrical equipment and mechanical equipment. The electrical equipment includes the computer, the display and the receipt printer. This study involves a certain service station chain that had its own field engineering force to fix and maintain their gas pump equipment. One of the routine tasks that would be conducted

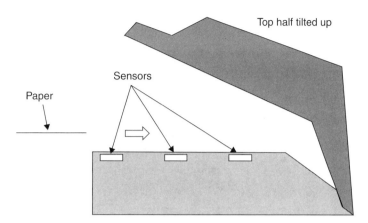

Figure 13-3 Receipt Printer Preventative Maintenance
In the diagram above, the sensors on the lower half of the printer tends to get covered by paper dust that is generated from the serration particles of the receipt paper. By having the sensor on the bottom half of the design, this basically ensures that regular cleaning will need to be performed. In this particular case, this was not anticipated or planned for by the design engineers.

by the engineering department was in the area of scheduling regular preventative maintenance on the receipt printer.

The receipt printer was the type that relied on a group of sensors that were located in the bottom half of the printer unit to detect the receipt paper as it traveled through the printer. Unfortunately, because the sensors were located on the bottom half of the printer, they were subjected to paper dust that covered them after many receipts were printed (typically after 5,000 to 10,000 receipts). See Figure 13–3. When the sensors were blocked, an error was flagged indicating an undefined state situation. This error would prevent the printer from working and printing out a receipt to the customer.

Thus, a preventative maintenance program was needed where the printers were cleaned on a regular basis, depending on the volume. The engineering department then kept a spreadsheet that listed all the sites and what day the cleaning was done. By cleaning the sensors on a regular basis before this error occurs, both failures and downtime were reduced significantly (over 90 percent reduction in failures).

This simple example shows how a preventative maintenance program can be implemented and still be tremendously effective. This particular case study shows how the use of preventative maintenance can help alleviate the impact of a design issue or fault (the locations of the sensors in the bottom half of the printer). Also, if a good maintainability review was conducted during the design stage, the manufacturer may have come up with an alternate design in which the sensors were not located in such a vulnerable location.

13.6.2 Preventative Maintenance Case Study #2

Item: Aircraft access doors

Since the evolution of the aircraft, access doors that are located on the airframe overall have always been a key ingredient for airframe design. These doors and panels are needed

A) The F-105 fighter started production in the late 1950s and a number of slotted screws that were spring-loaded were used for each of the quick-opening access doors as shown with this door pictured here for hydraulic servicing.

B) The A-10 attack aircraft started production in the mid-1970s and used the updated concept of quick-release latches on several of its access doors such as the one shown here for ground refueling.

Figure 13-4   Aircraft Access Doors for Periodic Maintenance

to allow the ground crews to get to items that need regular maintenance either after each flight or after a number of flights have been accumulated. The ideal design of an access door is to have some sort of quick-release fastener that allows the ground personnel to get in quickly, yet is strong enough that it does not fall off during flight.

Early designs of access doors feature simple Phillips head screws to secure the door or panel to the airframe. This has evolved into a quick-release design where the screw is spring-loaded. With two turns of the screw, the head is pushed out while being retained by the spring to the panel. Later access panel designs saw the use of quick-release latches, particularly on doors that are opened frequently by the ground crews, such as the refueling panel access. Figures 13–4A and B show the two types of particular access doors discussed here. It is noted that with the use of two quick-release latches on the newer A-10 aircraft design (compared to the older F-105 access screw design), significant time is saved.

Access time is part of repair time that makes up the value for MTTR. Thus, by keeping this time to a minimum through better and quicker access, the overall repair time and MTTR value will be minimized also. Airlines have tight schedules for each individual aircraft in their fleet, and thus, the maintenance turn around time has to be low. For military aircraft during peacetime, maintenance crews are trained to achieve periodic and preventative task efficiency. Thus, during wartime situations, the crews are equal to the task because of their training. But more importantly, wartime situations, such as battle damage repair, becomes an additional task. Figure 13–5 shows a situation where battle damage repair is being performed by the maintenance crew.

The concept of access doors is not unique to aircraft, as they are used on many other products such as printers, copy machines, automobiles, and trains. It is important that this aspect of the design be addressed during the initial stages of product development.

13.7 DESIGNING FOR MAINTAINABILITY

Contrary to what some design engineers may think, the concept of designing for ease of maintenance has to be done up front on a new design. If this is not thought of during the design stage, it is not going to be implemented, and it may be virtually impossible to incorporate additional access panels later on after production has started. Thus, the maintenance concept has to be understood fully right from the start and features promoting the ease of maintenance have to be planned as part of the design.

Unfortunately, manufacturers do not always pay attention to the concepts of easy access or maintenance-friendly designs. Witness the case where a car manufacturer puts the grease fittings for the front tie rods too close to the wheel such that a mechanic cannot put the grease gun attachment to the grease fittings unless the front wheel is taken off. Another case is a manufacturer who solders electrical components (that may be updated in the future), such as EPROMs into circuit boards in lieu of using sockets for these components. Oversights like these by the manufacturer will add access tasks and minutes to a routine maintenance task. This is primarily what the concept of designing for maintainability is about.

Figure 13-5 Aircraft Maintenance During War
During wartime conditions, aircraft maintenance crews have additional repair tasks because of damage sustained to the aircraft during combat mission. In the photo, a maintenance technician is applying a metal patch to an area of damaged skin surface on the fuselage of an A-10 close air support aircraft. Furthermore, maintenance crews often have to improvise to effect certain repairs such as these, particularly structural items or equipment not available at temporary bases set up near the front lines.

13.7.1 Designing for Maintainability Case Study #1

Item : 14-inch CRT display

Cathode ray terminals (CRTs) are fairly heavy devices that required high voltage and a number of adjustments to get the screen size and focus right. These adjustments are performed using various types of potentiometers.

A CRT manufacturer designed a 14-inch CRT display that could be used in a multitude of product applications, including use in personal computer monitors, customer service terminals, video games, and for ATMs. The electronics for the CRT were housed in a sheet metal cover that was secured to the chassis of the CRT with a dozen screws.

A number of units were shipped for evaluation to the engineering department of a company using the CRT for a customer service terminal that was used to issue subway and bus tickets. The company had its own field service organization and repair depot, so it was important

Maintainability Concepts 277

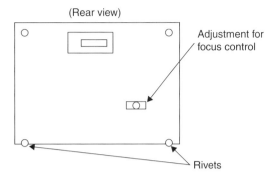

Figure 13-6 CRT Adjustment

that the CRT was reasonably easy to install, maintain, and repair. As these CRTs were being installed into a fixed structure, all adjustments were performed prior to final installation.

However, upon powering up the CRT display, the evaluation units were all out of focus. A call to the CRT manufacturer's engineering staff revealed that the focus was controlled by a potentiometer that was located in the rear of the unit. It had a slot knob that could only be accessed through a slot in the cover (see Figure 13–6). The root of the focus problem was due to the fact that the CRT manufacturer's did not secure the potentiometer with locking adhesive. By not doing this, the potentiometer moved during cross-country shipping. What affected the maintainability aspect of the design it was that it was extremely difficult to apply the adhesive after the unit was assembled with the cover. It could not be applied into the small slot, and the cover had to be completely removed.

The irony is that the CRT manufacturer should have made a smaller panel that could be removed just by using screws in lieu of rivets on the rear portion of the cover assembly. The manufacturer did not have an engineering staff and the mechanical design was sent out to a consulting firm. By the way, this was not a small company, but the idea of not having some sort of internal engineering represents small-company thinking.

This example shows the pitfalls of dealing with companies that do not have its own engineering staff as well as them not having the insight to incorporate reasonable maintainability features into the product. The burden fell on the procuring company's engineering staff to spot these difficulties because it would affect how their field engineering and depot staff would be able to work with these CRTs.

13.7.2 Designing for Maintainability Case Study #2

Item: Automobile water pump design

A number of items on an automobile require regularly scheduled maintenance. For example as mentioned earlier, a car must go through regular oil changes, typically between 3000 and 4,000 miles of service. In addition, the oil filter must be replaced on a regular basis with the replacement schedule based on the various parameters of the engine and type of filter.

One of the things that many prospective new car buyers do not often consider (particularly if they will do the maintenance themselves) is to take a close-up look at the engine and see where the oil filter is situated. Those filters that are located in the front part of the engine typically allow for easy maintenance while those located in the rear of the engine next to the fire wall can lead to hard-to-perform maintenance.

The same principle can be applied to other automobile components, some of which do not require regular maintenance but can occasionally fail and require replacement. One of these items that can tell a lot about an automobile manufacturer's approach to the ease of maintenance is the physical location of the water pump used for engine cooling.

An automobile water pump may fail at least once during the life of a car, with the main failure mode being that it is leaking engine coolant. (The other possible failure mode is where the pump seizes and this was covered in a case study that was presented in Chapter 9). When the water pump fails in the mode of leaking, it has to be replaced as this would eventually cause a significant loss of coolant and ultimately lead to more serious failures. What is profoundly interesting in looking at different models of cars is that the time to remove and replace a water pump can vary significantly to as little as one hour of total labor on some car models to as much as seven hours of total labor on other car models.

For example, on a medium-size or large-size car, the water pump may be mounted on a bracket that is located on the side of the engine and driven by a rubber belt off of the engine pulley system. On the other hand, many compact car models have the water pump embedded underneath a number of other components such as the engine cover, exhaust manifold and frame. In the worst case, not only is the water pump under other components, but it is engaged in the metal timing chain and replacement of the water pump also requires removal of the timing chain and subsequent retiming. This is why a removal of the water pump can take up to seven hours of labor (as documented in the car manufacturer's service manual). Thus a simple $100 water pump as shown in Figure 13–7 can cost over $400 of labor to install!

The obvious solution would be not to bury the water pump under too many other parts. An even more reasonable solution that some manufacturers use is to have a water pump that can disengage from the timing chain in some easy manner. This is sometimes made possible by having the gear and slotted shaft remain in the timing chain while the water pump can be pulled off the slotted shaft of the gear.

Why should ease of maintenance be a consideration to the car manufacturer and ultimately the customer? If one relegates the repair to a garage mechanic, not only does the cost become a concern but whether the mechanic will have the capability to even do the repair in the shop. The mechanic may even pass on the job and recommend that the automobile dealer perform the repair.

13.7.3 Designing for Maintainability Case Study #3

Item: Portable card embosser

Card embossers are machines that stamp and print out letters on credit cards, bank ATM cards, casino ID cards, cruise ship ID cards, and insurance ID cards. In addition to the stamped lettering, vital customer information is encoded on one or two of the magnetic tracks on the magnetic strip located on the rear of the card. Most credit cards and similar cards are made in mass quantities in large facilities with printing press-like machines. However, portable type embossers are needed for printing cards on the spot for hospitals, banks, cruise ships, colleges and casinos. Even though classified as portable, these machines can typically weigh over 75 pounds and generally require two people to move.

Figure 13-7 Automobile Water Pump

The design and location of some automobile components will have a major impact on maintainability. In the specific case of the water pump that is pictured here, the water pump is not optimally designed, nor is it located ideally for easy maintenance. Because it is buried underneath many other parts, such as the exhaust manifold, and is located in a limited access area between the engine and the firewall, it takes over seven hours to perform a removal and replacement. One part of the problem is that the gear on top of the pump is enmeshed in the timing chain of the engine, and this requires removal of the chain and other components. Some water pump designs have a removable gear, allowing for easier removal. Water pump design is a case of how a car manufacturer approaches the ease of maintenance concept, and it is one that consumers should look at carefully.

A field service organization provided intermediate-level repair for a particular brand of embosser that was used by several banks and hospitals. In addition, the field service organization provided a facility for depot-level repair. However, this particular model embosser that was used proved to be very difficult to maintain, as well as having other undesirable aspects related to the way that it was designed.

For example, the embosser was not able to be transported from the warehouse to a branch location or hospital because it would lose its internal adjustments during the transport. This required that a special transport cart be made in order to minimize this risk. The embosser had many adjustments and it would take over two hours to do a complete setup.

To make matters worse, the manufacturer of the embosser did not follow the concept of commonality of hardware. For example, there were seven different sizes of Allen-head screws that were used in the design and this would make a repair technician have to go back and forth between different tool sizes when making adjustments and this increased the MTTR values for embosser repair. Simply put, this brand embosser was a maintenance-unfriendly machine.

Eventually, it was realized that significant amounts of man-hours were expended in the repair of the embosser (on all three levels of maintenance—organizational at the bank or hospital, intermediate by the field service, and at the field service's repair depot). The next version of the embosser proposed by the manufacturer did not have significant enough improvements over the original model in terms of ease of maintenance. After several years of service, the original embosser was wearing out and the replacement proposed by the manufacturer meant more of the same type of maintenance issues.

Eventually, the banks and hospitals decided to eliminate the need for an embossed card, and decided that a printed card that was encoded was suitable for their needs. There were many printer/encoder units on the market and a design was chosen that was easier to maintain as the unit was much smaller and lighter. The original embosser either sold for low prices or was relegated to the role of a boat anchor (i.e., scrap). It is difficult for an older design to endure the changes in the market trends, particularly if the end user has difficulty maintaining the product.

13.7.4 Designing for Maintainability Case Study #4

Item: Computer server equipment

Companies that have many PCs in their workplace which are linked into a local or system network, generally require a computer server in order to accomplish this. A large organization was evaluating a number of off-the-shelf server designs that would be used to run their network for several workplace locations. Of four designs that were evaluated, one design became the clear favorite and eventual winner. The main reason, in addition to cost, was the aspect of maintainability taken by the company on the design.

Maintainability features were clearly designed into the server unit. The covers for the units snapped into place with the minimal number of screws. Slots that held the various memory and coding were used for the connectors of wires for the keyboard input, mouse input and monitor input, both on the jacks of the server and on the plugs for the devices.

The server featured three redundant power supplies that were easy to remove and replace. There was a handle on each power supply and it snapped into place. Each power supply could be removed and replaced while the server was powered on ("hot plug-in"). In addition, each power supply had indicator LEDs for good or fail status.

Indicator LEDs were used elsewhere in the server to show either faults or "working OK" status of individual assemblies such as drivers and circuit board assemblies. The server had a number of self-diagnosis tests that could be viewed using the monitor either during installation or during troubleshooting.

The winning company also provided training classes and repair documentation that explained repair procedures to the purchasing organization's field engineering and repair depot personnel. The company also provided on-site support for their product. The winning company had a top-to-bottom approach in all phases of maintainability. It is a real pleasure when a reliability engineer sees this approach taken by a vendor, as it can make life much easier.

13.8 LIFE CYCLE COST

An area that a reliability or maintainability engineer may be involved in is figuring out the life cycle costs (LCC) for a product. As the name implies, this involves the calculation of all costs used to support a product throughout its life. A very simplistic LCC model is:

$$LCC = \text{Equipment non-recurring costs} + \text{equipment recurring costs}$$

Non-recurring costs involve items such as training, acquisition, qualification and reliability testing, research and development.

A finer resolution model is:

$$LCC = R\&D \text{ costs} + \text{Production costs} + \text{Operation costs} + \text{Retirement costs}$$
(where R & D is research and development)

Many other versions and refinements of life cycle cost models exist, but the basic models have been shown here in order to give the reader a general idea of the model. It would not be unexpected that this would be a task that a reliability engineer would perform sometime during the course of a career.

Indeed, the issue of Life Cycle Cost management for a product line may be bigger than the issues that are typically addressed by the reliability or maintainability engineer. Company management may have to consider issues as to whether it is economically feasible to maintain a repair depot to fix field returns or whether to maintain a sustaining engineering staff to address field issues. In addition, decisions regarding vendor management may come into play. Vendor issues such as warranties, part repairs and part obsolescence are all factors that have to be examined by company management. LCC has been covered in significant detail as a subject by itself in other textbooks.

13.9 SUMMARY

This chapter has provided an overview of some of the important maintainability concepts and terms along with how they relate to the field of reliability engineering. It is important for any engineer involved in the reliability field to also have knowledge of maintainability issues because there is overlap of effort between the two areas. Reliability and maintainability may often be grouped into the same department of a company or perhaps maintainability may be put in separately in the logistics area of a company. Sometimes the reliability engineer may end up wearing two hats as it is during the early stages of a design where reliability and maintainability issues are often best addressed. In any event, the reliability engineer cannot be too focused or overly specialized in just performing reliability tasks. A basic knowledge of maintainability is required.

Maintainability-type analysis tends to be more straightforward then reliability analysis. A large part of the analysis relies on making realistic estimates with regards to repair times or in the issue of developing sufficient quantity of spares to support a product that is in field service.

13.10 EXERCISES

1. **A.** What is the overall MTTR for the VHF radio based on the following repairs as listed below on the summary sheet (time listed in terms of hours)?

ITEM	Preparation	Localization	Isolation	Disassembly	Interchange	Reassembly	Alignment	Check-out
VHF Radio	.01	.1	.1	.2	.1	.2	.1	.3
VHF Radio	.01	.1	.2	.2	.1	.2	.1	.3
VHF Radio	.01	.2	.1	.2	.1	.2	.1	.3
VHF Radio	.01	.1	.25	.2	.1	.2	.1	.3
VHF Radio	.01	.15	.15	.2	.1	.2	.1	.3

B. If the radio has a MTBF of 100 hours, what is the inherent availability?

2. What is the mean downtime (MDT) for two ticket vending machines that for a one-month period had a total of seven repair actions with a total downtime (for response and repair time) totaling 98 hours.
3. A specific brand of printer is used in a chain of stores where a population of 100 printers saw a total of 157 down hours for a given week. Calculate the operational availability of the printer for this week in terms of hours.
4. What are the mission capable rates (MC, FMC, NMC, and PMC) for a military fighter aircraft squadron that has accrued the following: 3,451 missions flown, 21 missions canceled because of failed equipment on the MEL, and 9 missions allowed to go forward with equipment failures not on the MEL. (Hint: use the example in Section 13.1.6 as a guide.)
5. With the junk appliance that was used in the exercise in Chapter 5, make an estimate towards which components will have to be stored in spares in significant quantities.
6. A railroad line has the requirement of canceling a scheduled train if any mechanical delays should exceed 25 minutes. For a given month, the railroad line had 650 scheduled trains scheduled that saw 11 cancellations, along with 37 mechanical delays. What is the dispatch reliability?
7. List some of the preventative maintenance tasks that would be performed on a car over its lifetime.
8. Perform a maintainability demonstration on the junk unit used in the exercises in Chapter 5. Make a list of items for potential areas of improvement.
9. Describe the concept of designing for maintainability. Look at some of the everyday products and appliances that you may use and see examples of where design for maintainability is either employed or not employed properly.

CHAPTER 14

Reliability Evaluations and Prototypes

INTRODUCTION

It is extremely beneficial for a company to perform a reliability evaluation of a new product, whether it would be an internal product made by the company or an external product being considered for purchase by the company. Evaluations can reveal potential issues with the new product in the area of reliability performance or in the maintenance of the product. These evaluations should be performed on the prototype or first production model, before full-scale production begins, so that any reliability issue that arise out of the evaluation can be addressed by the manufacturer of the product.

Thus, if it is possible, the task of building a prototype unit for conducting a qualitative reliability evaluation or assessment should be done. The prototype is important for verifying the functionality of the unit along with being available as a test asset for some initial testing. A reliability evaluation of a new product that is being contemplated for use by a company is based on the reliability engineer's experience in which a checklist of important items to look for is followed.

14.1 GENERAL GUIDELINES FOR A RELIABILITY EVALUATION

Product reviews are usually performed on major expense items or on items for which large quantities may be purchased. With an organization such as a bank or an aircraft company, this purchase can be in the form of new equipment in quantities of several hundred units that will be installed into a branch or into an aircraft. Thus, it is very important to conduct a preliminary review of this equipment with respect to reliability parameters.

The reliability engineer has to play the role of a product reviewer, as if writing for a magazine or newspaper. This review has to focus on the pluses and minuses of the product for the application for which it is planned to be used. In most cases, the review will have an end product in the form of a trade study, white paper, or product evaluation summary that is generated and forwarded to management to aid them in their decision to purchase or not.

Qualitative reviews can often be conducted on the equipment with power off. The major areas for review of the design are:

- Mechanical aspects
- Electrical aspects
- Miscellaneous items (such as proposed software functionality)

14.2 QUALITATIVE RELIABILTY OF MECHANICAL DESIGN

The reliability engineer must review several items with respect to the mechanical aspects of the design. The unit does not have to be powered on or activated when performing a mechanical review. Basically a set of tools for taking the unit apart will be needed. Many of the mechanical items listed here have been touched upon in Chapter 5 and in Chapter 13, but we will reemphasize some of them here.

14.2.1 Mechanical Reliability Evaluation Checklist

The following are typical questions that should be asked when reviewing the basic mechanical aspect of a design, particularly in the area of electrical units:

- Does the design present a general modular concept?
- Are the circuit boards and other modules easy to remove and replace?
- Is there a motherboard into which the other circuit boards plug?
- If there is no motherboard, how do the circuit boards connect or plug into each other?
- Is there sufficient clearance between modules and circuit boards when the unit is assembled?
- Is there sufficient area between circuit boards with regards to cooling air for components that will run hot?
- Are common hardware sizes used?
- Is there room for component expansion on the circuit board growth for design upgrade?
- Does the design make use of smaller Surface Mount Technology (SMT) components on circuit boards?
- Are there a minimal number of sizes used with regards to Allen keys and similar-type hardware?
- Are lock washers used in areas where movement is expected and slippage has to be prevented?
- Are connectors, circuit board slots, and input jacks labeled or color-coded?

Again, this reliability review of a company's mechanical design has to be done up front when the design concept is being formulated or when the prototype is built. For a company that is reviewing another vendor's product, this review should be performed on the first unit that is built and submitted by the vendor.

It is important to realize that the effort expended to manufacture a product can reveal a lot about the company, in terms of its procedures and approach to quality. Inside a product are clues about the history of the company, particularly in the area of design or manufacturing. Both the highlights and the warts can be reflected in a design, and these can be discovered during a reliability evaluation.

14.3 STRUCTURAL ISSUES

In addition to the basic mechanical issues to review, the structural aspects of the product should be examined for a number of features that are important to overall reliability. While there might be a tendency to leave this to the mechanical engineer, the reliability engineer can play an important role in the decision-making process by conducting a reliability review in this area.

14.3.1 Sharp Corners

An area that is often overlooked in design is the incorporation of radius into structural areas, such as mounting and pivoting. When no radius is used in a corner, there is a tendency for more stress to be placed in the corner, which can result in cracks in this area. There is stress placed on mounting brackets when a unit is installed, and a radius will reduce this stress. Another area of concern for stress is if a design has a pivot (such as a bracket that surrounds a peg). This would also benefit from a radius in the corners (see Figure 14-1).

14.3.2 Material

Weight savings may be accomplished by using lightweight materials, such as plastics and certain metals. While the goal of weight savings is a worthy goal, it is important not to sacrifice structural integrity or the possibility of introducing more failures into the normal handling process of the product. Some plastic structures are capable of handling a certain amount of weight when properly molded. This is the case with CRT monitors used with computers.

However, for aircraft applications, most avionics and electronic equipment are housed in aluminum because of the high-stress environment, both in handling and during flight conditions. The reliability engineer has to keep this aspect of the design in mind when conducting an evaluation with regards to durability attributes of the product.

14.4 QUALITATIVE RELIABILITY REVIEW OF ELECTRICAL DESIGN

For any piece of equipment that uses electrical circuitry, the reliability engineer should look for a number of preferred features to see if these features are incorporated, whether it be

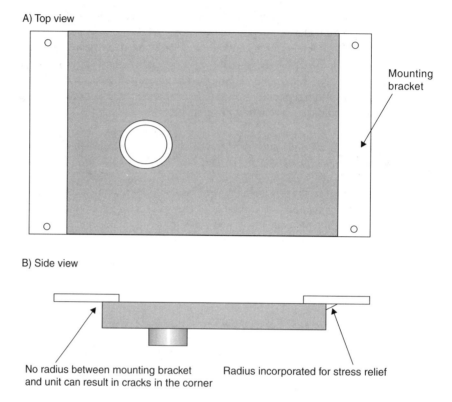

Figure 14-1 Structural Evaluation Issues—Use of sharp corners in mounting brackets.

equipment manufactured by the engineer's company or newly developed vendor equipment. The following is a generalized list of items and questions for the reliability engineer to review with regard to new electrical equipment. Most of these items can be reviewed by looking at the unit's circuit board layout or reviewing schematics. As in the mechanical review, power does not need to be on when conducting this aspect of the reliability evaluation.

14.4.1 Wiring

The way in which wiring is secured or routed can reveal a lot about the housekeeping approach that was taken by the manufacturer in terms of the product's design. Poor wiring practices can lead to wires getting tangled or caught on other components, and this can eventually lead to damage or failure. A manufacturer that applies good housekeeping principles for wiring in a new design is showing concern for the maintenance of the product and for the service personnel who have to do periodic maintenance or repair.

There is no real cost savings if no additional effort is given in making sure wire bundles are routed neatly and protected from catching on adjacent devices that make up a system in a product. Also, use of quick disconnect-type connectors at convenient places is a great aid to the maintenance personnel. Perhaps there is a small cost savings initially, but then the cost of ownership in the area of maintenance is made higher. Imagine the frustration of maintenance personnel that has to deal with poorly secured wire bundles when servicing either an automobile, an aircraft, or an ATM. In several cases, the company that designs a system or vehicle is also involved in providing the maintenance, so there is an incentive to do the right thing. Good housekeeping in a product will help the reliability.

Proper wiring is particularly important for assemblies that may be extended from the unit by the use of rails for better access for normal servicing. For example, the cash dispenser or the receipt printer of most ATM designs is able to be pulled out of the ATM on rails for better branch and maintenance access. Thus, the wire harness that powers this device has to be routed to allow the extension of the device, as well as being routed such that it does not get caught on adjacent components. Some companies use the concept of plastic tracks that neatly unroll the bundle as the device is pulled out on rails or in similar fashion (see Figure 14-2).

All wiring should be routed and secured properly inside the unit or piece of equipment using tie wraps or similar means of fastening every few inches. Some companies will use

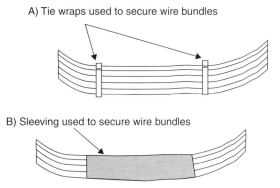

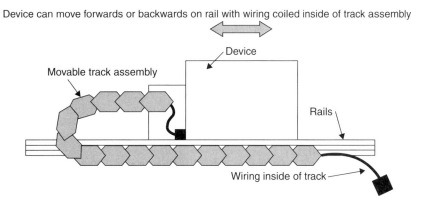

Figure 14-2 Wiring Routing Examples

plastic tubing or sleeving to place the wire harness inside. Not only should wire bundles be secured, but also single wires to items should be examined for proper routing. Another example in this category are magnetic heads used in reading cards, in which wiring comes from the head as discussed in Chapter 5. Is the wiring routed in such a way so that there is no pivoting on any spot on the wire?

14.4.2 Noise Bypass Protection

One would expect to see bypass capacitors on any voltage line coming into a circuit board from an external voltage source (such as a power supply input). These capacitors are designed to filter out low- and high-frequency noise that can come on the line. Figure 14-3 shows a typical example of how these capacitors are installed.

The filtering of noise is a major consideration in the age of computers. Certain levels of noise may actually be erroneously read as data on digital circuits. Some noise may actually affect signal lines such that the watchdog circuits of the computer may be triggered and cause the computer to reboot.

The use of bypass capacitors would be considered a standard part that should be employed in any new electronic product and be looked for during an evaluation.

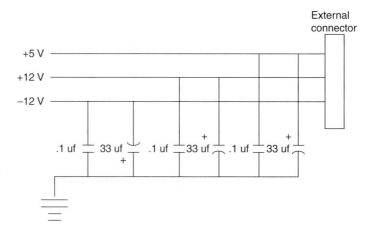

Figure 14-3 Bypass Capacitors in Electrical Design

In the schematic diagram, the higher-value electrolytic or tantalum capacitors (33 uf) are used to filter out low-frequency noise and act as storage for the circuit while the lower value ceramic capacitors (1 uf) are used to filter out high-frequency noise. A ceramic cap has a much higher resonant frequency than an aluminum electrolytic, so the ceramic is better at filtering out high frequencies. Unfortunately, a ceramic cap has limited capacitance values. On the other hand, electrolytic caps offer larger capacitance in smaller packages than ceramic caps, but they have much lower resonant frequencies. Large-value capacitors offer local energy storage and filtering of low frequencies. This is why both types of capacitors will often be used to filter voltage lines of electrical circuits.

14.4.3 Isolation or Protection Circuitry

This is a major feature to look for with regards to the environment that the equipment will be subjected to and is generally a requirement for most equipment installed on military and commercial aircraft. This involves using protection components, such as diodes and resistors, that are installed on the signal lines between the external connector and the internal circuitry. The reasoning is to reduce the effects of external problems, such as voltage spikes caused by lightning strikes or the external failure of other equipment in the system that are hooked up to the same voltage bus.

Unfortunately, the practice of using protective circuitry is not uniformly used in commercial equipment as it is in military aircraft equipment, and this difference shows up in the results. Often commercial-type equipment, such as ATMs or computers, are subjected to voltage spikes, lightning strikes, system failure of other units on the same electrical bus, or even hot "plug-ins" that can cause various types of circuit damage. Special concern is given to those pieces of electrical equipment that use different voltages such as those that have both analog ($+12$ and -12 volt based linear ICs) or digital ($+5$ volts) or those equipment pieces that have digital and motor circuits (which may be on the order of 28 volts). These types of equipment may experience more potential situations of voltage spikes or improperly applied voltage during hot plug-ins. Figure 14-4 shows an example of protection diodes that were added to a circuit to protect against high-voltage spikes that could be caused by lightning on the lines.

When an internal company review is being conducted, it is desirable for the reliability engineer to try to get the company to have this protection feature incorporated in the early stages of a new program, as it is impossible to make such changes in the circuit board later on. It may be hard to convince management, particularly if it holds the philosophy that unless a cost savings is realized, this type of fix won't be incorporated. Again, it is up to the reliability engineer to sell this feature at the beginning of a new program, since the cost of isolation components are minimal compared to the cost of damage components that occur during field use because there was no protection. This may be accomplished through the use of a trade study or white paper in which the need for protection on the product is analyzed.

14.4.4 Printed Circuit Board Design

The printed circuit board design has a number of issues that must be examined during any evaluation. Of primary concern is the thickness of the circuit board. Is the thickness of the circuit board sufficient enough to properly support the components that are installed on it, as well as the proposed environment that the product will see?

Manufacturers tend to make the thickness of the glass epoxy material used for the main printed circuit board on the thin side (less than .100-inch thick) and this is a major concern, particularly when heavy transformers are attached to the boards. The thin boards are not durable when it comes to shipping or repair. Figure 14-5 shows why this is a concern. Also, refer to Chapter 5 for examples of how a large component can induce damage to a thin circuit board such as a CRT.

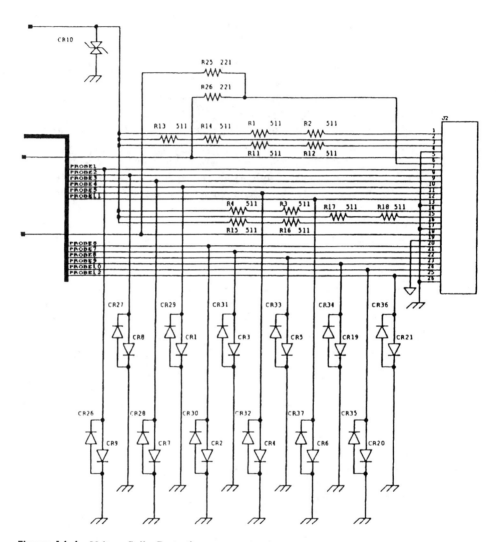

Figure 14-4 Voltage Spike Protection

The schematic is the input portion of an electronic module that takes voltage signal inputs from aircraft fuel tank probes and processes the signal into data that can be displayed on a cockpit display unit. The dual diode combination is added to each signal line to prevent any excess voltage over the rating of the diodes to enter into the circuit and cause possible damage. There are some low-value resistors that are on many of the lines to isolate the lines against electrostatic damage (ESD) and other external failures to the unit such as an electrical fuel probe short.

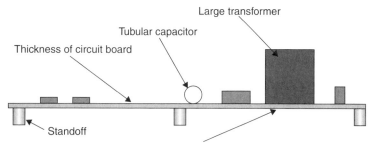

Figure 14-5 Circuit Board Thickness and Component Placement

Along with the thickness of the circuit board, there has to be adequate support for the circuit board inside the packaging. In addition, the evaluator should look for an even distribution of components on the circuit board where not all of the large components are located in only one area, causing uneven weight and potential vibration-related failures. As discussed in Chapter 5, individual large components such as tubular capacitors and transformers may require additional accommodations if the product is to endure a high-vibration profile in the field. These additional accommodations may be in the form of adhesive and other supports.

Likewise, if the product is going to be used in an outdoor environment in some way, the circuit board will probably require some sort of protection. This protection may be in the form of conformal coating or even silicon coating.

If the evaluator is examining a product built by his or her company, it is possible to get more deeply involved with the processes that are used to construct the circuit board. Multilayer boards are of concern with their internal construction processes. As discussed in previous chapters, the processes used to make the circuit board are critical, especially in the area where the through-hole is connected to the individual traces.

Also in the area of circuit boards is the use of flexible printed circuit boards in new product design. The processes used to make these boards have to be on the highest level, as would be used in the making of a glass epoxy printed circuit board. One reason for using the flexible circuit board is because some bending and routing will be required inside the packaging of the product, making this is a major area for review. The amount of bending used in a flexible circuit board is always of concern, and this has to be carefully evaluated by the reviewer to see whether damage could occur.

14.4.5 Part Selection

For electrical design, part selections must be made as per parts requirement and for the temperature environment that the item will see. Again, the reliability can be an important part of this process with the calculation of stress ratios as discussed in Chapter 3. The case study in Chapter 6 concerning driver chips that were not able to perform properly over the temperature range shows the importance of proper parts selection.

The reliability engineer should look out for short cuts that may be taken on circuit board assemblies, particularly in the area of commercial products. For example, some electronic units designed by manufacturers may use cheaper parts with lower current or power ratings installed in circuit boards when higher-rated components are required. A specific case is a controller board of a record printer where there are high current motors and solenoids that need to be driven. Yet, in many areas of the circuit design, the manufacturer has used small-signal diodes in the role of blocking diodes. This diode is only rated for .45-amp peak current, but typical motor circuit current can run on the order of 2 to 5 amps. If a motor should fail in a stuck-on position, the current may back up into the circuits where this blocking diode is used, and it may burn it up. In a sense, the diode acts like a fuse, but it is very easy for other components in the circuit to be damaged with this particular design mentality. Questions must be asked by the engineer who conducts the evaluation if there is a suspicion of shortcuts in the part selection or circuit design.

Commercial circuit board assembly manufacturers may often use components in circuit designs where they are immediately overstressed. For example, a bypass capacitor that is rated for 10 volts may be inserted on a 12-volt line, meaning that this capacitor is stressed at 120 percent. This is a blatant overstress condition that would never be allowed on any military program or on most commercial aircraft programs. Yet, in the area of commercial products, such as ATMs, printers, appliances, and commercial displays, this type of overstressing is not all that rare. Why is it done? Sometimes, savings of pennies may be realized across a large production run, but often the reason is ignorance. Engineers who have a background in fields where there are rigid requirements (such as military electronics) are often surprised at the liberty taken in the commercial product area. It is still bad practice to overstress components in excess of 80 percent as a general rule of design. Please refer to the chart in Chapter 3 that lists the maximum recommended stress ratio for electrical components.

14.5 MISCELLANEOUS DESIGN REVIEW ITEMS

A number of items fall in the miscellaneous category and needs to be reviewed during a new design. These are not technically in the reliability engineering area, but they can affect reliability performance if they are not implemented correctly.

14.5.1 Software

If the design makes use of software, there are a number of questions that should be asked up front. Does the unit employ software that displays messages on a LCD-type screen? Are the messages useful to the user as well as being user-friendly in fixing common problems? Are the diagnostics in the software useful? Chapter 12 references some common things to look for.

14.5.2 Continuity of Design

What is the concept of continuity of design, you may ask? Well, it is not unusual for some new products to have changes in the design concept because of changes in the engineering staff. Each engineer has an approach that follows personal preferences. Imagine what happens to a design if an engineer is removed from the program and another engineer takes over the effort. Also, there may be a need for engineering support throughout the life of the program. That is why a company must not only make the commitment in the quality but in the area of sufficient engineering support being available when certain types of problems are identified. This support that may be required will often be on the level beyond a help desk support, and a product support group has to be available for these types of problems.

14.6 THE STRAW MAN CONCEPT FOR A NEW DESIGN

Eventually something in the way of a new design has to be put on paper, as well as some type of mockup or prototype constructed. This initial attempt might be known as setting up a straw man in order to get something down on paper. It is recognized that this straw man design will be modified further as the program moves forward. The first task that will take place is a computer model or story board design on paper.

14.6.1 Computer Modeling or Virtual Prototyping

Virtual prototyping of a new design can be performed using computer tools such as computer-aided design (CAD) programs or finite element modeling (FEM). These tools allow multiple iterations to be performed on the design as it progresses. Different concepts or alternate versions can be saved on a disk or computer network file with hard copies being generated at various stages of the design for the purpose of presenting at regular design review meetings.

14.7 CONSTRUCTION OF A PROTOTYPE UNIT

Perhaps the smartest thing that a company can do is to build a prototype unit prior to making preproduction units for tests or prior to full-scale production. This is especially important for electrical equipment with regards to the mechanical aspects, as discussed in the checklist in Section 14.2. The following are some of the benefits that come out of building a prototype:

- Manufacturing processes
- General layout of the assembly
- Test unit for initial testing such as thermal survey
- Initial software and electronic functionality

The prototype is the end result of all processes that the company uses. Any difficulties in obtaining parts or in the manufacturing process can usually be discovered during the making of a prototype unit.

In addition, the prototype is the first attempt of actually laying out the physical arrangement of the internal circuit boards and attaching hardware for an electrical piece of equipment. Initial layout will generally be done on a breadboard-type setup as shown in Figure 14-6. The use of a prototype breadboard lends itself to orderly layout, as well as allowing for the proper selection of component values and rating. After the breadboard layout is finished, initial circuit boards can then be manufactured. However, it is not uncommon for

294 Chapter 14

Figure 14-6 Example of Prototype Circuit Board
The photo is a breadboard prototype of a crystal-controlled amateur radio transmitter for 3.5 MHz. Once the layout has been completed on the breadboard, a permanent circuit board design can be determined and subsequently etched for larger quantities to be made.

many of the circuit boards to start to accumulate jumper wires and tacked-on components as revisions are made during the early design stages.

Perhaps more importantly, the prototype is the means by which a formal thermal survey is conducted on the design (as described in Chapter 6). The paperwork task of thermal analysis may be verified by conducting a thermal survey on the prototype unit.

Prototyping extends beyond making individual equipment, it also is used in entire systems and even aircraft. This can range from a nonfunctional mockup of an aircraft or system that is constructed to an actual working aircraft prototype or working system.

The use of prototype units is discussed in two of the following case studies.

14.8 CASE STUDIES

The following are various case studies associated with different types of reliability-oriented reviews and shows some of the different aspects and approaches used to conduct reviews. In addition, the role of the prototype in the early design evaluation is investigated.

14.8.1 Engineering, Reliability and Maintainability Review (ER & MR)

Item: Electronic module for commercial aircraft company

A vendor had been making a series of electronic units for a commercial aircraft when a new program came along to update the equipment on revised aircraft that were based on the original aircraft line. As part of the new requirements, the manufacturer imposed a data item called Engineering, Reliability and Maintainability Review (ER & MR). This item was a qualitative review at the customer's facility in which a prototype of each of the electronic units would be examined by all engineering disciplines of the company for each major piece of equipment to be installed on the aircraft.

The design of each unit represented significant circuit board and electronic layout changes from the original design made ten years earlier. This was made possible by newer technology and increased functionality of ICs, as well as the use of flexible circuits in lieu of wire bundles. One of the units saw a weight savings of almost ten pounds from the original unit, and the weight savings were significant in that the carrying handles could be eliminated. Changes in another unit saw reductions in components and wiring that made the assembly process easier, and thus, less workmanship-related failures were introduced (see Figure 14-7).

The review of each unit-type also focused on the maintainability aspect of the design in the area of ease of maintenance where the unit was disassembled down to the individual circuit board levels. The ease of disassembly was aided by the use of quick disconnect type connectors and the use of standoffs to connect the circuit boards to the chassis. A list of action items was generated at the review for the vendor to work on in improving the design further.

After the review, the prototype was taken to the aircraft flight line where the commercial aircraft model that used this unit was being assembled. The prototype was handed to a shop technician to install into the aircraft's cargo bay area where the original design was installed. He noticed the difference in the weight right away and this made it much easier to install the unit while it still had the same mounting footprint that the original unit had.

Both the review and the trial installation were excellent things to accomplish prior to the start of full scale production and before the design was finalized. The value of these simple tasks to improving a design cannot be emphasized enough.

14.8.2 Reliability and Maintenance Evaluation of New Product Case Study

Item: Compact size ATM

An engineer for a field service organization was asked by an ATM manufacturer to evaluate the prototype of a new compact-size ATM for the purpose of identifying potential reliability and maintenance issues, along with recommending any potential improvements. The field organization would be doing the maintenance on the ATM, so their inputs for potential improvements could be implemented prior to the beginning of full-scale production of the design.

The basic design of the ATM was similar to a conventional-size ATM, however some features were put into place to make the design more compact with a smaller overall footprint to fit into smaller areas. One of the main features was a hinged assembly that opened the entire front of the ATM unit so that it would lift up and allow servicing and replenishment. However, this feature, along with other items in the machine, had several issues come up in the area of reliability, maintainability, durability, and safety that definitely concerned the engineer performing the review.

In the area of reliability, the evaluation uncovered some concern about the use of certain components that had poor reliability history. These included the two air pistons that were used to open the front assembly, as they had a high failure rate that would degrade over time and require periodic replacement. This was also a major safety item because if one or both of the pistons holding the front end fail, injury to the servicing personnel may result. There was no

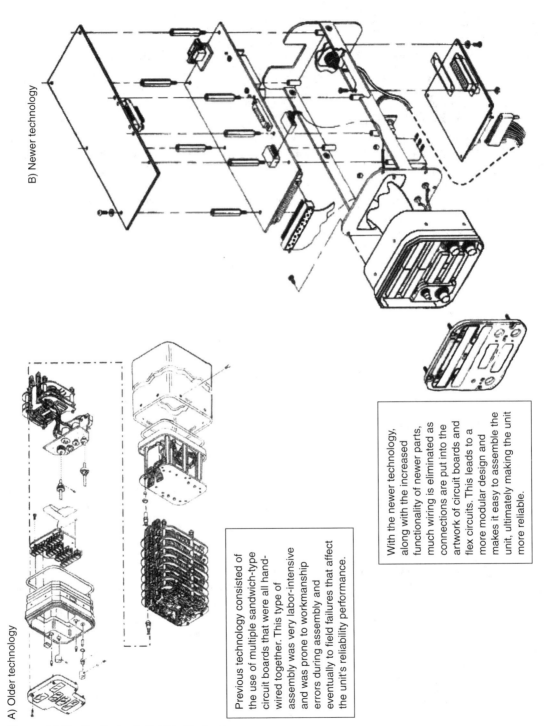

Figure 14-7 Qualitative Reliability Evaluation

thought toward implementing a safety latch or notch in the design to hold the front end. In addition, there was insufficient clearance with the front assembly in terms of human factors where most service personnel would have to bend down and be aware of the structure.

A number of maintenance issues came up, particularly in the area of gaining reasonable access to replacing components, especially those components that wear out and have to be replaced on a regular basis. Another concern involving reasonable access was in the area of the computer. Several screws held the computer into the ATM, and some of them were not readily accessible. Removal of the computer unit from the compact ATM was more time-consuming than from the current ATM design that the field service was servicing.

Issues with durability included the brackets for the outside door and safe door not being rugged and perhaps being unable to endure medium- to high-volume usage and breaking often. The bracket assemblies are projected to be high failure-rate items, requiring replacement and eventually an engineering change to reinforce the bracket. Also, there were concerns involving the use of plastic bearings in lieu of metal bearings on shafts that saw high levels of stress.

There were insufficient clearances inside the ATM especially between the cash dispenser, and exposed and unsecured wiring was on one side of the machine, similar to issues discussed in Section 14.4.1. It was noted that when the cash dispenser was undocked for replenishment, it was very easy for damage to occur to this wiring. Here the engineer saw that the wiring would be a candidate for track-style wire harnessing as shown in Figure 14-2C.

The review of this compact ATM design shows that the unit in the form that it was not robust enough to meet the high-volume requirements when deployed, as well as meeting the requirements of simple maintenance. A number of standard maintenance tasks would take longer than for similar tasks for the existing ATMs that were currently deployed in the bank. The longer duration of these maintenance tasks will have an impact on customers, particularly in the area of downtime. In addition, there are concerns of safety and durability as detailed in the text.

14.8.3 Reliability Evaluation of a Replacement Technology

Item: Use of flat-screen LCD monitors to replace CRT monitors

A large financial institution was in the position of upgrading computer equipment used by their customer representatives and teller personnel. Among the equipment considered for upgrade was a seventeen-inch diagonal cathode ray tube (CRT) monitor. Initially, the thinking was to replace these CRT monitors with a new CRT model. But the engineering department's interface with the customer representatives showed that there was a preference for the new flat-screen LCD monitor. At that point, the engineering department conducted a product evaluation of the flat-screen LCD monitor and compared it with CRT monitors. Among the advantages of the flat-screen LCD monitor over the CRT monitor were the following:

- Smaller base footprint
- Less weight
- Smaller overall size for warehouse storage
- Less power consumption
- Swivel capability for moving the screen

The latter function was considered particularly useful by the customer representatives because they could show the customer what was on the screen by simpler movement. The only negative was the unit price, as it was double the cost of the CRT. But the institution was able to justify this upgrade by buying in large quantities to reduce the cost along with the reduction of power consumption and longer display life.

14.8.4 Competition between Prototypes—Case Study

Item: Military aircraft prototypes in fly-off competition

Building a complete aircraft prototype in order to participate in a fly-off competition for a new military aircraft program is a fairly common practice that the U.S. Air Force follows. A recent example is the Advanced Tactical Fighter (ATF) program. Here, the YF-22 prototype had a fly-off with an opposing design, the YF-23, which was made by an opposing team headed by McDonnell Aircraft in the late 1980s. During fly-off competition, not only is flight-performance data measured, but also all of reliability and maintainability characteristics are tracked and studied. Failures of components or major systems are similarly looked at with scrutiny.

Sometimes fly-off competition between two teams show two completely different concepts in addressing the requirement set forth by the military agency. These concepts may not only vary in the basic design but also in the area of reliability and maintainability as well. This was the case in the fly-off competition for the U.S. Air Force close air support program.

In 1972, two competing companies submitted two prototype aircraft each of their design in a fly-off competition for this program. The A-10 prototypes were made by the Fairchild Republic Company, and the A-9 prototypes were made by Northrop. The fly-off competition took over six months to complete with the two prototypes from each company accruing over 300 flight hours for each design. Besides the unusual shape of the A-10 (as described in Chapter 9), compared with the more conventional looking A-9, some aspects of the design eventually gave the A-10 the edge in the competition.

While the A-10 had better handling characteristics, the A-10 had the edge in maintainability and, most important of all, in survivability. The A-10's radical design, featuring the high placement of the engines, was the key to the edge in survivability. A very important point was the fact that the A-10 prototype was very close to being the final production model. The A-9 design had difficulties in many key areas that hurt its chances during the competition. Cracks were discovered in the wing spar during flight testing, the tail design did not work well with the wing causing stall, and the engine inlets were too close to the gun, resulting in engine flameout problems during gun firing. Additionally, the engine used in the A-9 design could only generate 7,500 pounds of thrust as opposed to 9,000 pounds of thrust by the A-10 engines. The maintenance technicians that worked on the two aircraft praised the ease of maintenance that the A-10 featured in comparison to other aircraft.

The fly-off was conducted in a fair manner. The A-10 and the A-9 prototypes would be flown in mixed pairs so both planes saw the same weather conditions, such as wind or temperature. During flight testing, each plane would take turns flying lead or wing position. The A-10 came out the winner of the fly-off, and the aircraft went into production beginning in 1975. The company learned the valuable lesson of having a prototype aircraft and would construct a prototype aircraft during its Air Force trainer proposal in 1982.

Fly-off competitions give a chance to test out two different design concepts under the same environmental conditions and the same mission profiles. Also, reliability and maintainability characteristics are measured during the flight-testing.

14.9 EVALUATION TECHNIQUES USED IN MAKING A MORE ROBUST DESIGN

A company can use a number of evaluation processes to evaluate its own design, as well as designs made by other manufacturers. Many of these processes are under the auspices of the area of the quality engineering field and sometimes covers issues of reliability.

Product improvements to a design are an iterative process. There are a number of other techniques that allow an honest evaluation of a product during the early design stages. By doing this evaluation early enough, it is possible for product improvements to be incorporated early into the design.

This section will provide a general review and some introductory details on some of these techniques and processes that are used. It will also provide some of the quality textbook references where these processes are fully explained.

14.9.1 Taguchi's Loss Function

Taguchi provides a methodology for quality improvement and a process for improving robustness. The basic axiom of the approach is:

1. Quality should be measured with respect to deviation from a specified target value, not by conformance to preset tolerance limits.
2. Quality can only be ensured by having it built-in through the design of the product and its processes, not by inspection and rework.

As opposed to statistical process control (SPC) that emphasizes the attainment of a parameter within a tolerance range, the methods used by Taguchi emphasize obtaining a specified target value and eliminating variation. Only through the proper design of a product can the product and the processes involved in its manufacture be made insensitive to variations. By reducing the effect of noise factors that cause variation, costly rework and repair can be avoided.

The actual application of Taguchi's functions uses an orthogonal matrix to show the effect of environmental parameters on a product design. The actual implementation of this matrix is beyond the scope of this book and is discussed in detail in *Total Quality Management* by Besterfield (Prentice Hall).

14.9.2 Design of Experiments (DOE)

The design of experiments, or DOE, treats the development of a product design as a controlled experiment. Through this process of experimentation, changes are intentionally introduced into the process in order to observe the effect on the performance of the product. Using a statistical approach is the most efficient way to conduct this process of change. DOE involves a number of factors such as:

- Sequential experimentation to model process behavior
- Prediction of future process behavior through use of empirical models

- Investigation and isolation of factors that cause variations from the mean
- Selection of experimental design to identify interactions of factors that affect a product

This subject is also covered in significant detail in *Total Quality Management* by Besterfield (Prentice Hall).

14.9.3 Evolutionary Operation (EVOP)

Engineers have a tendency to treat product design as an unchanging process in which everything is kept under control at some fixed set of operating parameters. There is no attempt to experiment with this type of mentality. Nothing new is ever learned as no experimentation is done.

With the concept of evolutionary operation (EVOP), the concept is different than stated in the previous paragraph. With this approach, experimental conditions are changed very slightly. By making these small changes, the product could still be operating within its specified values. By monitoring the changes that came from the slight changes in experimental conditions, additional information could be gained about the product that may be used to improve the product.

Some aspects of EVOP were demonstrated in Chapter 4, where environmental stresses were increased on undergoing reliability testing in order to gain more knowledge about the specific failure mechanisms. The primary lesson here is not to be afraid to experiment within the limits of the test or the design process.

14.9.4 Benchmarking

An organization will often compare its product to similar products that already exist on the market made by competitors. This process is known as benchmarking. The various attributes are compared to competitor's products in a pros and cons table. A company that does an honest evaluation of its products will gain much from this exercise. A detailed look at benchmarking is provided in the book *Quality*, by Donna Sommers (Prentice Hall).

14.9.5 Best Practices

This is a qualitative exercise in which a company can improve its internal processes to help in its overall manufacturing reliability. For example, a company may explore the entire routing of parts and process used to assemble an individual product. Among the things that are examined for improvement are:

- Simplification of manufacturing instructions
- Co-location of assembly areas and processes
- Reduction of parts travel

The best practice process is a very subtle aspect of reliability. It can be reasoned that an excessive amount of handling, too much physical movement, and extra processes can allow for more opportunity for damage to occur to the part along the way. For example, it may make sense to limit the number of acceptance tests to be performed to just key points during the production of a new product, as there is always a possibility of a failure being introduced when a test procedure is performed by the technician. Excessive part movement may result in more potential of an accident, such as a part being dropped.

Thus, the more efficient the overall manufacturing process for the part, the better chance for higher reliability with regards to manufacturing and workmanship. This sums up the principle behind the best practice approach.

14.10 SUMMARY

This chapter has demonstrated to the reader the importance of conducting a timely reliability evaluation of a new product. This review will generally be a qualitative review, wherein the product is examined for potential situations that can impact the product's reliability, as well as issues involving ease of maintenance. Many of the items that are looked for involve common sense type problems, as well as using the evaluator's previous experience with certain design features and components.

It was shown that the best time to conduct a reliability evaluation of a new product is at the beginning of the program, prior to a major purchase of a new product. By conducting this review at the beginning, there is still a chance for the manufacturer of the product to incorporate any recommended changes prior to beginning full-scale production.

It is important to realize the value of conducting this exercise and to relay any issues back to the design group. Companies that do not take heed of their customer's feedback and concerns when designing a product will eventually experience some fallout as a result of this attitude. When customers detect a nonresponsive attitude on the part of a company with regards to issues that are raised, they may look to other sources who manufacture similar products as a potential replacement.

14.11 EXERCISES

1. When is the best time to conduct a reliability-focused design evaluation on a product? Give the reasons why.
2. Describe the approach that you would take in evaluating a new printer that will be used to work with the computers utilized by your organization. What features would be most important to examine with regards to reliability performance and ease of maintenance?
3. What structural items of a new product would you examine during an evaluation? Why would some structural items be important to this review?
4. Select one of the case studies presented in each of Chapter 4 (reliability testing), Chapter 8 (failure analysis) and Chapter 9 (aspects of design). Describe how a reliability review might have been beneficial in either eliminating or reducing the impact of the failure that occurred in the specific case study being reviewed. Could the failure condition have been an item to list on a design review checklist?
5. Take a junk item from the trash, such as a monitor, computer, or electronic appliance, and perform a reliability evaluation. Make note of any damage that might have occurred to the unit during its lifetime. List the various ways that you could make the design better and address any of the common failure modes.

CHAPTER 15

Reliability Management

INTRODUCTION

Throughout the previous chapters of this book, there have been several references for the need to have a definitive approach to managing reliability for new designs. In this chapter, we will formally introduce aspects of integrated reliability management. Taking an aggressive approach to reliability management will lead to the timely completion of reliability program tasks. We will discuss how this is accomplished through discussion of the following topics:

- The benefits of having a reliability program
- How reliability program tasks are managed
- Decisions to be made for performing reliability tasks

A successful reliability management program cannot be put into place too long after a program is started. Rather, it has to start right at the beginning, at the design or concept stage of a new product or piece of equipment. A reliable design has to be a principle that management agrees to and implements as part of the program.

15.1 BENEFITS OF AN INTEGRATED RELIABILITY PROGRAM

The benefits of having an integrated management approach to reliability is primarily in the area of preventing major failures and problems from occurring at any time during the product design cycle. The fallout of not having a successful reliability program has already been illustrated in the failures that occurred in the numerous case studies that were provided in the previous chapters of this book. The following were typically the results when there is insufficient reliability incorporated into a design:

- Excessive engineering time in redesign efforts
- Excessive field returns or high amount of customer returns of the product
- Expensive retrofits or product recalls
- Lawsuits
- Loss of confidence in the company by the customer

Lawsuits involving product liability have become a major issue in the present age of technology. It is not uncommon to hear of a case in the news where a product manufacturer has been named as a co-defendant in a civil case and sometimes even in criminal court. One case in point may involve an automobile manufacturer that used a defective part in the car that caused the car to crash or catch on fire resulting in injury or death. In a previous chapter, we discussed the importance of maintaining high reliability in the manufacturer of medical equipment in order to prevent the possibility of injury or death. The old adage of an ounce of prevention is worth a pound of cure is very much appropriate in the area of product liability. Just by implementing an integrated reliability management program, thousands of dollars may be saved in court costs and product redesign costs as well as the prevention of negative publicity.

A lesser-known fact in our society which is often not publicized is that often the widows and widowers of U.S. military pilots killed in aircraft crashes may sue the manufacturer of the aircraft for damages in civil court. This is because the federal government can not be sued according to U.S. law, so the only recourse of the party seeking damage is going after the aircraft manufacturer for a perceived or real defect in the aircraft. The manufacturer may have to bring their engineers into court, either from the reliability department or safety group to testify to the aircraft's safety or reliability, as well as whether any aspect of the design played a part in the crash of the aircraft. These company experts will be subjected to cross-examination by the plaintiff's attorney, so it would seem imperative that there has to be a belief by them in the reliability and the safety of the aircraft in order to present a reasonable defense.

The aspect of negative publicity may occur in a lesser degree even when there is much less money or injury involved. How often are we as consumers affected by the opinions of others when we look to buy a product? Word of mouth can travel quickly through many forms of media. Sometimes a company may obtain a bad reputation because of poor reliability experienced by consumers but not necessarily documented in the news media. Many of the examples provided in Chapter 9 regarding automobiles were major repairs occurring because of poor reliability design that cost the consumer much money and ultimately remembered for the next new car purchased.

All of us should agree that having reliability integrated into the design process of a product is a good thing. But how can this be accomplished at a reasonable price?

15.2 MANAGING RELIABILITY PROGRAM TASKS

The size of the company will often determine who will perform the reliability tasks, as well as how many people will be assigned the task of integrating reliability into a product. Also, depending on the size and type of organization involved, the reliability engineer could end up being assigned to any of a number of different groups. These may include the following departments:

- Engineering
- Integrated logistic support
- Quality engineering
- Product support
- Configuration management
- Test engineering department

There is no absolute rule throughout industry regarding the optimal setup and where to place the reliability engineer (as well as the reliability section) within the organizational structure. Regardless of which manager is charged with overseeing reliability tasks, it is critical for that person to realize the importance of ensuring that the reliability program tasks that were defined in Chapter 3 and any reliability testing (as described in Chapter 4) are performed in a timely manner.

The manager responsible for overseeing the completion of reliability tasks, such that the program schedule is not adversely impacted, is able to do this by:

- Conducting regular meetings with the engineers assigned to the work
- Monitoring the reliability engineer's activity and participation with design
- Providing appropriate resources for the reliability engineer to complete tasks
- Participating in the program stage when manpower is determined
- Having the engineers participate in preliminary design review
- Convincing the company as to the importance of conducting reliability test or piloting a product in a field environment

The manager should be aware of the formal reliability tasks that must be performed, as they are spelled out in the contract. Often, the type of tasks that have to be completed are known for some time prior to the formal contract being awarded, during the proposal stage as requirements are spelled out by the customer, whether it be military or commercial.

As mentioned in Chapter 4, the management of reliability becomes very important in the case of a reliability warranty program (RWP). This program is mandated by the customer in which the MTBF values are tracked for production units of a product over a specific time frame in the field environment. If a company does not meet the specified MTBF during the measured time period, the customer may have the option of requesting free spares from the company as a penalty for not meeting performance levels. There may be free retrofit to all existing units for any corrective actions developed during this time period. It is critical for company management to help the reliability engineer by making sure that the appropriate material and engineering resources are available to address failures when they occur during the warranty period.

The manager who oversees the efforts of the reliability engineer during the RWP, reliability tests, or pilot programs, must be somewhat familiar with the functionality of the unit along with a general understanding of any failures that could occur as explained by the engineer. The manager has to keep in mind the potential cause and effect of each failure on the product integrity as well as on the program schedule.

Reliability tasks are associated with both new programs and existing designs. The following sections address the issues associated with both.

15.2.1 Reliability Management for New Programs

A number of things need to fall into place for a successfully managed reliability program. A major part of the job for the reliability engineer is in the area of interfacing with other engineering departments in the company in order to get the necessary information to complete each reliability task. In the beginning of the program, a certain amount of information from design engineering is needed in order to start work on the basic reliability tasks. The reliability engineer needs to be involved on a daily or weekly basis with regards to obtaining the documentation or preliminary drawings that are evolving out of engineering. Often, this requires the reliability engineer's presence at weekly design review meetings where these issues are discussed. The reliability engineer needs to take a fairly aggressive role in obtaining this information, and the manager helps ensure that this information flow is there for the task to be completed. It is important to maintain a teamwork atmosphere with the other groups in this effort.

Table 15–1 shows some of the typical engineering documents needed for the reliability engineer to start and complete work on reliability tasks.

Some of the major tasks are described in detail in Chapter 3 of this book. As seen in the figure, several of the reliability tasks are performed in parallel as the information for each task is used for parts of the other reliability tasks. For example, information that is obtained when performing electronic stress analysis is fed into the formula for predictions of electrical parts using the MIL-HDBK-217 methodology or similar software program.

The final stage is the reliability test program. Chapter 4 described the benefits of completing reliability testing in a timely manner. Some of the reasons included the lessening of the impact of costly retrofit in production units. Again, this task has to be watched carefully by the reliability manager and any issues that can impact the effective running of the test has to be addressed and brought to the attention of the appropriate departments.

Table 15–1 A) Items Needed for Reliability Analysis of Electrical Equipment

Items	Stress Analysis	Predictions	FMEA/FMECA
1) Functional Block Diagram		X	X
2) Schematic Diagrams	X	X	X
3) Parts List	X	X	

Table 15–1 B) Items Needed for Reliability Analysis of Mechanical Equipment

Items	Predictions	FMEA/FMECA
1) Functional Block Diagram	X	X
2) Schematic Diagrams	X	X
3) Parts List	X	
4) Historical Failure Data	X	

One of the real hard-sell tasks by anyone associated with reliability is trying to convince one's company for the need to perform a reliability test even when not required by the customer. There have been cases in which a company had convinced the customer to drop the reliability test requirement, perhaps citing similarity with another product line as a reason not to perform the test. A company may do this in order to reduce program costs by not running an expensive test, however costs may be higher later. A case in point where this happened ended up haunting an aerospace company that made electronic equipment for a commercial aircraft when catastrophic failures appeared during the vibration portion of ESS (burn-in) testing. A design fault that would have been picked up by the reliability test showed up during the ESS, which as a production test, is perhaps the worst time to get involved with major redesign issues.

15.2.2 Reliability Management for Mature Programs

Much emphasis has been placed on the reliability tasks of a new program. However, with regards to sustaining engineering functions of an existing program, there are less issues involving paperwork-type analysis and more issues in other areas. Some issues regarding a program that is deep into production or is at maturity are:

- Appearance of certain failure modes caused by wearout or parts quality problems
- Obsolescence issues

The latter item can be referred to in a fashion as "a fighting retreat," where an existing product line is prolonged as much as possible until a new product is rolled out. The reliability engineer may be involved in troubleshooting failures and asked to come up with solutions for keeping the existing product operating at a reliable level until it is eventually retired from service.

A case in point involves the power supply that is used to power the different devices in an envelope depositor used by a bank. It costs a lot of money for a bank to buy new equipment for all of their branches, so often considerable effort may be expended to maintain the existing product line for as long as possible. Thus, the reliability engineer may be called to investigate failures that occur on the current product. With a mature product, failures involve parts wear-out or damage inflicted by the users.

15.3 DECISIONS REGARDING RELIABILITY TASK MANAGEMENT

Early in a new program or design, a number of things must be considered by management regarding the "who" and the "how" in performing reliability tasks. The following represents a short shopping list of some things that must be considered.

15.3.1 Cost Analysis

Cost analysis involves estimating the number of hours needed to complete a task. Because of the frugal environment created by intense competition, there may not be an abundance of hours allocated for completing reliability tasks, so inventive ways will be needed to do them. For example, combination of tasks wherever possible should be done to save labor hours. Realistic estimates are needed to ensure that sufficient hours will be available throughout the program.

15.3.2 Manpower and Schedule

Manpower not only includes the number of people needed but also the matching of the proper skill set to complete the specific reliability task. In a small company, one person may wear many hats and perform all the tasks. Some companies may have a dedicated test engineering department that can handle most of the day-to-day running of a reliability test while another engineer is assigned to perform the analysis-oriented tasks.

Schedule is the most critical item for any program. Program managers want to see tasks completed on time; otherwise delinquent items have to be reported to higher management while tracked by the customer. But schedule in the area of reliability should be considered important not only for the fact that the task is completed but also the design has been properly reviewed in terms of reliability.

Companies must understand the importance of providing sufficient time for the reliability engineer to complete important reliability tasks. The withholding of sufficient manhours to complete a task will only result in that task not being adequately completed. By the same token, it is important for the reliability engineer to realize that each task is not meant to be a lifetime sustaining charge, so a realistic estimate has to be worked out between management and the engineers involved. Of course, any way that can save time or combine assignments are very helpful to the process.

It is very, very important for companies to free up sufficient manhours early in the program for reliability engineers to start performing the necessary analysis for reliability tasks. For example, a FMEA (failure mode and effects analysis) can be completed in preliminary form based on what is expected of the product, even before a unit is built. For complex systems, it is very useful for such work to be done as soon as possible so that when details are filled in on the FMEA worksheet, concerns can be raised immediately. For example, if a failure of a part in the system becomes undetectable, there may be a chance to correct this while the design is still ongoing. This is also true for the task of electrical stress analysis where overstressed components in a circuit can be identified right away and a change can take place in using a higher-rated component in the circuit.

Reliability testing is one area that may sometimes not be fully appreciated by the program manager with regards to the impact on design or schedule. It would be wrong to think of reliability testing as merely another contractual obligation that has to be satisfied. To the eyes of the customer and the company reliability engineer, this testing is of greater importance than just a paperwork requirement. It verifies that the design can work for temperature extremes, as well as vibration levels that it may be subjected to in the field environment. Also, it brings up any failures that should have corrective action developed and implemented.

It would be wise to ensure the timely completion of this test right after some qualification tests are completed and before the delivery of too many production units, primarily because of the risk of retrofit of corrective action into any units that have already been built.

15.3.3 Outsourcing Reliability Tasks

Smaller companies may not find it feasible to maintain a full-time reliability engineer because of the amount of reliability-related work that is performed over the course of a year or the life of a new program. Thus, it may be better to explore options such as hiring a consultant or job shopper to come in for the period of time that the reliability work needs to be performed. These individuals typically work on an hourly rate or sometimes an overall price for each task can be worked out. Sometimes the work load may be too much for an existing company reliability engineer and some of the work can be farmed out.

Another option is to outsource the work to a firm that specializes in performing reliability engineer tasks. The areas where this is done is during the calculation of reliability predictions, or the performing of an FMEA. Another area where this is commonly done is when the reliability testing is performed in a test lab that uses their equipment and their technicians. The company will usually provide the engineer to oversee the work being done by the lab. Again, this cost could be charged on an hourly, daily, or task rate.

15.3.4 Tools Selections

Various tools are available to perform reliability predictions and FMEA worksheet analysis, usually in the form of software packages or spreadsheets. One of the issues facing management is whether it is useful to purchase a software package or to construct a simple spreadsheet program with built-in macros and subroutines to do reliability predictions. The answer is governed by the complexity of the design and the amount of work that has to be performed.

Is it better to buy a software package that costs $10,000 and complete the analysis in two days, or is it better to have a spreadsheet analysis constructed by an engineer that takes two weeks? These are the types of questions that have to be considered with regards to selecting the right tools for the job.

15.4 RELIABILITY PROGRAM CASE STUDY

Item: Organize a reliability program for an electronic control unit that drives the rudder actuator for the rudder flight control surface of a commercial aircraft

The product will be installed in the cargo area of the aircraft and will see moderately high levels of vibration during field service. The product is a single box that has external connectors and has five circuit boards built into it and consists of approximately 200 electronic components for the entire unit. The customer has requested that all of the major reliability tasks be performed that were described in Chapter 3.

1. Tasks to Be Completed Prior to PDR (Preliminary Design Review) The manpower estimate for completing the reliability program can be based on the work that was done for previous plans written for other programs, thus an update of an existing file can be completed with moderate effort by the engineer. The program plan lists each of the major reliability tasks that are to be completed for the program along with the methodology that will be used to complete each task. A new program plan may take up to thirty hours to complete while an update of an existing file may take less than twenty hours of editing work.

The unit used in this example is of moderate complexity, so a parts count prediction and initial FMEA should be straightforward for an experienced reliability engineer to complete. The parts count method can use a simple spreadsheet methodology in which the base failure rate, the quality factor, and the quantity are listed into the appropriate column. As the parts count prediction is used during the early stages of the design, it is generally not required to know the exact part number for many of the electrical components, just the general type of the component. For example, an exact value resistor that is drawn in a preliminary electrical schematic may not be known, but the type that would be used is known (i.e., carbon resistor).

Likewise for the initial FMEA, the general outputs of a functional block may be known, even though all of the components may not be identified at this time. In this case, the outputs are fairly well defined with regards to the control of the rudder actuator. It is generally understood that as details become available, they will be used in a more refined version of the FMEA that will be done later on.

Depending on program constraints, a general estimate would be about thirty to forty hours for each task (parts count reliability prediction and initial FMEA).

2. Tasks to Be Completed Prior to CDR (Critical Design Review) At this stage in the program, the design is starting to be finalized where the actual parts are known. The electrical schematic is taking shape at this point where specific values are known. A working prototype may actually be constructed at this stage by design engineering.

At this stage in the program, a stress analysis is required to verify that the electronic components are not being overstressed in the circuit where premature failure could result. As the reliability engineer examines each component in the circuit and performs stress analysis as described in Chapter 3, any overstressed component must be brought to the attention of the design engineer for further action, usually in the area of upgrading the rating of the component used. For this example, an estimate of about fifteen hours per circuit board would yield an overall effort of about fifty hours for this task.

While stress analysis is being performed, the information can be fed into a formal reliability prediction program. The use of software in this area is essential as updates and revision are quite common at this stage. This task takes approximately the same amount of time that the parts stress analysis effort took, so fifty hours is a good estimate for this example.

The FMEA becomes more of a complete document at this stage from the initial report that may have been submitted for the PDR. The worksheet that is used for the FMEA is described in Chapter 3. More information is known as the design starts to become more defined. At this stage, with the reliability predictions being completed, this information can now be plugged into the criticality effort of the FMECA worksheet effort. Each of the efforts for the FMEA and FMECA involve a lot of typing in filling out the various fields of the worksheet, typically requiring forty hours for each task for a design of this complexity.

3. Task after CDR and Prior to Production Reliability engineers may be involved with two tasks after CDR. Typically these would be a reliability test and sustaining efforts during early production. As explained in Chapter 4, several types of reliability tests can be imposed by a customer. Since the equipment is to be installed on an aircraft, the test will most likely involve both thermal and vibration cycling. A reliability test may take a minimum of three months to complete, if all failures are addressed immediately or if no failures occur. If the reliability engineer is performing this task on a part-time basis (along with other tasks from other programs), it may require up to twenty hours of effort weekly. Thus, for fifteen weeks of effort, this would work out to be about 300 hours just for the test routine. Associated with the reliability test is the procedure and the final report, each of which could take up to forty hours of effort to complete. Thus, one could come up with a minimum of 500 hours of effort.

4. Production Support Management may elect to allocate some hours for the reliability engineer to provide sustaining support for production units that enter field service. The amount of hours that may be needed for reliability engineering to provide analysis of field failures will depend on the complexity of the unit and how well it is performing in the field. A reliability engineer will often aid in the troubleshooting of repetitive field failures and help develop corrective action. There may be a tendency for management not to include hours for this sustaining effort, but they should be included if the company has had a history of problems with field units in previous programs. For this example, an estimate can be on the order of 5 hours per week for one year of support or approximately 250 hours.

5. Summary of Reliability Manpower for the Control Unit Program For the example used, we come up with the following shopping list of reliability tasks along with manpower estimates:

I PDR Tasks
    1) Reliability Program Plan   20 hours
    2) Parts count reliability prediction   30 hours
    3) Initial FMEA   30 hours

II CDR Tasks
    1) Parts Stress Analysis   50 hours
    2) Reliability Predictions   50 hours
    3) FMEA and FMECA   80 hours

III Pre-Production Tasks
    1) Reliability Test   500 hours

IV Production Tasks
    1) Sustaining product support   250 hours

The total hours needed for reliability manpower in this program is 1,000 hours.

This is one example of how to estimate hours for a reliability program. This example is meant to demonstrate the basic template that may be used for listing the various tasks and a typical methodology that can be used towards estimating hours. Methods may vary between different companies. It is also understood that the number of hours can vary significantly depending on the complexity of the product, as well as additional customer demands that may be requested.

15.5 SUMMARY

The need for an organized reliability management approach benefits the company in all aspects. It may be a hard sell for this approach for a smaller company but as explained in this chapter, there are a number of ways that this can be accomplished. The philosophy of invest now rather than pay later is very much the philosophy that has to be pushed. As demonstrated in the case history that was presented in Section 15.4, about 1,000 hours of reliability manpower is needed for performing the reliability tasks for a typical aircraft program. This effort may be less for many commercial products, as certain tasks may not be required.

But even at 1,000 hours of effort and at an engineering labor rate of $60 an hour, the expenditure for manpower will be less than $100,000. Compared to the amount of money that would be needed to defend against a lawsuit and payment for subsequent damages, this is a worthwhile investment!

15.6 EXERCISES

1. Examine a design program in which you are involved. If the program does not already have a reliability engineer assigned to it, describe how the program would benefit from having such an engineer. What reliability tasks could be performed and when should they be performed in this program?
2. State some of the risks and liabilities that a company may face when certain reliability tasks are not performed.
3. A major issue for many companies is the cost of performing reliability analysis. What are some ways to address this for a simple product line (less than fifty parts)? How would one address ways to save costs for performing reliability task for complex systems?

CHAPTER 16

Epilogue: Other Useful Skills

INTRODUCTION

In recent years, it has become more imperative for the reliability engineer to have other skills beyond analytical and engineering skills. Some of these are intangible skills that include so-called "people" skills, such as the ability to be able to communicate effectively, either by speaking or writing. It becomes critical in many aspects of the reliability engineer's job, such as design review meetings or in the area of regular customer interface. A more worldly approach, being aware of the big picture for a program, is required of the reliability engineer with special emphasis on being flexible enough in order to deal with unusual or difficult situations that will inevitably arise.

This chapter will cover this unusual aspect of the reliability engineering profession that is often not covered in textbooks, by using some of the personal experiences of the author and others in the reliability field.

16.1 NON-ENGINEERING SKILLS

Over the course of an engineer's career, a number of non-engineering skills will become a necessary part of his or her toolkit in dealing with various situations. For example, an engineer running a reliability test program that uses a company's environmental lab facilities will need to obtain the full cooperation from the lab manager and technicians. Often engineers will run into the situation where production equipment needing burn-in takes priority of thermal chambers and vibration tables over any reliability testing. Sometimes a compromise can be reached in which a uniform thermal profile fits the needs of both production units and test units, and thus a thermal chamber can be shared with both sets of units.

This is one aspect of teamwork building. Another aspect of teamwork is when engineers of various disciplines work together during the beginning of a new design program. They share their knowledge, skills, and common experiences toward developing the best product possible. By having a synergy of different talents, the team as a whole will be greater than the sum of its parts. The experienced reliability engineer will be a major part of this by pointing out aspects of the design that could experience failures if not corrected.

Over time, it is possible for the engineer to develop some talent and skills in these other areas. The following are some of intangible skills that can help the reliability engineer become a more complete engineer:

1. Knowing what the customer wants
2. Ability to multitask (do several jobs at once)
3. Maintaining a distrustful nature
4. Expecting the unexpected
5. Having a natural curiosity in other areas
6. Writing skills
7. Not being afraid to break things while testing
8. A realization that there is no one "magic bullet" to solve a problem
9. Dispelling the myth: "anyone can build these parts (or do this job)"

16.2 CASE STUDIES AND DISCUSSIONS

An explanation of the different concepts listed in Section 16.1 will be explored further in the following case studies. Several of these case studies are based on the author's firsthand experience.

16.2.1 Knowing What the Customer Wants

Item: A year in the life of a reliability engineer

The customer is the end product for a reliability program and this fact should always be remembered. The author, along with many other engineers, got a big dose of this lesson in a single year—the year of 1987.

At the beginning of 1987, the author was a senior reliability engineer for the Fairchild Republic Company in Farmingdale, Long Island, and was working primarily on the new T-46 jet trainer program that the company was building for the U.S. Air Force. The company had just completed the production run of its successful A-10 aircraft program of 713 aircraft in early 1984, and the T-46A program was to be the next major program. This new aircraft was to be the primary jet trainer for the U.S. Air Force, and it featured side-by-side seating for optimal student and instructor arrangement. The company had won this contract against two other companies in late 1984 through their technical proposal as well as their low bid. They had also helped their efforts by constructing a three-quarter-scale flying prototype of their design (as shown in Figure 16–1) and this impressed the air force. In addition to the trainer version of this aircraft, there was also another version of this aircraft being developed that could be used as an attack aircraft with excellent potential for foreign sales. However, things would not work out.

Unfortunately, so many issues came up with this program that ranged from parts delay to design and reliability issues, that the air force was beginning to lose confidence with the

Figure 16-1 As part of the proposal effort for the air force trainer, the Fairchild Republic Company constructed a three-quarter-scale flying prototype (as shown) in order to demonstrate flight characteristics. This was instrumental in the company winning the bid. However, other issues such as cost overruns later created severe problems for the company.

company being able to meet milestones in the T-46 program. Problems started right from the beginning for Fairchild Republic because of their low bid for the program, resulting in insufficient funds to drive the effort. Because it was a fixed-price contract, the company was forced to pump additional money from its own resources. However, even this was insufficient to fund design engineering as well as buying equipment from outside vendors. With the lack of money, the company sought to solicit data items and products from vendors at no cost using the carrot of production sales as the incentive to do so.

The lack of money for the vendors became a serious problem in the world of reliability when vendors were slow to submit the various reliability data items of stress analysis, predictions, and FMECAs. Even worse was the fact that reliability tests for over two dozen new pieces of electronic equipment were not being started because there was little promise of money for the vendors to be paid for running these tests. Thus, parts were not being delivered because they were not finished or even entered into any sort of qualification testing. It was hard to enforce any kind of effort to get testing done, and thus the reliability department, along with the other disciplines, became a paper tiger in trying to get tasks completed.

The original aircraft MTBF requirement of 10 hours to be reached at maturity did not look like it was going to be achieved as a number of issues came up during the design stage. Indeed, the 10-hour aircraft MTBF requirement was derived incorrectly by Air Force analysts. While they used fairly conservative estimates for individual pieces of equipment that were to be installed on the aircraft, they forgot to take into account that there were two sets of some of the avionics equipment in the cockpit: one for the student and the other for the instructor. This would have reduced the overall MTBF value for the aircraft. However, despite being alerted to this arithmetic error, it was too late in the program for the Air Force to change the 10-hour aircraft MTBF requirement, thus further handicapping the efforts by the company. Given the complexity of cockpit equipment, a more realistic aircraft MTBF requirement should have been on the order of 8 hours overall for the aircraft. The issues regarding reliability were typical of the many design and cost issues with this program.

The turning point in the program came in February of 1985 when the company presented an incomplete aircraft at its facility during a rollout ceremony in front of many major Air Force officials. The company was under pressure to meet this rollout date even though they were not getting cooperation from several vendors in the delivery of key parts. The Air Force was angry after it had found out afterwards that the plane was incomplete with several parts missing. In retaliation, a major contractor review was conducted in June of 1985; the company failed in several areas, and as a result, all funding was stopped for a period of time.

Over the next two years, a lot of effort was expended by the company to try to win back the confidence of the customer but to no avail. Flight testing did begin in 1986 as shown in Figure 16–2. There was even a brief holdup in the U.S. budget because of congressional debate in the Senate to try to save this program. The end finally came, appropriately enough, on Friday the 13th in March of 1987 when the Air Force formally terminated the T-46A program and 500 employees (including the author) were let go after only three aircraft were built. The Air Force hurt itself by canceling the contract, as it would now have to make do with aging T-37 trainer aircraft as their primary trainer from this point on. It would be several years before new trainers would be available, but these would be foreign-designed propeller aircraft, not jet engines like the T-46A design! Perhaps some sort of compromise by both the Air Force and manufacturer at the time would have prevented this result.

A secondary (and perhaps even more valuable) lesson is that several different things outside of the actual aspect of engineering can undermine even a good design concept. These things include politics, lack of supplier cooperation, and customer errors (such as the MTBF calculation) that are beyond the control of the design engineers. The T-46A story shows how bad things can become once the program starts on the wrong foot. Thus, once a situation reaches a point of no return and out of the hands of the engineers and in the hands of non-engineers (such as politicians), it can get pretty ugly, resulting in the loss of jobs. Some of the Fairchild Republic employees were able to find jobs in the New York area but generally with smaller companies. A new attitude would have to be adopted by some of these people in order to make the adjustments in working for these smaller companies through their understanding of the T-46A experience. They now had experience from the T-46 situation in knowing what was the type of product or data that customers wanted and were able to apply this knowledge in their new companies.

Finally, it is important to note that paper analysis alone does not guarantee that a product will meet the specifications. Witness the error that was made by the Air Force during the estimated MTBF prediction. The fact that testing was not going well also showed some of the weakness in the paper analysis. Sometimes the customer is willing to work through issues, and sometimes the customer can be very firm and not bend on the requirements that they specified.

For the reliability engineer working on a new design program, it becomes necessary to build working relationships with the reliability counterpart from the customer in order to make sure what was expected in the way of content for data items that were to be submitted. Sit-down face-to-face meetings may be part of this process of customer interface. Once a mutual trust and understanding is established, it becomes easier to address issues when they come up

316   Chapter 16

Figure 16-2 In the photo, the second pre-production T-46A aircraft is taking off from Republic field in Farmingdale, Long Island, in July of 1986. Within nine months, the program would be cancelled. The irony about the cancellation of this program was that this was a working aircraft, but politics and funding entered into the situation. These are things that reliability engineers have to go through despite any good work that they may do.

(such as the situation of the erroneous MTBF calculation). It is all part of understanding exactly what the customer wants. The worst situation that a company can get into is the loss of confidence by the customer in the company's ability to make or deliver a product or service.

Companies come and go as the result of events that occur during a program. It is important to always remember one's coworkers from a former company that exhibited excellent skills in the area of design, as these former coworkers may become useful if one should end up in another company at some point in the future.

16.2.2 Ability to Multi-Task

Item: Multiple programs and multiple reliability tasks

Some years ago, the way to work was to complete one task at a time and then go to the next task. This old way of serial thinking is obsolete in recent times where there is demand for quick turnaround of task completion. With the advent of many tasks to be performed, it is impossible to do things in a serial manner because there are tight schedule constraints. A better way to manage time is needed, and thus parallel-type thinking or multitasking is the current trend. Basically, as the name implies, several tasks would be worked on during the course of day. No one task can afford to be neglected in this type of setup. More time during the day may be devoted to priority items that are due sooner, however each task that is due for completion in the near future must have some attention paid to it either on a daily or a weekly basis.

From the author's experience, he found that during his time with an electronics company that made aircraft equipment, he learned that the busiest time was when he was in the reliability test phase of one program at the same time that another program was starting up. During start-up of a new program, usually a whole slew of reliability tasks have to be performed and submitted to the customer within a short period of time after the contract has been awarded. Yet, neither the test program, nor the new program starting up was any more important than the other was where a priority could be set. So, at this point, multitasking was the only possible way to deal with this situation.

A situation that describes this point was when the author was conducting reliability testing on fuel-gauging equipment at the same time that he had to give a reliability presentation during a preliminary design review (PDR) with the customer for a display unit on another program that his company had built. It so happened that on the day of the PDR, the author was able to get access to the vibration table for the reliability test on the same day. So here was the author in a formal suit, setting up a unit on the vibration in the lab that was located in the basement. A half-hour later, he was upstairs in a conference room giving a talk on the various reliability tasks that had to be performed in a PDR presentation. The author uses this example whenever he is asked a question during an interview with regards to the ability of being able to handle several tasks at once.

Every person has been exposed to some aspect of multitasking at an early stage of life. It actually begins in school and is especially emphasized during high school and college where students have several homework assignments at once. And thus the trend continues for us on a daily basis in our professional lives. The wrong answer for an engineer to say to management is "I can't do that right now," rather than, "I will try to get some work done on this today." Serial thinking is not going to be acceptable to most managers in business today, and multitasking is the way it is in today's world.

16.2.3 Maintaining a Distrustful Nature

Item: A small manufacturer attempts to go beyond their experience

It serves the reliability engineer well to have a natural distrust in dealing with vendors that supply parts to the company. The golden rule to follow is not to trust any vendor until the work is satisfactorily completed. It would be wrong to use a stereotype such as "all vendors lie," but if one has the experience of having been burned by one or two vendors, one learns to maintain a distrust until the trust is gained.

This is a story of a reliability engineer who had the experience of working for a small company on the east coast of the United States that made survival kits for military aircraft as well as various aircraft seats that were made to drawings supplied by the contractors. This company was also the second-source vendor for an ejection seat for various navy aircraft, with the prime source being a well-known ejection seat company that was located in the United Kingdom. The company was a family-owned business that employed less than 100 people and was looking at a major contract in-house that could put the company over the edge toward becoming bigger in size.

An opportunity arose when a number of seat contracts came up on an Air Force transport aircraft. There were four different seat designs for this plane: a simple pullout seat for the troops, a bench seat, a load master seat, and a complex seat for the pilot and copilot. The latter seat was complex and similar to that of a barber shop seat with the pilot/copilot seat having all sorts of controls as well as riding on a track. Actually the vendor was only going to bid on the first two seat designs, which were relatively simple, but the vendor encouraged by the contractor to bid on the other two as well. At the time that the reliability engineer was working on the proposal for the company, he could see that this was going to be a lot of design-related work, something to which the company had not been previously exposed. There was going to be a ton of data items, including the complete spectrum for reliability data items. Basically, he saw this contract would be a make-or-break situation for this small company.

The engineer had another opportunity during this time and left the company as he did not have a trust or healthy feeling about the long-term prospects for this company making it on a higher level. Shortly after the engineer left the company, they won all four seat contracts. Now the fun would begin as it was wondered how this company was going to develop new designs from scratch as well as completing all of the reliability data items with a very small engineering staff.

The company was in the hole right from the start. The amount of work required at least twenty to twenty-five experienced engineers, not the staff of eight that they had in the company. They scrambled to pull in consultants to do some of the engineering tasks, as well as the reliability data items. This approach was like applying a bandage to fix a major cut.

The aircraft manufacturer, or the prime contractor, was located on the west coast and should have been more suspicious about the vendor's capability right from the start. The signs were all there. People from the prime contractor who visited the facility should have noticed that it was a small shop with a very small engineering department. They should have applied a sort of confidence test in their mind about the vendor's capability, and the warning flags should have gone up. The vendor had little previous experience in the area of designing seats from scratch on other programs. This situation is akin to hiring someone without references. A vendor with little background in the area provides very little reassurances to the main contractor that they are capable of doing the job. This was true in this case.

After the contract awards were made to the seat vendor, several data items had to be submitted almost immediately. This caught the vendor off guard because they did not have the necessary manpower or the expertise at the time to complete these data items. So in a relatively short period of time, the vendor fell behind in submitting the required data items, especially the reliability tasks.

The situation got progressively worst over the next few months, not only for the completion of the data items but in the design of the seats themselves. The vendor was not able to get a finished design on paper or even a working prototype concept ready in time for the critical design review. A working prototype was in order for this program, especially for the pilot seat because of the complex swivel and cable arrangements. The vendor was not spending any upfront money that was needed to properly develop the new design. This came from a lack of experience of what was needed to complete the job.

It was inevitable that the vendor would come under increased scrutiny by the contractor with numerous phone calls and visits being made by the contractor. The hiring of several consultants allowed for some reliability data items to be completed, but the company was not able to pay the consultants at all, and thus, many data items remained unfinished. The four designs remained unfinished and the vendor was formally cited for non-compliance of the contract.

There was a point of no return when not a single prototype was submitted of each seat type for the test aircraft. When it became apparent that no seats were going to be made, the prime contractor's representatives scheduled a trip to the east coast of the United States to visit a number of vendors on the aircraft program, with the last vendor that was visited being the seat vendor. When the prime contractor made this last stop, they formally told the vendor that they were off of the program as they were in default in accordance with the contract agreement.

The prime contractor had to quickly find a replacement seat manufacturer to take over this program. They were successful as the reliability engineer would later find out when he had a chance to visit the aircraft manufacturer's factory some three years later, as an employee for another company that made equipment for the same aircraft. He saw that all four seat designs were completed and saw the name plate of another company. It must have been quite an effort for the aircraft manufacturer to get this second company on board.

The lesson is clear here. Maintain a healthy distrustful nature, particularly with new vendors having no recorded experience. The deficiencies presented here could have been seen by the contractor in this story had they probed a little bit beyond the surface of what was being presented. A contractor can gauge the situation sometimes by seeing whether the company has a dedicated reliability engineer, particularly if there are many reliability tasks to be performed such as FMEA and FTA. The contractor has the right to ask how these data items will be completed if there is no dedicated reliability engineer in the company. A hired consultant may not be sufficient. There is a discernable line between the minor leagues, where smaller companies are situated, and the major leagues, where larger organizations deal with complex reliability tasks.

Also illustrated in this example is the fact that many small companies fall into the trap of "small thinking," even if the potential for growth exists when winning a major contract. This can lead to a conflict not only between companies but also with the goals of the employees of the company as they strive for optimal performance.

16.2.4 Expecting the Unexpected

Item: Circuit board failure on controller unit in test

These should be the key words that any reliability engineer should heed. These words can come true at any point of the design program, particularly during the test program or early field service. Many of the unexpected events come when assumptions are made. Just when veteran engineers think that they have seen it all, a new or unusual failure mode comes along.

This happened to a reliability engineer during a particularly difficult reliability development test that was being performed on a controller unit for an air force transport aircraft. The engineer was having a rough go of it with one of the two units used during testing. The older unit that was built had many failures that made it difficult to even complete one 8-hour thermal cycle. It was in this setting that the engineer was troubleshooting a strange failure mode. The unit had a two-line LED display and the lower display was not functioning. The engineer was able to isolate the failure to one circuit board which sent the data to this display but could not figure out why this board was causing the failure. When the board was installed inside the unit, the failure occurred, yet when the board was put on an extender card (see Figure 16–3), the display was functioning! Then by chance, the engineer touched one of the extractor assemblies on top of the board (the card extractor is a mechanical assembly that is used to pull the card in and out of the unit) and he felt a burning sensation. The engineer put a multimeter in the voltage scale on this point with reference to ground and found that 5 Volts was being read on the card extractor!

You can imagine how much fun the engineer had in showing the particular failure mode to the design engineers of the program. It was apparent that there was a fault with the circuit board itself. It was a multi-layered board, consisting of ten unique layers of traces connecting the various components with one layer dedicated to being a ground plane and another being an active voltage plane of 5 volts. The board layout artwork was examined and it was seen that on the 5-volt plane, the clearance between the hole for the extractor pin and the rest of the 5-volt plane was very marginal. It took only a few cycles of pulling the board in and out before the pin shorted with the surrounding 5-volt plane. This short caused the data line input from this board to go to ground and cause a blank display. An immediate corrective action was developed in the artwork where the nonconductive area around the pin hole would be increased (as shown in Figure 16–3).

The lesson learned here was that one should expect during testing and early production that any possible failure could occur. As shown in this case study, even a circuit board can be taken for granted where one would normally not expect a voltage plane failure. But some carelessness occurred when the board layout was being done and there was insufficient clearance between a voltage plane and potential grounding point (in this case, the pin holding the extractor). So unfortunately, there is no given in that even the simplest device is working properly.

Lest one thinks that artwork errors in electrical circuits are rare, the same engineer encountered a similar situation in a completely different setting. In this case, the piece of equipment was a printer for an ATM and the electrical short was the result of a circuit trace carrying 5 volts that was routed directly underneath a 30-volt fuse holder. As the coating over the 5-volt trace degraded over time, arcing occurred between the 5-volt and the 30-volt circuits which caused destruction of several 5-volt ICs. Indeed, from these examples, the layout of electrical circuitry is not something to be taken for granted and vigilance is required to prevent subtle errors from entering!

16.2.5 Having a Natural Curiosity in Other Areas

Item: The aurora borealis and the interview process

It is very important to have a natural curiosity, not only as an engineer in your professional career but also in life in general. On occasion, one's outside interests or hobbies can overlap with a special work assignment. A true story involves a reliability engineer who was able to use his amateur radio background when dealing with RF problems that come up occasionally in his job. However, this additional interest would later play an even bigger part during a critical stage in his life and is perhaps one of the strangest examples of having other interests besides your main field.

It occurred for this reliability engineer when he had gotten laid off from an aerospace company and was looking for another job. Fortunately for him, he was able to get an extension from the company where he could continue to work until he could find something. A friend of his pointed out an interesting ad in the weekday paper which simply said, "Reliability/Engineer, develop failure rates." He called immediately, and it turned out that the ad was placed by an employment agency that was fronting for the company that had the position. He was not told who the company was, just that it was a bank that had a lot of equipment, both inside the branch and in the form of ATMs.

It took over a month, but he finally got the call for an initial interview, and it turned out to be in the building next to the building where the engineer was already working! So it was

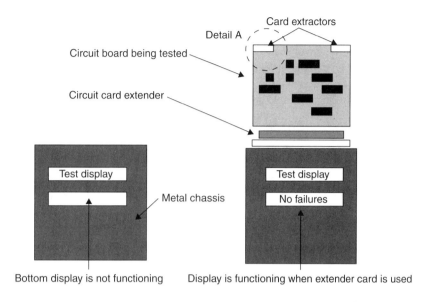

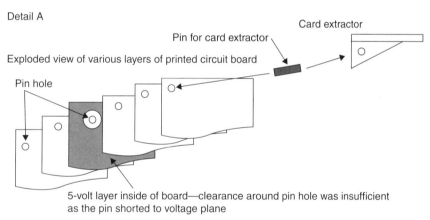

Figure 16-3 Description of Circuit Board Voltage Plane Failure

just a short walk across the parking lot to the interview. The engineer spoke with two people during the first interview that lasted over two hours. Then a second interview was scheduled a month later and lasted just as long with over twenty scenario questions being thrown at the engineer. The engineer's previous experience had told him that usually by the second interview, one has an idea if they will be hired or not. In this case, he was told that he was tied with one other candidate who was being considered for the job. This was particularly frustrating, as the interviews were quite extensive, and he was not getting a good feeling that he was making significant headway in being the one that they wanted.

A third interview was set up a month later, where he was to be interviewed by the higher management of the organization. After speaking with one manager, the engineer was brought to the office of the leader of the engineering organization. He was a very friendly person, and he focused not on his experience but on an item that was listed in the engineer's resume.

The engineer had been writing for a few years for amateur radio publications about various types of radio propagation. He listed a few of these articles in his resume under the section of published works. One of these articles that he had listed had caught the attention of this manager, an article on aurora propagation. (Aurora propagation is where radio waves can be reflected off of the aurora borealis, also known as the northern lights.) As it turned out, this manager had an intimate knowledge of this phenomenon from his military service, where he had spent two years in Fort Churchill in Manitoba, Canada, helping with the scientific studies that launched rockets into active auroras that were seen up there. The engineer had known about this launching of rockets that was conducted at Fort Churchill based on research conducted for the article. This was a most unusual connection to have with the interviewer!

So for the next hour, this was all that was talked about—the aurora borealis and information that was shared on this amazing phenomenon. At the end of an hour, the manager felt he had to ask a job-related question, and he asked why would the author want to work for a bank. The engineer gave him a straightforward answer about bringing the skills that he had picked up from the aerospace field and using it for the bank. This answer satisfied him, and they went to lunch. He promised an answer within a week, which was good considering that the process had already taken for three months and time was running out in the current job. The engineer was finally happy to get the phone call a week later telling him that he had the job. After he joined the company, he had some additional conversations not only about the job but also about the aurora.

The important point being presented here is that it is not how much you know but what different things you know. These different things become important, especially when a connection is made with an interviewer who has experienced the same thing. It could have been something else besides the aurora borealis, but the point is that a resume that reflects knowledge or written works in other areas besides your immediate profession is a real plus for getting a job. Knowledge of subjects off of the beaten path of mainstream subjects reflects another aspect of your personality. Management looks for people who think out of the box and an inclination or curiosity toward learning about other areas of knowledge is important. In this case study, the engineer's knowledge about the aurora borealis was a major part toward him obtaining this job, even though it had nothing to do with the job itself!

16.2.6 Writing Skills

Item: Writing test reports and procedures

Writing is a very important skill for anyone to have, particularly for engineers. This is because engineers will be called upon to write many procedures, data items and reports during the course of a new product being developed. For reliability engineers, there will be many items to write up during the beginning of a new program, such as reliability predictions, FMEAs and test procedures. Towards the end of product development, the reliability engineer will be writing up test reports and occasionally, white papers that discuss a failure investigation.

The interesting thing that occurs after writing many procedures and reports is that you start to develop an organized approach to writing. Many reliability reports will follow the same basic format as follows:

- *Cover Page:* This contains the title of the document, data item reference and the signatures of the author and managers.
- *Section 1:* This section is an introduction where the purpose of the report or procedure is stated.
- *Section 2:* This section may provide a listing of applicable documents that are used as a guideline for what the particular data item or report should cover.
- *Section 3:* This is the main body of the report and it covers the aspects of the test to be performed or how the reliability analysis was performed.
- *Section 4:* For most reports, this will be the last section of the report where a summary of the results are provided.
- *Appendix:* For test reports, this will be a depository for data sheets and test charts. For reliability data items such as predictions or FMEA, this may be the area to provide the actual failure rate prediction analysis or FMEA work sheets.

The tendency is that because of the high amount of technical matter presented, many of these reports can be quite dry so anything that can help save the report from being totally boring will be a testament to one's writing skills. Unfortunately, because of the heavy emphasis on engineering courses that are needed for an engineering degree, many people in the engineering field may not have had the benefit of creative writing courses during college. But in the professional world, writing is usually going to be a major part of the job. For an engineer involved with reliability testing, it would not be uncommon to write on the order of forty to fifty reports and procedures during a given year. So organization becomes very important.

It becomes possible to improve one's organization by writing often, and this also will show up in other presentations (both written and verbal). It can also lead to improved writing in other areas of one's life, as well as outside of work where small writing projects may lead to larger writing challenges such as a book. Each step in writing that is taken is progress towards being able to take on bigger projects.

A person cannot help but get better after writing these types of reports on a regular basis. Organization of ideas in a presentation gets better with each report that is written and will make it easier for management to approve the report during the review process prior to it being sent to the customer. One sign of success that the report was written well is when the customer approves the report with minimal or no comments.

Being able to write well or present one's ideas clearly in any type of presentation is a power that can be used during the course of a career. The skill of being able to write well can be obtained over time by continuous writing or significant amounts of reading and, in some cases, by taking a course in creative writing.

16.2.7 Not Being Afraid to Break Things During Testing

Although this may seem like a strange thing, it actually helps to have a natural clumsiness when you are a reliability engineer. People who have a natural talent for taking things apart and correctly putting things together right all the time, actually make lousy test engineers. Nothing will be discovered about the durability of the product if the item is not tested either to or beyond the breaking point. The product will have to endure all kinds of handling when in field service, so it is so important to be able to recreate real-life situations when testing, whether it is a reliability test, qualification test, or a maintainability test. Think of it as a legitimate excuse to break something on the job and not get fired for it!

This particular talent for clumsiness is a good thing to have when product testing. The author has had the opportunity to review two or three amateur radio products a year and write up the review for various amateur radio magazines. Part of the testing involves running the products in different environments and testing for durability. Durability is an area where many products fail.

It is proper to test products to their limits as part of the product review. This is exactly the same approach that a reliability engineer should take when reviewing a new product whether it be an aircraft manufacturer evaluating vendor equipment that goes on aircraft or a bank reviewing a new ATM design for possible use.

It is important to focus on the areas that see the most user interface as well as any possible weakness in the design that could show up during field service. For example, a lever that opens a printer or copier machine may be the first piece to break because of normal use by those who service the machine to replenish with paper, ribbons or toner cartridges. By identifying any potential weakness early enough, it may be possible for the manufacturer to reinforce the design of weak components. Also, a potential high use spare item is identified if a component is susceptible to regular breakage during normal field use.

16.2.8 A Realization That There Is No One Magic Bullet

It is a natural human tendency for engineers and their managers to jump at finding the one solution that will solve a reliability-related problem. Management in particular loves the concept of a "magic bullet" that will solve a difficult problem. They may say, "Good, that's the solution!" and "Now we can continue with the program without any more delays!" This may be possible on occasion, but quite often there may be a series of solutions that will be required to solve a reliability problem. Also, sometimes it may be necessary that an interim solution be developed and be put into action right away while a long-term solution has to be worked on.

One of the reasons why a "magic bullet" may not exist for a problem is because the failure may be the result of a complex series of events and there may be many steps needed to solving a problem. A case in point that was discussed earlier in this book described how analog circuits are particularly prone to the effects of water. Conformal coating of an entire analog circuit board may be viewed as the magic bullet, but in reality, other things may be required such as covering up analog test points on a board or changing the impedance of some of the input circuits to make it less susceptible to moisture. Thus, the overall solution to a problem may become a chain of smaller solutions that are implemented together.

It can be particularly difficult when engineers are couched in a non-technical organization and have to explain to their management that the one solution that is proposed may not be the complete answer. The engineer who conducts an exhaustive investigation may see that more work has to be done in addition to an initial solution that may have been found. Thus, the engineer has to caution management or provide a caveat with regards to expectations when solutions are developed for problems.

16.2.9 Dispelling the Myth: "Anyone Can Build These Parts (or Do This Job)"

The myth that anyone can build a certain product or do a certain job is perhaps the biggest myth in the field of engineering and manufacturing. Most of the time, non-technical types such as marketing, contracts, purchasing or upper management, who may not have the technical knowledge to know better, carry this attitude. Unfortunately, the burden of dispelling this lie, or the task of picking up the pieces after a seemingly incorrect decision has been made, often falls on the reliability engineer. This will be illustrated through two different case histories.

The first involves a situation where during the manufacture of a military aircraft, a number of machined parts were sent to outside vendors for production. After these parts were made, they were then sent to the receiving inspection facility of the aircraft's prime contractor for verification that the dimensions were correct. One part that was sent out was a large flange fitting that had four decimal-place dimensions on the inside diameter of the flange fitting. A vendor made and then sent these parts for inspection and it was seen that there were significant cutter grooves on the inside diameter of the flange, thus causing the inspector to reject and send the parts back. Three weeks later, another large shipment of this flange came from the vendor. This time there was no quality defects on the part as was found in the first shipment, but it was found that the inside diameter was out of tolerance by a significant amount. It then became apparent that the vendor had merely taken the original order and machined out the cutter grooves! However, this repair then put the part outside of the specified dimension.

Further research revealed that the vendor was working out of a garage next to a house, and he was accepted conditionally to make these parts by the prime contractor's source inspectors. The conditional acceptance was granted because this was a new vendor, and there were political reasons (as required by the government) to award certain amounts of work to small companies. Needless to say, this vendor was pulled from the job, and the contract was given to a more-proven outfit. Thus, not just anyone could make this part.

The second case demonstrates this situation where a manufacturer of displays for a commercial aircraft had developed and built a number of units that passed qualification testing for a subcontractor of one of the systems for the aircraft. The display manufacturer had a significant amount of experience in the field as well as developing an efficient and rugged design for the display, particularly in the area of the backlighting. Because of a change in the subcontractor's contract management, and against the advice of its engineering department, the contract manager decided that "anyone else could build this display." They then proceeded to use another manufacturer to make these displays for large production orders for reasons involving politics. Unfortunately, this second manufacturer did not have the experience or the same approach as the original manufacturer. And since the backlighting design was proprietary in the original design, the second manufacturer had to come up with a different design. Also, the second manufacturer did not use pre-screened parts that could meet the temperature range requirements in their unit.

As a result, the first production lot of two dozen units only saw three units passing the burn-in test (the burn-in test consisted of thermal and vibration testing and was still a necessary requirement). It is interesting to note that even though the second display manufacturer met many of the production quality requirements (such as the ISO-9000 quality requirements used worldwide), it did not mean that the manufacturer was technically capable of building this unit. It is so important to realize the distinction between quality process requirements and engineering capabilities—they are two different areas!

Both delays and significant cost impact resulted, yet after six months the second manufacturer still could not make the required number of units. Now the contractor was in a deep pickle because they were missing orders for installation on the aircraft along with cost overruns and was now forced to go to the original manufacturer to make more units. Fortunately, the original display manufacturer was somewhat forgiving and a compromise was worked out. Needless to say, the contractor's engineering department had the "I told you so" attitude with the contracts department.

These two cases, like many others that the reliability engineer will experience during the course of a career, demonstrate that the issues of "deep pockets and politics" may override sound engineering choices to the detriment of a program. The idea that "anyone can build these parts" is a myth that has to be dispelled by the engineer as much as possible.

Likewise, there may be a tendency by some companies to say that anyone can do the job of a reliability engineer, even if they have not had some training or experience in the field. This, too, is a myth and may result in oversights during the critical early stages of a design or improperly performed analysis that is submitted to the customer. There is no shortcut to adequate training and experience!

16.3 SUMMARY OF BOOK

It is brought to the reader's attention that although there has been many changes in the reliability field in recent years, there will always be products that will continue to require some reliability analysis and some development testing. Product lines that will benefit from the combination for reliability analysis and reliability testing are:

- Aerospace products
- Commercial products
- Medical products

It is important to note that many opportunities continue to develop for engineers with reliability experience in these fields, often in new product lines not previously associated with reliability tasks being performed. Regulating organizations for these products may still insist on the major reliability tasks, such as FMEA and reliability testing, be performed in order to get better confidence for the projected performance of the product in the field.

It may seem like many examples of poor reliability were presented in this book, many of which were experienced by the author on a firsthand basis. But to the author, the negative experience of a failure may actually become a positive experience in that a valuable lesson was learned. Each problem can be a valuable learning tool and can be presented later to a much larger audience in textbook form. There is a challenge in dealing with the great unknown that is present when new and unprecedented problems appear. New technologies will beget new failure modes that require an engineer with an open and intuitive mind to get arms around the problem and solve it. A textbook can serve as a guide, but the engineer has to provide the rest in terms of ability to want to solve the problem.

The field of reliability is a dynamic field that changes regularly as new methods and findings are developed. Because of this fact, it has been difficult for one book to be written on the subject that is able to remain current with new developments as they occur. This has been seen with previous books that focused heavily on reliability component prediction techniques that has been committed to print.

It is for this reason that the focus of this textbook has been placed on methodology and ways of thinking, as opposed to the mechanical plug-in type methods that can change. Thus, the emphasis on mathematical tools and statistical techniques have been presented in much detail as to how they can be applied. These tools will always be useful regardless of technological changes in the reliability field.

Indeed a key proponent of this type of thinking for reliability analysis was W. J. Willoughby, Jr., who was the Secretary of the Navy for Logistics during the 1980s. He was a major force in instituting many reforms in the area of product reliability for all major systems and equipment purchased the Navy. He was particularly concerned with the knowledge that new engineers were obtaining from engineering schools. He noted that while engineering schools were very good in teaching theory and analysis, more effort was needed for engineers to gain knowledge in technical fundamentals as well as putting personal integrity into the design. Willoughby did not believe in the concept that product integrity and design quality could be managed into a product. Rather he had the approach that a successful product design followed a complete process from the initial concept to the final delivery.

During his tenure as secretary, he had many concerns with the way elaborate contracts were constructed to ensure successful delivery, even though there may be issues with the contractor's design procedures or engineering department's abilities. Thus he strove to incorporate changes in the way contracts were constructed to emphasis the concepts of best practices and other key aspects of product reliability. This is the same spirit in which this book was written where the focus is on the practical side of reliability.

From the many examples provided in this book, it is shown that the problems cited by Willoughby still exist and show up in many situations that the reliability engineer will have to face when working on new product designs. It is hoped from the approach used in this book that the student has picked up valuable tools, both quantitative and qualitative, that can be used not only in academic studies but throughout the course of one's professional career.

Much of reliability was developed through the imposition of government requirements, both in the military equipment field (for all four services of the U.S. armed forces) and in the commercial aircraft field (F.A.A., J.A.R., etc). Despite the decline of mandatory government requirements in recent years, many aspects of reliability have filtered down to other commercial fields, ranging from bank equipment to environmental gauging equipment. With less guidelines passed down from the customer, more and more emphasis will be placed on the individual reliability engineer's judgement and experience.

In the upcoming years, fewer ads may appear in the classified section for engineers that specifically ask for a reliability engineer. In comparison to the peak of the aerospace era in the 1980s, current ads will more likely list positions such as product support engineer, quality specialist, or performance engineer to handle reliability engineering functions. The field of reliability has not vanished; it has been renamed and called other things. Sometimes companies may try to define the position in writing by using ambiguous terms, but they really need an engineer with a reliability background.

Often, a person with a reliability background can see the opportunity to apply reliability expertise even when the company is not clear on what it really needs. It is important not to be deterred by any such ambiguity, rather the prospective candidate must seize the opportunity and describe how one's reliability expertise will help the company in its quest to make more reliable products in order to maintain customer satisfaction.

It is important for many companies to recognize the need for a reliability engineer or for someone in the company to perform reliability tasks. If for no other reason, it should be done for the goal of preventing costly product recalls or avoiding more serious litigation. Cases appear in the news on a frequent basis where such recalls occur. A recent case involved a major automobile manufacturer having to pay or replace faulty ignitions on 23 million cars and trucks at a cost of $2 billion! The problem was caused by the poor location of the ignition module next to the hottest point of the engine and this would cause the vehicle to stall. Certainly the cost of a reliability engineer performing a durability test would appear to be a good investment compared to billions in replacement costs!

Thus, a case has been made for the need for a reliability engineer in the commercial product field as well as the military field. It is hoped that the reader, whether a student or practicing engineer, has been able to obtain useful tools from this book towards the solving of reliability problems and will continue to use this book for future reference.

16.4 EXERCISES

1. Examine the classified ads in the newspaper and list the jobs that may be looking for an engineer with reliability skills. Describe the approach that you would take in selling your skill set to the company listing the position. What other skills beside engineering would be useful to perform the job? Does your resume include an area of outside interest that would be useful to the prospective employer?
2. In your current or previous job, list the non-engineering skills that you may have employed to enhance the way that you perform reliability (or similar field) functions.
3. Review your current job with respect to how much writing you perform on a weekly or monthly basis. Is it organized enough to meet the requirements for acceptance by either management or the customer? How can you improve your writing skills?

Glossary of Reliability Terms

Confidence limits A method of incorporating a level of confidence into raw reliability data. This is accomplished through the use of statistical methods using multipliers selected from tables of values that are based on statistical functions such as the chi-square function.

Electronic Part Stress Analysis Analysis that is performed during the design stages to verify that parts are not overstressed. The stress is calculated as the applied value over the rated value for the component.

Failure Modes and Effects analysis (FMEA) A qualitative analysis that is arranged in the form of a worksheet that addresses all possible failure modes and effects for a device or product. The numerical portion of the FMEA is the Failure Modes, Effects and Criticality Analysis (FMECA)

Failure rate A reliability measurement that is calculated by the taking the ratio for the number of failures to the number of operating hours.

Fault-Tree Analysis (FTA) A flowchart analysis that graphically illustrates the faults of a component or system as identified by the FMEA. From this analysis, minimum cut sets, i.e., single point of failure paths, can be identified.

Mean Time Between Failures (MTBF) A reliability measurement that is calculated by taking the ratio of the number of operating hours to the number of failures.

Mean Time to Repair (MTTR) A maintainability measurement that is calculated by taking the cumulative totals of active repair times during a specific period of time and dividing it by the total number of malfunctions performed during this period of time.

Reliability The definition of reliability is the probability of a device performing its purpose adequately for the period of time intended under the specified conditions encountered.

Reliability distribution A distribution that tracks reliability through mathematical expressions that are based on probability functions. There are two types of reliability distributions: continuous (time-based) and discrete (event-based). Examples of continuous reliability distributions include the exponential and the Weibull while examples of the discrete reliability distribution include the binomial and Poisson distributions.

Reliability model The reliability model is the configuration of how components in the system are hooked up in terms of reliability paths. There are several types of reliability models, including serial and parallel models.

Reliability testing The reliability test is a defined test that subject test units to environmental tests that simulate some of the conditions that will be seen during field service. Testing may include both thermal and vibration testing for specified period of time and using specific profiles as defined by the customer or environment that the product will see during field service. There are many types of reliability tests that can be performed that are known by different names.

APPENDIX A

Statistical Tables

Table 1 Upper Percentage Points of the Chi-Square (χ^2) Distribution

Deg of Freed	.995	.990	.975	.950	.900	.800	.750	.700	.600	.500	.400	.300	.250	.200	.100	.050	.025	.010	.005
1	.000	.000	.000	.000	.016	.064	.101	.148	.275	.455	.708	1.07	1.32	1.64	2.71	3.80	5.02	6.63	7.88
2	.010	.020	.051	.103	.211	.446	.575	.713	1.02	1.39	1.83	2.41	2.77	3.22	4.60	5.99	7.38	9.21	10.6
3	.072	.115	.216	.352	.584	1.00	1.21	1.42	1.87	2.37	2.95	3.67	4.11	4.64	6.25	7.81	9.35	11.3	12.8
4	.207	.297	.484	.711	1.06	1.65	1.92	2.19	2.75	3.36	4.04	4.88	5.38	5.99	7.78	9.49	11.1	13.3	14.9
5	.412	.554	.831	1.14	1.61	2.34	2.67	3.00	3.66	4.35	5.13	6.06	6.63	7.29	9.24	11.1	12.8	15.1	16.7
6	.676	.872	1.24	1.63	2.20	3.07	3.45	3.83	4.57	5.35	6.21	7.23	7.84	8.56	10.6	12.6	14.4	16.8	18.5
7	.989	1.29	1.69	2.17	2.83	3.82	4.25	4.67	5.49	6.35	7.28	8.38	9.04	9.80	12.0	14.1	16.0	18.5	20.3
8	1.34	1.65	2.18	2.73	3.49	4.59	5.07	5.53	6.42	7.34	8.35	9.52	10.2	11.0	13.4	15.5	17.5	20.1	22.0
9	1.74	2.09	2.70	3.32	4.17	5.38	5.90	6.39	7.36	8.34	9.41	10.7	11.4	12.2	14.7	16.9	19.0	21.7	23.6
10	2.16	2.56	3.25	3.94	4.86	6.18	6.74	7.27	8.30	9.34	10.5	11.8	12.5	13.4	16.0	18.3	20.5	23.2	25.2
11	2.60	3.05	3.82	4.57	5.58	6.99	7.58	8.15	9.24	10.3	11.5	12.9	13.7	14.6	17.3	19.7	21.9	24.7	26.8
12	3.07	3.57	4.40	5.23	6.30	7.81	8.44	9.03	10.2	11.3	12.6	14.0	14.8	15.8	18.5	21.0	23.3	26.2	28.3
13	3.56	4.11	5.01	5.89	7.04	8.63	9.30	9.93	11.1	12.3	13.6	15.1	16.0	17.0	19.8	22.4	24.7	27.7	29.8
14	4.07	4.66	5.63	6.57	7.79	9.47	10.2	10.8	12.1	13.3	14.7	16.2	17.1	18.2	21.1	23.7	26.1	29.1	31.3
15	4.60	5.23	6.26	7.26	8.55	10.3	11.0	11.7	13.0	14.3	15.7	17.3	18.2	19.3	22.3	25.0	27.5	30.6	32.8
16	5.14	5.81	6.91	7.96	9.31	11.2	11.9	12.6	14.0	15.3	16.8	18.4	19.4	20.5	23.5	26.3	28.8	32.0	34.3
17	5.70	6.41	7.56	8.67	10.1	12.0	12.8	13.5	14.9	16.3	17.8	19.5	20.5	21.6	24.8	27.6	30.2	33.4	35.7
18	6.26	7.01	8.23	9.39	10.9	12.9	13.7	14.4	15.9	17.3	18.9	20.6	21.6	22.8	26.0	28.9	31.5	34.8	37.2
19	6.84	7.63	8.91	10.1	11.6	13.7	14.6	15.4	16.9	18.3	19.9	21.7	22.7	23.9	27.2	30.1	32.8	36.2	38.6
20	7.43	8.26	9.59	10.8	12.4	14.6	15.4	16.3	17.8	19.3	21.0	22.8	23.8	25.0	28.4	31.4	34.2	37.6	40.0
21	8.03	8.90	10.3	11.6	13.2	15.4	16.3	17.2	18.8	20.3	22.0	23.9	24.9	26.2	29.6	32.7	35.5	38.9	41.4
22	8.64	9.54	11.0	12.3	14.0	16.3	17.2	18.1	19.7	21.3	23.0	24.9	26.0	27.3	30.8	33.9	36.8	40.3	42.8
23	9.26	10.2	11.7	13.1	14.8	17.2	18.1	19.0	20.7	22.3	24.1	26.0	27.1	28.4	32.0	35.2	38.1	41.6	44.2
24	9.89	10.9	12.4	13.8	15.7	18.1	19.0	19.9	21.7	23.3	25.1	27.1	28.2	29.6	33.2	36.4	39.4	43.0	45.6
25	10.5	11.5	13.1	14.6	16.5	18.9	19.9	20.9	22.6	24.3	26.1	28.2	29.3	30.7	34.4	37.6	40.6	44.3	46.9
26	11.2	12.2	13.8	15.4	17.3	19.8	20.8	21.8	23.6	25.3	27.2	29.2	30.4	31.8	35.6	38.9	41.9	45.6	48.3
27	11.8	12.9	14.6	16.1	18.1	20.7	21.7	22.7	24.5	26.3	28.2	30.3	31.5	32.9	36.7	40.1	43.2	47.0	49.6
28	12.5	13.6	15.3	16.9	18.9	21.6	22.7	23.6	25.5	27.3	29.2	31.4	32.6	34.0	37.9	41.3	44.5	48.3	51.0
29	13.1	14.3	16.0	17.7	19.8	22.5	23.6	24.6	26.5	28.3	30.3	32.5	33.7	35.1	39.1	42.6	45.7	49.6	52.3
30	13.8	14.9	16.8	18.5	20.6	23.4	24.5	25.5	27.4	29.3	31.3	33.5	34.8	36.3	40.3	43.8	47.0	50.9	53.7
40	20.7	22.2	24.4	26.5	29.0	32.3	33.7	34.9	37.1	39.3	41.6	44.2	45.6	47.3	51.8	55.8	59.3	63.7	66.8
50	28.0	29.7	32.4	34.8	37.7	41.4	42.9	44.3	46.9	49.3	51.9	54.7	56.3	58.2	63.2	67.5	71.4	76.1	79.5
100	67.3	70.1	74.2	77.9	82.4	87.9	90.1	92.1	95.8	99.33	103	107	109	112	118	124	130	136	140

Percentage

Table 2 Critical Values for the F Distribution $P(F) = .95$ ($\alpha = 0.05$)

Numerator (K)

L \ K	1	2	3	4	5	6	7	8	9	10	12	15	20	24	30	40	60	120	500	∞
1	161.45	199.50	215.71	224.58	230.16	233.99	236.77	238.88	240.54	241.88	243.91	245.95	248.01	249.05	250.09	251.14	252.20	253.25	254	254.32
2	18.51	19.00	19.16	19.25	19.30	19.33	19.35	19.37	19.38	19.40	19.41	19.43	19.45	19.45	19.46	19.47	19.48	19.49	19.50	19.50
3	10.13	9.55	9.28	9.12	9.01	8.94	8.89	8.85	8.81	8.79	8.74	8.70	8.66	8.64	8.62	8.59	8.57	8.55	8.54	8.53
4	7.71	6.94	6.59	6.39	6.26	6.16	6.09	6.04	6.00	5.96	5.91	5.86	5.80	5.77	5.75	5.72	5.69	5.66	5.64	5.63
5	6.61	5.79	5.41	5.19	5.05	4.95	4.88	4.82	4.77	4.74	4.68	4.62	4.56	4.53	4.50	4.46	4.43	4.40	4.37	4.36
6	5.99	5.14	4.76	4.53	4.39	4.28	4.21	4.15	4.10	4.06	4.00	3.94	3.87	3.84	3.81	3.77	3.74	3.70	3.68	3.67
7	5.59	4.74	4.35	4.12	3.97	3.87	3.79	3.73	3.68	3.64	3.57	3.51	3.44	3.41	3.38	3.34	3.30	3.27	3.24	3.23
8	5.32	4.46	4.07	3.84	3.69	3.58	3.50	3.44	3.39	3.35	3.28	3.22	3.15	3.12	3.08	3.04	3.01	2.97	2.94	2.93
9	5.12	4.26	3.86	3.63	3.48	3.37	3.29	3.23	3.18	3.14	3.07	3.01	2.94	2.90	2.86	2.83	2.79	2.75	2.72	2.71
10	4.96	4.10	3.71	3.48	3.33	3.22	3.14	3.07	3.02	2.98	2.91	2.84	2.77	2.74	2.70	2.66	2.62	2.58	2.55	2.54
11	4.84	3.98	3.59	3.36	3.20	3.09	3.01	2.95	2.90	2.85	2.79	2.72	2.65	2.61	2.57	2.53	2.49	2.45	2.41	2.40
12	4.75	3.89	3.49	3.26	3.11	3.00	2.91	2.85	2.80	2.75	2.69	2.62	2.54	2.51	2.47	2.43	2.38	2.34	2.31	2.30
13	4.67	3.81	3.41	3.18	3.03	2.92	2.83	2.77	2.71	2.67	2.60	2.53	2.46	2.42	2.38	2.34	2.30	2.25	2.22	2.21
14	4.60	3.74	3.34	3.11	2.96	2.85	2.76	2.70	2.65	2.60	2.53	2.46	2.39	2.35	2.31	2.27	2.22	2.18	2.14	2.13
15	4.54	3.68	3.29	3.06	2.90	2.79	2.71	2.64	2.59	2.54	2.48	2.40	2.33	2.29	2.25	2.20	2.16	2.11	2.08	2.07
16	4.49	3.63	3.24	3.01	2.85	2.74	2.66	2.59	2.54	2.49	2.42	2.35	2.28	2.24	2.19	2.15	2.11	2.08	2.02	2.01
17	4.45	3.59	3.20	2.96	2.81	2.70	2.61	2.55	2.49	2.45	2.38	2.31	2.23	2.19	2.15	2.10	2.06	2.01	1.97	1.96
18	4.41	3.55	3.16	2.93	2.77	2.66	2.58	2.51	2.46	2.41	2.34	2.27	2.19	2.15	2.11	2.06	2.02	1.97	1.93	1.92
19	4.38	3.52	3.13	2.90	2.74	2.63	2.54	2.48	2.42	2.38	2.31	2.23	2.16	2.11	2.07	2.03	1.98	1.93	1.90	1.88
20	4.35	3.49	3.10	2.87	2.71	2.60	2.51	2.45	2.39	2.35	2.28	2.20	2.12	2.08	2.04	1.99	1.95	1.90	1.85	1.84
21	4.32	3.47	3.07	2.84	2.68	2.57	2.49	2.42	2.37	2.32	2.25	2.18	2.10	2.05	2.01	1.96	1.92	1.87	1.82	1.81
22	4.30	3.44	3.05	2.82	2.66	2.55	2.46	2.40	2.34	2.30	2.23	2.15	2.07	2.03	1.98	1.94	1.89	1.84	1.80	1.78
23	4.28	3.42	3.03	2.80	2.64	2.53	2.44	2.37	2.32	2.27	2.20	2.13	2.05	2.00	1.96	1.91	1.86	1.81	1.77	1.76
24	4.26	3.40	3.01	2.78	2.62	2.51	2.42	2.36	2.30	2.25	2.18	2.11	2.03	1.98	1.94	1.89	1.84	1.79	1.74	1.73
25	4.24	3.39	2.99	2.76	2.60	2.49	2.40	2.34	2.28	2.24	2.16	2.09	2.01	1.96	1.92	1.87	1.82	1.77	1.72	1.71
26	4.23	3.37	2.98	2.74	2.59	2.47	2.39	2.32	2.27	2.22	2.15	2.07	1.99	1.95	1.90	1.85	1.80	1.75	1.70	1.69
27	4.21	3.35	2.96	2.73	2.57	2.46	2.37	2.31	2.25	2.20	2.13	2.06	1.97	1.93	1.88	1.84	1.79	1.73	1.68	1.67
28	4.20	3.34	2.95	2.71	2.56	2.45	2.36	2.29	2.24	2.19	2.12	2.04	1.96	1.91	1.87	1.82	1.77	1.71	1.67	1.65
29	4.18	3.33	2.93	2.70	2.55	2.43	2.35	2.28	2.22	2.18	2.10	2.03	1.94	1.90	1.85	1.81	1.75	1.70	1.65	1.64
30	4.17	3.32	2.92	2.69	2.53	2.42	2.33	2.27	2.21	2.16	2.09	2.01	1.93	1.89	1.84	1.79	1.74	1.68	1.64	1.62
40	4.08	3.23	2.84	2.61	2.45	2.34	2.25	2.18	2.12	2.08	2.00	1.92	1.84	1.79	1.74	1.69	1.64	1.58	1.53	1.51
50	4.03	3.18	2.79	2.56	2.40	2.29	2.20	2.13	2.07	2.03	1.96	1.87	1.79	1.75	1.71	1.65	1.58	1.51	1.46	1.44
60	4.00	3.15	2.76	2.53	2.37	2.25	2.17	2.10	2.04	1.99	1.92	1.84	1.75	1.70	1.65	1.59	1.53	1.47	1.41	1.39
120	3.92	3.07	2.68	2.45	2.29	2.18	2.09	2.02	1.96	1.91	1.83	1.75	1.66	1.61	1.55	1.50	1.43	1.35	1.27	1.25
150	3.91	3.06	2.67	2.43	2.27	2.16	2.07	2.00	1.94	1.89	1.81	1.73	1.64	1.59	1.54	1.47	1.41	1.33	1.25	1.22
200	3.89	3.04	2.65	2.41	2.26	2.14	2.05	1.98	1.92	1.87	1.80	1.71	1.62	1.57	1.52	1.45	1.40	1.32	1.22	1.19
400	3.86	3.02	2.62	2.39	2.23	2.12	2.03	1.96	1.90	1.85	1.78	1.70	1.60	1.54	1.49	1.42	1.36	1.28	1.16	1.13
1000	3.85	3.00	2.60	2.38	2.22	2.10	2.01	1.95	1.88	1.84	1.76	1.68	1.58	1.53	1.47	1.41	1.32	1.23	1.13	1.08
∞	3.84	3.00	2.60	2.37	2.21	2.10	2.01	1.94	1.88	1.83	1.75	1.67	1.57	1.52	1.46	1.39	1.32	1.22	1.11	1.00

Denominator (L)

Table 3 Critical Values for the F Distribution $P(F) = .99$ ($\alpha = 0.01$)

L \ K	1	2	3	4	5	6	7	8	9	10	12	15	20	24	30	40	60	120	500	∞
1	4052.2	4999.5	5403.3	5624.6	5763.7	5859.0	5928.3	5981.6	6022.5	6055.8	6106.3	6106.3	6208.7	6234.6	6260.7	6286.8	6313.0	6339.4	6361	6366.0
2	98.50	99.00	99.17	99.25	99.30	99.33	99.36	99.37	99.39	99.40	99.42	99.43	99.45	99.46	99.47	99.47	99.48	99.49	99.50	99.50
3	34.12	30.82	29.46	28.71	28.24	27.91	27.49	27.34	27.34	27.23	27.05	26.87	26.69	26.60	26.50	26.41	26.32	26.22	26.14	26.12
4	21.20	18.00	16.69	15.98	15.52	15.21	14.98	14.80	14.66	14.55	14.37	14.20	14.02	13.93	13.84	13.74	13.65	13.56	13.48	13.46
5	16.26	13.27	12.06	11.39	10.97	10.67	10.46	10.29	10.16	10.05	9.89	9.72	9.55	9.47	9.38	9.29	9.20	9.11	9.04	9.02
6	13.74	10.92	9.78	9.15	8.75	8.47	8.26	8.10	7.98	7.87	7.72	7.56	7.40	7.31	7.23	7.14	7.06	6.97	6.90	6.88
7	12.25	9.55	8.45	7.85	7.46	7.19	6.99	6.84	6.72	6.62	6.47	6.31	6.16	6.07	5.99	5.91	5.82	5.74	5.67	5.65
8	11.26	8.65	7.59	7.01	6.63	6.37	6.18	6.03	5.91	5.81	5.67	5.52	5.36	5.28	5.20	5.12	5.03	4.95	4.88	4.86
9	10.56	8.02	6.99	6.42	6.06	5.80	5.61	5.47	5.35	5.26	5.11	4.96	4.81	4.73	4.65	4.57	4.48	4.40	4.33	4.31
10	10.04	7.56	6.55	5.99	5.64	5.39	5.20	5.06	4.94	4.85	4.71	4.56	4.41	4.33	4.25	4.17	4.08	4.00	3.93	3.91
11	9.65	7.21	6.22	5.67	5.32	5.07	4.89	4.74	4.63	4.54	4.40	4.25	4.10	4.02	3.94	3.86	3.78	3.69	3.62	3.60
12	9.33	6.93	5.95	5.41	5.06	4.82	4.64	4.50	4.39	4.30	4.16	4.01	3.86	3.78	3.70	3.62	3.54	3.45	3.38	3.36
13	9.07	6.70	5.74	5.12	4.86	4.62	4.44	4.30	4.19	4.10	3.96	3.82	3.66	3.59	3.51	3.43	3.34	3.25	3.18	3.17
14	8.86	6.51	5.56	5.04	4.70	4.46	4.28	4.14	4.03	3.94	3.80	3.66	3.51	3.43	3.35	3.27	3.18	3.09	3.02	3.00
15	8.68	6.36	5.42	4.86	4.62	4.32	4.14	4.00	3.89	3.80	3.67	3.52	3.37	3.29	3.21	3.13	3.05	2.96	2.89	2.87
16	8.53	6.23	5.29	4.77	4.32	4.20	4.03	3.89	3.78	3.69	3.55	3.41	3.26	3.18	3.10	3.02	2.93	2.84	2.77	2.75
17	8.40	6.11	5.18	4.67	4.34	4.10	3.93	3.79	3.68	3.59	3.46	3.31	3.16	3.08	3.00	2.92	2.83	2.75	2.67	2.65
18	8.29	6.01	5.09	4.58	4.25	4.01	3.84	3.71	3.60	3.51	3.37	3.23	3.08	3.00	2.92	2.84	2.75	2.66	2.59	2.57
19	8.18	5.93	5.01	4.50	4.17	3.94	3.77	3.63	3.52	3.43	3.30	3.15	3.00	2.92	2.84	2.76	2.67	2.58	2.51	2.49
20	8.10	5.85	4.94	4.43	4.10	3.87	3.70	3.56	3.46	3.37	3.23	3.09	2.94	2.86	2.78	2.69	2.61	2.52	2.44	2.42
21	8.02	5.78	4.87	4.37	4.04	3.81	3.64	3.51	3.40	3.31	3.17	3.03	2.88	2.80	2.72	2.64	2.55	2.46	2.38	2.36
22	7.95	5.72	4.82	4.31	3.99	3.76	3.59	3.45	3.35	3.26	3.12	2.98	2.83	2.75	2.67	2.58	2.50	2.40	2.33	2.31
23	7.88	5.66	4.76	4.26	3.94	3.71	3.54	3.41	3.30	3/21	3.07	2.93	2.78	2.70	2.62	2.54	2.45	2.35	2.28	2.26
24	7.82	5.61	4.72	4.22	3.90	3.67	3.50	3.36	3.26	3.17	3.03	2.89	2.74	2.66	2.58	2.49	2.40	2.31	2.23	2.21
25	7.77	5.57	4.68	4.18	3.86	3.63	3.46	3.32	3.22	3.13	2.99	2.85	2.70	2.62	2.54	2.45	2.36	2.27	2.19	2.17
26	7.72	5.53	4.64	4.14	3.82	3.59	3.42	3.29	3.18	3.09	2.96	2.82	2.66	2.58	2.50	2.42	2.33	2.23	2.15	2.13
27	7.68	5.49	4.60	4.11	3.78	3.56	3.39	3.26	3.15	3.06	2.93	2.78	2.63	2.55	2.47	2.38	2.29	2.20	2.12	2.10
28	7.64	5.45	4.57	4.07	3.75	3.53	3.36	3.23	3.12	3.03	2.90	2.75	2.60	2.52	2.44	2.35	2.26	2.17	2.09	2.06
29	7.60	5.42	4.54	4.04	3.73	3.50	3.33	3.20	3.09	3.00	2.87	2.73	2.57	2.49	2.41	2.33	2.23	2.14	2.06	2.03
30	7.56	5.39	4.51	4.02	3.70	3.47	3.30	3.17	3.07	2.98	2.84	2.70	2.55	2.47	2.39	2.30	2.21	2.11	2.03	2.01
40	7.31	5.18	4.31	3.83	3.52	3.29	3.12	2.99	2.89	2.80	2.66	2.52	2.37	2.29	2.20	2.11	2.02	1.92	1.84	1.81
50	7.17	5.06	4.20	3.72	3.41	3.18	3.02	2.88	2.78	2.70	2.56	2.40	2.26	2.18	2.10	2.00	1.90	1.85	1.73	1.70
60	7.08	4.98	4.13	3.65	3.34	3.12	2.95	2.83	2.72	2.63	2.50	2.35	2.20	2.12	2.03	1.94	1.84	1.73	1.63	1.60
120	6.85	4.79	3.95	3.48	3.17	2.96	2.79	2.66	2.56	2.47	2.34	2.19	2.03	1.95	1.86	1.76	1.66	1.53	1.41	1.38
150	6.81	4.75	3.91	3.44	3.14	2.92	2.75	2.62	2.53	2.44	2.30	2.16	2.00	1.91	1.83	1.72	1.60	1.50	1.37	1.33
200	6.76	4.71	3.88	3.41	3.11	2.90	2.73	2.60	2.50	2.41	2.28	2.11	1.97	1.88	1.79	1.69	1.58	1.47	1.33	1.28
400	6.70	4.65	3.83	3.36	3.06	2.85	2.69	2.55	2.46	2.37	2.23	2.09	1.92	1.84	1.74	1.64	1.52	1.41	1.24	1.19
1000	6.66	4.62	3.80	3.34	3.04	2.82	2.66	2.53	2.43	2.34	2.20	2.06	1.90	1.81	1.72	1.61	1.48	1.34	1.19	1.11
∞	6.63	4.61	3.78	3.32	3.02	2.80	2.64	2.51	2.41	2.32	2.18	2.04	1.88	1.79	1.70	1.59	1.47	1.32	1.15	1.00

Numerator (K) / Denominator (L)

APPENDIX B

References and Source Material

RELIABILITY RESOURCES

The following are government standards which are used as guides only:

Military Standards

MIL-STD-721C	Definition of Terms for R& M
MIL-STD-756B Notice 1	Reliability Modeling & Prediction
MIL-STD-781D	Reliability Testing for Engineering Development, Qualification & Production
MIL-STD-785B Notice 1	Reliability Program for Systems & Equipment Development & Production
MIL-STD-790D Notice 1	Reliability Assurance Program for Electronic Parts Specifications
MIL-STD-1543A (USAF)	Reliability Program Requirements for Space & Missile Systems
MIL-STD-1629A	Procedures for Performing a Failure Mode, Effects and Criticality Analysis
MIL-STD-2155(AS)	Failure Reporting, Analysis & Corrective Action System (FRACAS)
MIL-STD-2164(EC)	Environmental Stress Screening Process for Electronic Equipment

Handbooks

DOD-HDBK-344	Environmental Stress Screening of Electronic Equipment
MIL-HDBK-189	Reliability Growth Management
MIL-HDBK-217F	Reliability Prediction of Electronic Equipment
MIL-HDBK-251	Reliability Design Thermal Applications
MIL-HDBK-338 Vols I & II	Electronic Reliability Design Handbook
MIL-HDBK-781D	Reliability Test Methods, Plans and Environments for Engineering Development, Qualification and Production

OTHER RELIABILITY TEXTBOOKS

The following are existing reliability textbooks that cover other areas of reliability:

RADC Reliability Engineer's Toolkit (RADC 1988)

Reliability Theory and Practice (Bazovsky, Prentice-Hall 1961)

Reliability for Technology, Engineering and Management (Kales, Prentice-Hall 1998)

Reliability Engineering in Systems Design and Operation (Dhillon, Van Nostrand 1983)

RELATED SUBJECT TEXTBOOKS

The following textbooks are of interest and cover some aspects of failure analysis or engineering design processes:

To Engineer is Human (Petroski, Vintage 1992)

Differential Equations and Their Application (Braun, Springer-Verlag 1975)

Statistical Analysis (Parsons, Harper and Row 1978)

Quality, Second Edition (Sommers, Prentice Hall 2003)

Total Quality Management, Second Edition (Besterfield et al., Prentice Hall 2003)

INDEX

A-10 Close Air Support aircraft, 185–193, 271–272, 274–275, 276
Ability to break test articles, 323
Accelerated Life Test, 76, 85–86
Access doors (for maintainability)
Admundsen, race to South Pole, 193–194
Amateur radio software issues, 257
Application software, definition, 244
Automatic Teller Machine (ATM), 247
Availability, 263

Bathtub curve, 2
Bazovsky, Igor, vi
Benchmarking, 300
Best practices, 300
Binomial confidence limits, 36–37
Binomial reliability distribution, 7–8
BIT demonstration, 270
Bulletproofing of design, 176, 180–183
Burn-in test, 91

Capacitors, failure modes of, 101–103
Cathode ray terminal, (CRT), 276–277
Chi-square distribution, 31–35
Chi-square distribution, 31–36
Chi-square table, 331

Chi-square test of independence, 213–235
Circuit board design, 290–291
Circuit board, failure modes of, 105–108
Circuit board, flexible, 109
Circuit board support, 119
Combination reliability model, 21–22
Component maintenance manual, 267
Component spares, 267
Confidence Limits:
 One-sided, lower, 28–29
 One-sided, upper, 28–29
 Two-sided, 29, 31, 33
Continuity of design, 292
Continuous distribution, 5
Connectors, failure modes of, 110–113
Corrosion, effects of, 140, 146–147
Cumulative MTBF graph, 196–199
Current technology, issues, 175, 179–180
Cut sets, 60

Depot Level Maintenance, 266
Design Concepts, 175–193
Design of Experiment (DOE), 299–300
Design Pitfalls, 175–176
Design management, approaches, 193–194

Designing for maintainability, 275–281
Discrepancy Reports, 266
Discrete distributions, 5
Dispatch reliability, 268–269
Downtime, 262
Duane reliability growth curve, 204–217
Durability investigation test, 76, 81–84

Electrical component failure modes, 54
Electronic stress analysis, 41–45
Embedded software, definition, 244
Environmental testing, 66
Environmental Stress Screening (ESS), 91
EPROM, 244–246, 247–248
Evolutionary operation (EVOP), 200
Exponential reliability distribution, description, 5 -7
 graph, 6

F distribution tables, 332–333
F test of significance, 221–223
Failure modes and effects analysis (FMEA), 49
Failure modes, effects and criticality analysis (FMECA), 56–57
Failure rate (λ), definition of ,2
Failures, types of,
 Intermittent, 162–172

Latent, 159–161
Repetitive, 172–173
Secondary, 161–162
Failure rate predictions, 45–48
Failure report form, 75
Fans,
    failure modes of, 114
    use for thermal relief, 134
Fault isolation technique, 158
Fault Tree Analysis, 57, 60–62
Fully Mission Capable (FMC), 264

Graphs, reliability, 196–204
Growth rate, 215–217
GPS units, 258

Highly Accelerated life test (HALT), 89–90
Heat sinks, 131–133
Hypothesis testing, 221

Ice, 150
Inherent availability, 263
Instantaneous MTBF graph, 196–199
Integrated circuits (ICs), failure modes of, 112, 129, 134
Intermediate Level Maintenance, 266
Isolation resistor, 310

K of N configuration reliability model, 19–20

Lamps, failure modes of, 102–103, 119
LCDs, failure modes of, 127–128, 149, 179–180
Least-square model, 206
Life History curve (bathtub curve), 2
Linear regression, 206

Magic bullet, 324
Magnetic head reader, reliability issues, 167–168, 180–183
Maintainability demonstration, 269–270

Maintenance Concept, 265–266
Maintainability, definition of, 261
Margin of error in design, 185–193
Markov model, 22
Mean, definition of, 26
Mean downtime, (MDT), 262
Mean time between failures, (MTBF), 3
Mean time between maintenance, (MTBM), 263
Mean time to repair, (MTTR), 261–262
Medical equipment reliability issues, 255–256
Meter movement, 79
MTBF (see Mean Time Between Failures)
MTBF graphs, 196–204
MTBF field demonstration test, 87
Mission Capable Rate (MC), 264–265
MIL-HDBK-217, 45–47
MIL-STD-1629, 51
Multi-tasking, 316–317

Noise bypass circuit protection, 288
Not mission capable (NMC), 264
Null hypothesis, 221

O-rings, 128
Ohm's law, 43
Operational availability, 263
Organizational Level Maintenance, 265

Parallel reliability model, 14–15
Part selections, 291–292
Part derating guidelines, 42
Partially mission capable (PMC)
Plastic parts, 128, 136
Potentionmeters, failure modes of, 98–100
Poisson reliability distribution, 7–9
Power dissipation, 43–44
Power supply, 41, 81, 115, 117
Preventative maintenance, 271–275
Printer, reliability issues, 160–161

Maintainability issues, 267–268, 272–273
Production reliability tests, 90–93
Production reliability acceptance test (PRAT), 92
Protection support, 310
Protection circuitry, 289
Prototype, 293–294

Q (term for unreliability), 6–9
Quality, 299

R (term for reliability), 5–9
Relays, failure modes of, 113–114
Reliability, definition of, 2
Reliability apportionment, 48–49
Reliability Development Test (RDT), 76, 78–81
Reliability distributions, 4–10
Reliability Enhancement Test (RET), 89
Reliability evaluations, 283–284
Reliability graphs, 196–204
Reliability life curve, 2
Reliability predictions, 45–48
Reliability program, 40–41, 302–311
Reliability tests, 66–95
Reliability test failure troubleshooting, 92–93
Reliability test guidelines, 93
Reliability test procedure, 74
Reliability test report, 76
Reliability Warranty Program (RWP), 76, 87–88
Resistor networks, failure modes of, 100–101
Rolling average, 198, 200

Safety engineering tasks, 57
Secondary failures, 161–162
Serial reliability model, 11–14
Software reliability, 243–259
Stabilization, thermal, 125–127
Standard deviation, definition of, 26
Standby reliability model, 16–18
Structural issues in design, 285
Support software, definition, 244

Switches, failure modes of, 113–114
System reliability models, 10–20

T-46A aircraft, 86–87, 210–212, 313–316
Taguchi's loss function, 299
Temperature ratings, 124
Terminal post installation, 121
Test articles, 72
Test documentation, 72
Test point analysis, 270–271
Test report, 76
Thermal analysis, 124
Thermal chamber, 68, 73
Thermal lag, 125
Thermal profile, 69

Thermal survey, 125–127
Thermal cycling, 67–69
Thermal rise, 124, 126
Thermocouples, 125–127
Transformers, failure modes of, 102, 104
Trend analysis, reliability, 196, 198, 202–203
Troubleshooting, of failures, 154–164

Unproven design concept, 175, 176–178

Variance, definition of, 26
Vendor, relationships, 317–319

Vibration cycling, 70–72
Vibration profile, 71
Vibration table, 70, 73
Voltage, calculation in Ohm's Law, 43
Voltage spike protection, 290
Voltage plane failure, 320–321

Water, effects on reliability, 139–152
Water pump, automobile, 277–279
Weibull reliability distribution, 9–10
Wiring, issues in design, 120, 286–288
Wiloughby, W. J., 327
Writing skills, 322–323

WITHDRAWN

TA 169 .N48 2004
Neubeck, Ken.
Practical reliability
 analysis   11/2008